Lecture Notes in Statistics 148

Edited by P. Bickel, P. Diggle, S. Fienberg, K. Krickeberg, I. Olkin, N. Wermuth, S. Zeger

Springer
New York
Berlin
Heidelberg
Barcelona
Hong Kong
London
Milan
Paris
Singapore
Tokyo

S.E. Ahmed
N. Reid (Editors)

Empirical Bayes and Likelihood Inference

Springer

S.E. Ahmed
Department of Mathematics
University of Regina
Regina, Saskatchewan S4S 0A2
Canada

N. Reid
Department of Statistics
University of Toronto
Toronto, Ontario M5S 3G3
Canada

Library of Congress Cataloging-in-Publication Data

Empirical Bayes and likelihood inference / editors, S.E. Ahmed, N. Reid.
p. cm.—(Lecture notes in statistics; 148)
Includes bibliographical references.
ISBN 0-387-95018-4 (softcover: alk. paper)
1. Bayesian statistical decision theory. 2. Estimation theory. I. Ahmed, S.E. (Syed Ejaz), 1957– . II. Reid, N. III. Lecture notes in statistics (Springer-Verlag); v. 148.

QA279.5 .E47 2001
519.5'.42—dc21 00-059550

Printed on acid-free paper.

Camera-ready copy provided by the Centre de Recherches Mathématiques.
Printed and bound by Sheridan Books, Ann Arbor, MI.
Printed in the United States of America.

9 8 7 6 5 4 3 2 1

ISBN 0-387-95018-4 SPIN 10761462

Springer-Verlag New York Berlin Heidelberg
A member of BertelsmannSpringer Science+Business Media GmbH

Preface

In November, 1997, the Centre de Recherches Mathématiques hosted a one-week workshop called "Empirical Bayes and Likelihood Inference" as part of a year long program in statistics. This workshop was attended by about fifty researchers, featured eighteen speakers and a lively discussion. In January, 1995, in the newsletter of the Royal Statistical Society, Efron predicted that shrinkage and empirical Bayes methodology would be a major area of statistical research for the early 21st century. Shrinkage and empirical Bayes methods provide useful techniques for combining data from various sources, and recent asymptotic theory has advanced understanding of the fundamental role of the likelihood function for much the same purpose. The goal of the workshop was to explore and create common ground in likelihood based inference, Bayesian inference and empirical Bayes methods. This last topic has until fairly recently emphasized point estimation, while the first two are typically focussed on a distributional assessment. The thirteen papers presented in this volume cover a range of interesting practical problems, and illustrate some common themes as well as some distinct differences in approach.

Empirical Bayes methods were originally developed for random effects models, which have a natural hierarchical structure that can also be treated by Bayesian methods. Random effects models are playing an increasingly prominent role in applications, possibly because they provide a way to model population heterogeneity that is easy to describe.

Several papers (Louis, Guttman, Rao, Sen) consider versions of random effects models and aspects of inference in these models. Louis uses random effects to model Poisson overdispersion and spatial heterogeneity, with the "triple goals" of estimating the random effects, estimating their ranks, and estimating their distribution. He notes that in fields such as outcomes research, where an end goal may well be to rank several units, it is important to be aware of these competing goals and of the fact that the performance of estimators can depend heavily on the goal. Guttman draws attention to the fact that in normal theory random effects models the behaviour of posterior point estimates depends on whether these are computed from the marginal or joint posterior, and considers this phenomenon in relation to the EM algorithm. Rao uses random effects to link model-based and sampling-based estimation methods, to address the important topic in survey sampling of small area estimation.

Several papers consider the construction of point estimates and their comparison by either Bayes risk, mean squared error, or asymptotic versions of this. Sen, Ahnsanullah and Ahmed, Singh, and Ahmed all investigate various versions of this problem. Both Sen and Singh emphasize the extension to (partially) nonparametric settings. Ahsanullah and Ahmed and Ahmed study estimation in problems of interest in survival data studies, the latter incorporating type II censoring.

Doksum and Normand and O'Rourke are more explicitly concerned with applications in medical studies. Doksum and Normand consider the the estimation of aspects of the prior distribution of an unknown changepoint, the changepoint corresponding to infection time with the HIV virus. Accumulated responses on CD4 counts and covariates measured at study entry are used to estimate the distribution of the change point. O'Rourke considers the relationship of empirical Bayes methods to the important area of meta-analysis.

Estimation of the variance or mean squared error of empirical Bayes estimators is an area needing much further study. These shrinkage type estimators are often sufficiently complicated that this is a difficult problem. In contrast, fully Bayesian and likelihood approaches typically provide distributional assessments, usually asymptotic, and point estimates with standard errors are at most a handy summary of these. Thus the Bayesian and likelihood approach may have something to offer to the empirical Bayes approach, most likely by considering in more detail random effects models, a research area of current interest. These approaches raise various problems of their own, of course, often best addressed through detailed study of special cases. Yin and Ghosh investigate the construction of so-called matching priors in in the problem of inference about a ratio of means, and Sprott considers likelihood methods for the same problem. Fraser and Reid show that ancillary statistics needed for higher order asymptotic likelihood theory can be constructed in multiparameter problems. Hu and Zidek consider a new approach to likelihood that weights components differentially. Interestingly, this leads to shrinkage type estimators that also feature prominently in empirical Bayes methodology.

Several directions for inference were highlighted by the talks and the discussion, and we hope that this volume conveys some of the surprises, puzzles and successes of theoretical statistics.

We would like to express our thanks to the superb staff at the Centre de Recherches Mathématiques for the encouragement and support for the organization of the theme year in statistics and this workshop in particular. Marty Goldstein provided leadership and encouragement through the planning stages for the theme year and for the production of this book, Louis Pelletier and Josée Laferrière provided superb local arrangements for the workshop, and Louise Letendre and Diane Poulin outstanding technical support for the production of this proceedings. Funding for the theme year was provided by CRM, NSERC (National Science and Engineering

Research Council of Canada), Fonds FCAR (Le Fonds pour la Formation de Chercheurs et l'Aide à la Recherche) of Quebec, and NATO.

S. Ejaz Ahmed
Nancy Reid
May, 2000

Contents

List of Contributors

S.E. Ahmed, Department of Mathematics and Statistics, University of Regina, Regina, Sask., S4S 0A2, Canada
`ahmed@math.uregina.ca`

M. Ahsanullah, Department of Management Sciences, Rider University, Lawrenceville, NJ 08648-3099, USA
`ahsan@genius.rider.edu`

K. Doksum, Department of Statistics, University of California, Berkeley, CA 94720-3860, USA
`doksum@stat.berkeley.edu`

D.A.S. Fraser, Department of Statistics, University of Toronto, Toronto, On., M5S 3G3, Canada
`dfraser@utstat.toronto.edu`

M. Ghosh, Department of Statistics, University of Florida, P.O. Box 118545, Gainesville, FL 32611-8545, USA
`ghoshm@stat.ufl.edu`

I. Guttman, Department of Statistics, SUNY at Buffalo, Buffalo, NY 14214-3093, USA
`sttirwin@acsu.buffalo.edu`

F. Hu, Department of Mathematics, National University of Singapore, Singapore 119260, Singapore
`stahuff@stat.nus.edu.sg`

T.A. Louis, Biostatistics School of Public Health, University of Minnesota, Minneapolis, MN 55455, USA
`tom@biostat.umn.edu`

S.-L.T. Normand, Department of Health Care Policy, Harvard Medical School, 180 Longwood Avenue, Boston, MA02115, USA
`sharon@hcp.med.harvard.edu`

J.N.K. Rao, Department of Mathematics and Statistics, Carleton University, Ottawa, On., K1S 5B6, Canada
`jrao@math.carleton.ca`

N. Reid, Department of Statistics, University of Toronto, Toronto, On., M5S 3G3, Canada
reid@utstat.toronto.edu

K. O'Rourke, Loeb Health Research Institute, University of Ottawa/Ottawa Hospital, 1053 Carling Avenue Ottawa, On., K1Y 4E9, Canada
orourke@utstat.toronto.edu

P.K. Sen, Departments of Biostatistics and Statistics, University of North Carolina, Chapel Hill, NC 27599-7400, USA
pksen@bios.unc.edu

R.S. Singh, Department of Mathematics and Statistics, University of Guelph, Guelph, On., N1G 2W1, Canada
rsingh@msnet.mathstat.uoguelph.ca

D.A. Sprott, Department of Statistics and Actuarial Science, University of Waterloo, Waterloo, On., N2L 3G1, Canada
sprott@fractal.cimat.mx

M. Yin, Analytical Sciences Inc., Durham, NC 27713, USA
myin@asciences.com

J.V. Zidek, Department of Statistics, University of British Columbia, Vancouver, BC, V6T 1Z2, Canada
jim@stat.ubc.ca

1

Bayes/EB Ranking, Histogram and Parameter Estimation: Issues and Research Agenda

T.A. Louis

ABSTRACT We propose that methods of inference should be linked to an inferential goal via a loss function. For example, if unit-specific parameters are the feature of interest, under squared error loss their posterior means are the optimal estimates. If unit-specific ranks are the target feature (for example to be used in "league tables", ranking schools, hospitals, physicians or geographic regions), the conditional expected ranks or a discretized version of them are optimal. If the feature of interest is the histogram or empirical distribution function of the unit-specific parameters then the conditional expected edf or a discretized version of it is optimal.

No single set of estimates can simultaneously optimize the three inferential goals. However, in many policy settings communication and credibility will be enhanced by reporting a set of values with good performance for all three goals. This requirement leads to development of "triple-goal" estimates: those producing a histogram that is a good estimate of the parameter histogram, with induced ranks that are good estimates of the parameter ranks and with good performance in estimating unit-specific parameters. Using mathematical and simulation-based analyses, we compare three candidate triple-goal estimates for the two-stage hierarchical model: posterior means, constrained Bayes estimates and a new approach which optimizes estimation of the edf and the ranks.

1 Introduction

The beauty of the Bayesian approach is its ability to structure complicated models, inferential goals and analyses. The prior and likelihood produce the full joint posterior distribution which generates all inferences. To take full advantage of the structure, methods should be linked to an inferential goal via a loss function. For example, under squared-error loss the posterior mean of a function of parameters is optimal, but its optimality depends on the loss function. Reference Shen and Louis (1998) (hereafter referred to as S&L), Conlon and Louis (1999) and references thereof document the need to structure via loss functions in the two-stage, hierarchical compound sampling model. They show that the histogram of the posterior means of coordinate-specific parameters is under-dispersed as an estimate of the parameter histogram and that ranking these posterior means can produce sub-optimal estimates of the parameter ranks. Decision-theoretic structuring guides development of effective estimates of histograms and ranks.

Since estimates of coordinate-specific parameters are inappropriate for producing histograms and ranks, no single set of values can simultaneously optimize all three goals. However, in many policy settings communication and credibility is enhanced by reporting a single set of estimates with good performance for all three goals. To this end, S&L introduce "triple-goal" estimation criteria. Effective estimates are those that produce a histogram that is a high quality estimate of the parameter histogram, induced ranks that are high quality estimates of the parameter ranks and coordinate-specific parameter estimates that perform well. S&L evaluate and compare candidate estimates including the observed data, the posterior means, the constrained Bayes estimates (Ghosh, 1992; Louis, 1984) and the new S&L "G then Ranks" estimates. They evaluate these for exchangeable models with exponential family likelihoods.

In this report we summarize the status of this research and identify issues and research needs. Most mathematical details can be found in S&L. Areas requiring development include finite sample and asymptotic evaluation of models more general than those in S&L, additional study of empirical Bayes and frequentist performance, sensitivity to prior mis-specification and use of robust priors, use of loss functions other than squared-error, and inferences for multivariate coordinate-specific parameters.

2 Model and Inferential Goals

Consider the two-stage, hierarchical model with a continuous prior distribution:

$$\begin{aligned} \theta_k &\overset{\text{iid}}{\sim} G, \\ Y_k \mid \theta_k &\overset{\text{indep}}{\sim} l_k(Y_k \mid \theta_k), \end{aligned} \tag{1}$$

for $k = 1, \ldots, K$. Let g be the density function of G. Then the posterior distribution of θ_k is: $g_k(\theta_k \mid Y_k) = l_k(Y_k \mid \theta_k)g(\theta_k)/f_k(Y_k)$, where $f_k(Y_k) = \int l_k(Y_k \mid s)g(s)\,ds$ is the marginal density of Y_k. Our goals are to estimate the individual θ's, their ranks and the histogram associated with their empirical distribution function (EDF). To structure estimates in a mathematically convenient manner we use squared error loss (SEL) for the relevant estimation goal. Other loss functions are possibly more appropriate (e.g., absolute value producing the posterior median, 0/1 producing the posterior mode). Developing and evaluating these are part of the future research agenda.

Estimating the individual θ's

Under squared error loss, $(\hat{\theta} - \theta)^2$, the posterior means (PM) are optimal:

$$\hat{\theta}_k^{\mathrm{PM}} = \eta_k = E\big[\theta_k \mid Y_k\big] = \int \theta_k g_k(\theta_k \mid Y_k)\, d\theta_k.$$

These are the "traditional" parameter estimates. In this report estimates of all quantities of interest will be posterior means, but we reserve the "PM" label for estimates of the individual θ's.

Estimating the ranks

The ranks, $\mathbf{R} = (R_1, \ldots, R_K)$ are: $R_k = \mathrm{rank}(\theta_k) = \sum_{j=1}^{K} I_{\{\theta_k \geq \theta_j\}}$, where $I_{\{.\}}$ is the indicator function. The smallest θ has rank 1. The posterior mean rank $(\bar{R}_k)$ is optimal under squared error loss:

$$\bar{R}_k = E\big[R_k \mid \mathbf{Y}\big] = \sum_{j=1}^{K} P\big[\theta_k \geq \theta_j \mid \mathbf{Y}\big],$$

where $\mathbf{Y} = (Y_1, \ldots, Y_k)$. See Laird and Louis (1989) for the posterior covariance of the ranks.

Generally, the $\bar{R}_k$ are not integers. Though their mean $([K+1]/2)$ is the same as for integer ranks, their spread is smaller, frequently substantially smaller. For example, the largest $\bar{R}_k$ can be far smaller than K and the smallest can be far larger than 1. The noninteger feature of the $\bar{R}_k$ is attractive, because integer ranks can over-represent distance and under-represent uncertainty, but we will need integer ranks. To obtain them, rank the $\bar{R}_k$, producing:

$$\hat{R}_k = \mathrm{rank}(\bar{R}_k).$$

Estimating the EDF

In inferential contexts such as geographic risk assessment, subgroup analysis and finite-population inference, one is interested in the empirical distribution function of the θ's which operate in the current data set (denoted G_K) rather than the prior distribution which generated the θ's (G). This EDF can be represented as: $G_K(t; \mathbf{O}) = (1/K)\sum I_{\{\theta_k \leq t\}}, -\infty < t < \infty$. With $A(t)$ a candidate estimate of $G_K(t; \mathbf{O})$, under weighted integrated squared error loss, $\int w(t)\big[A(t) - G_K(t; \mathbf{O})\big]^2 dt$, the posterior mean is optimal:

$$\bar{G}_K(t) = E\big[G_K(t; \mathbf{O}) \mid \mathbf{Y}\big] = \frac{1}{K}\sum P(\theta_k \leq t \mid Y_k).$$

Estimating the θ's

For generic estimates $\widehat{\theta}_k$, ***posterior regret*** is the difference between their posterior risk and the optimal risk. For SEL, the optimal risk is the posterior variance and the regret is the squared posterior bias:

$$\begin{aligned}\text{regret }(\widehat{\Theta}) &= E\left[\frac{1}{K}\sum(\widehat{\theta}_k - \theta_k)^2 \middle| \mathbf{Y}\right] - E\left[\frac{1}{K}\sum(\eta_k - \theta_k)^2 \,\middle|\, \mathbf{Y}\right],\\ &= \frac{1}{K}\sum(\widehat{\theta}_k - \eta_k)^2.\end{aligned}$$

Performance of the CB *estimates*

Let $\mu = E(\theta)$, $\tau^2 = V(\theta)$ and $\delta^2 = V(\eta)$. Since $V(\theta) = E\big[V(\theta \mid Y)\big] + V\big[E(\theta \mid Y)\big] = E(v)+V(\eta)$, $\tau^2 = E(v)+\delta^2$. For large K, $(1/K)\sum \eta_k \to \mu$, $(1/K)\sum v_k \to (\tau^2 - \delta^2)$ and the sample variance of the ηs converges to δ^2. Therefore, $\lim_{K\to\infty} \widehat{\theta}_k^{\text{CB}} = \mu + (\eta_k - \mu)\tau/\delta$ and the ***preposterior regret*** is:

$$\text{regret}_{\text{CB}} = \tau^2 + \delta^2 - 2\tau\delta = (\tau - \delta)^2. \tag{4}$$

S&L provide additional details for exponential families.

4.0.1 Performance of the GR estimates

Let H be the CDF of η. For the exchangeable model, S&L show that the preposterior regret for the GR estimates converges to:

$$\begin{aligned}\text{regret}_{\text{GR}} &= \int \big\{\theta - H^{-1}\big[G(\theta)\big]\big\}^2 g(\theta)\, d\theta,\\ &= \tau^2 + \delta^2 - 2 \text{ cov } \big\{\Theta, H^{-1}\big[G(\Theta)\big]\big\}.\end{aligned} \tag{5}$$

The covariance in Eq. (5) is nonnegative, since $H^{-1}\big[G(\theta)\big]$ is monotone nondecreasing in θ.

4.0.2 GR versus CB

Using (4) and (5) we have that for all $l(Y \mid \theta)$, G, and for large K, $\text{regret}_{\text{GR}} \geq \text{regret}_{\text{CB}}$. In the limit as $K \to \infty$:

$$\begin{aligned}\frac{\text{regret}_{\text{GR}} - \text{regret}_{\text{CB}}}{2} &= \tau\delta - \text{cov}\big\{\Theta, H^{-1}\big[G(\Theta)\big]\big\},\\ &= \tau\delta\big[1 - \rho(G, H)\big] \geq 0,\end{aligned} \tag{6}$$

where $\rho(G, H)$ is the correlation between θ and η. The result follows from the Cauchy–Schwartz inequality (recall that $E(\eta) = E(\theta) = \mu$). The correlation $\rho(G, H)$ measures the probabilistic linearity of $H^{-1}\big[G(\theta)\big]$. With θ the mean parameter, the correlation is 1 for the Gaussian/Gaussian model and is close to 1 for other exponential family models. Since regret is the increment over the Bayes risk, (6) also computes the difference in risk between the GR and CB estimates.

Estimating G_K

To see that the EDF computed from the CB estimates is a poor estimate of G_K, we consider large sample (large K) performance. Since G_K is consistent for G, consistency or lack thereof to G implies the same performance in estimating G_K. As we have seen, neither the PMs nor the Ys are appropriate for estimating G_K or G. The CB estimates adjust the spread of the PMs so that the induced CDF (G_K^*) has the correct center and spread, but the estimate pays no attention to shape. As S&L show in the exchangeable case, the EDF of the CB estimates is consistent for G if and only if the marginal distribution of η differs from G only by a location/scale change. The Gaussian/Gaussian model with θ the mean parameter qualifies, but in general G_K^* will not be consistent for G though it will have the correct first and second moments. Importantly, consistency for G_K^* is θ transform-dependent.

For the GR approach, $\overline{G}_K$ is consistent for G whenever G is identifiable from the marginal distributions see (Teicher, 1961; Robbins, 1983; Lindsay, 1983). Consistency is transform invariant. Other than for the Gaussian/Gaussian model with equal sampling variance, $\widehat{G}_K$ will dominate G_K^*. Under mild regularity conditions, $\overline{G}_K$ and $\widehat{G}_K$ are consistent, while G_K^* generally is asymptotically biased. The integrated squared error for $\overline{G}_K$ and $\widehat{G}_K$ is $O(1/K)$, while in general that for G_K^* is $O(1)$.

Simulations

S&L simulated four models: Gaussian/Gaussian, Gaussian/Lognormal, Gamma/Inverse Gamma, and Gamma/Gamma. The first and third models investigate the relevant mean parameter; the second and fourth study a nonlinear transform of it. Results for $K = 50$ and $K = 20$ were qualitatively and quantitatively similar. As it must for estimating individual θ's under SEL, PM dominates the other approaches. However, both GR and CB increase SEL by no more than 32% in all simulation cases, and both outperform use of the MLE (the Y_k). Comparison of CB and GR was more complicated. Results for the Gaussian/Gaussian (where CB and GR are asymptotically equivalent) and Gamma/Gamma cases (where CB has a slightly better large-sample performance than GR) were as predicted by the S&L asymptotic analysis. However, for finite K simulations show that GR can perform better than CB for skewed or long-tailed prior distributions.

As it must, GR dominates other approaches to estimating G_K. The advantage of GR increases with sample size. The Gaussian/Gaussian model provides the most favorable setup for CB relative to GR, because G_K^* is a consistent estimate of G. Yet, the risk for CB exceeds GR by at least 39%. Performance for CB is worse in other situations, and in the Gaussian/Lognormal case CB is even worse than PM. Importantly, the GR estimates are transformation equivariant and their performance in estimating

G_K is transform invariant. These are very attractive properties for a triple-goal estimate.

For the simulated models, all approaches produce the same ranks. The percentage of simulations in which the coordinates associated with the extreme parameters were identified range from 22% to 43% for $K = 20$, and from 13% to 36% for $K = 50$.

Using Monte-Carlo output

For simple or complex hierarchical models, Monte-Carlo methods produce samples of parameters from the posterior distribution. These can be used to compute optimal inferences relative to a loss function, thus avoiding some mathematical developments. However, efficient production of the inferences and and both mathematical and simulation-based evaluations need to catch up with this ability to produce posterior distributions and to cater to nonstandard goals.

To see how samples from the posterior distribution enable straightforward computation of most estimates, assume that a Monte-Carlo has been run, for example, MCMC via BUGS, see Carlin and Louis (2000); Spiegelhalter et al. (1995) or Gilks et al. (1996), and that we have available an $I \times K$ array in which rows ($i = 1, \ldots, I$) index draws from the chain (draws from the posterior distribution) and columns ($k = 1, \ldots, K$) index region. If MCMC were used, the chains have been stripped of burn-in values. We have available the θ_{ik} and can compute a wide variety of summaries, including:

- The posterior means: $\theta_k^{\mathrm{PM}} = \theta_{\cdot k}$.
- The posterior expected ranks: $\overline{R}_k = R_{\cdot k}$, where R_{ik} is the rank of coordinate "k" among the θ_{ik}, $k = 1, \ldots K$. The standard error of $\overline{R}_k$ is the sample standard error of the R_{ik}. The $\overline{R}_k$ are not integers.
- $\widehat{R}_k$ = ranks of the $\overline{R}_k$.
- The posterior modal ranks, $\widetilde{R}_k$: These are computed as the mode of the R_{ik}, $i = 1, \ldots I$. A very large "I" will be needed to produce valid modes.
- $\widehat{G}_K$: The EDF of $U_1, \ldots, U_K$. To find the U's, pool all "IK" MCMC output values (all θ_{ik}) and for $\ell = 1, \ldots, K$ let U_ℓ = the $\left[(\ell - .5)I\right]$th smallest value in the pooled output. This is the $\left[100(\ell - .5)/K\right]$th percentile in the pooled output.
- The posterior modal G_K, $\widetilde{G}_K$: This can be computed as the modal $G_K^{(i)}$, where $G_K^{(i)}$ is the EDF of the ith row of the Monte-Carlo values. A very, very large "I" will be needed to produce valid modes and more efficient approaches are being developed.

5 Correlated θ's and Unequal l_k

We use spatial models to identify issues associated with unequal l_k, correlated parameters and a multi-level hierarchy. See Besag et al. (1991) and Conlon and Louis (1999) for details and references. Consider disease prevalence or incidence data that are available as summary counts or rates for a defined region such as a county, district or census tract for a single time period. Denote an observed count by y_k, with $k = 1, \ldots, K$ indexing regions. The observation y_k is a count generated from a population base of size n_k and a sampling model (likelihood) parameterized by a baseline rate and a region-specific relative risk ψ_k.

With $\mathbf{X} = (\mathbf{X}_1, \ldots, \mathbf{X}_K)'$ the matrix of region-specific covariate vectors and m_k the expected count under a baseline model, we use the hierarchy:

$$\boldsymbol{\eta} \sim h(\boldsymbol{\eta}),$$
$$\boldsymbol{\psi} \sim g(\boldsymbol{\psi} \mid \mathbf{X}, \boldsymbol{\eta}),$$
$$Y_k \mid \psi_k \sim \text{ Poisson } (m_k \psi_k),$$
$$l_k(y_k \mid \psi_k) = \frac{1}{y_k!}(m_k\psi_k)^{y_k} e^{-m_k\psi_k}.$$

Conditional on $\mathbf{X}$ and $\boldsymbol{\eta}$, ψ_k has the log-linear model:

$$\log(\psi_k) = \mathbf{X}_k\boldsymbol{\alpha} + \theta_k + \phi_k.$$

The θ_k are iid random effects that produce an exchangeable model with extra-Poisson variation; the ϕ_k are random effects that induce spatial correlation. The vector $\boldsymbol{\eta}$ includes $\boldsymbol{\alpha}$ and parameters specifying the distribution of the θ's and the ϕs. We focus on inferences for region-specific, covariate-adjusted risks relative to the baseline m_k: $\rho_k = e^{\theta_k + \phi_k}$.

Priors

The pure *exchangeable model* sets, $\phi_k \equiv 0$, $\theta_1, \ldots, \theta_K$ iid $N(0, \tau^2)$. The pure *spatial correlation model* sets $\theta_k \equiv 0$ and builds a correlation model for the ϕs, generally with correlations that decrease with geographic distance. Hybrid models are possible. The conditional auto regressive (CAR) is relatively easily implemented using BUGS and has proven effective. It builds the full joint distribution from complete conditional distributions for each ϕ_k given all others. The ϕs are Gaussian with conditional mean for ϕ_k given all other ϕs a weighted average of these others. The conditional variance depends on the weights. For weights $w_{kj}, k, j = 1, \ldots, K$:

$$\phi_k \mid \phi_{j \neq k} \sim N(\bar{\phi}_k, V_k), \quad i = 1, \ldots, K,$$

$$\bar{\phi}_k = \frac{\sum_{j \neq k} w_{kj} \phi_j}{\sum_{j \neq k} w_{kj}}$$

$$V_k = \frac{1}{\lambda \sum_{j \neq k} w_{kj}}.$$

The hyperparameter λ controls the strength of spatial similarity induced by the CAR prior; larger values of λ indicate stronger spatial correlations between neighboring regions. Setting $\lambda = \infty$ produces complete shrinkage to a common value; $\lambda = 0$ produces PM = ML. These situations are both special cases of the exchangeable model ($\tau^2 = 0$ and $\tau^2 = \infty$ respectively) and it may be difficult to choose between the two prior structures. Generally PMs from a CAR model will be more spread out than those from an exchangeable model.

The CAR weights can be very general so long as they are compatible with a valid joint distribution for the ϕs (in general, a nontrivial accomplishment!). In the *adjacency model* the weights are $w_{kj} = 1$ if areas k and j share a common boundary; $w_{kj} = 0$ otherwise.

Lip cancer data analysis

The data set, from International Agency for Research on Cancer (1985), includes information for the 56 counties of Scotland pooled over the years 1975–1980. The data set includes observed and expected male lip cancer cases, the male population-years-at-risk, a covariate measuring the fraction of the population engaged in agriculture, fishing or forestry (AFF), and the location of each county expressed as a list of adjacent counties. Expected cases are based on the male population count and the age distribution in each county, using internal standardization.

We report on the pure exchangeable and pure CAR/adjacency models:

$$\log(\psi_k) = \alpha_0 + \alpha_1 \mathrm{AFF}_k + \theta_k,$$

$$\log(\psi_k) = \qquad \alpha_1^* \mathrm{AFF}_k + \phi_k,$$

both implemented by BUGS. The CAR model includes a nonzero intercept and so $\alpha_0^* = 0$.

5.0.3 Posterior means

Figures 1 and 2 display the ML and PM estimates of relative risk and the SD of the ML for the exchangeable and CAR models respectively. In both figures, when the ML = 0 the estimated SD = 0. In Figure 1 shrinkage towards the overall value is more pronounced for regions with relatively unstable MLs than for regions with relatively stable MLs. A comparison of Figures 1 and 2 shows that the PMs from the CAR model are more spread out than are those for the exchangeable model. The CAR model preserves more of the local signal. Also, note that for both the exchangeable and CAR models the lines between the ML and PM axes cross, indicating that, unlike for the $l_k \equiv l$ model, ranks based on PMs are different from those based on MLs.

The regions with the 5th and 6th largest ML estimates provide a good comparison of the operation of exchangeable and CAR models, essentially of the influence of spatial correlation. These regions have approximately the same ML estimates (2.37 and 2.34 respectively) with approximately the same standard deviation. In Figure 1 their PMs are similar and are moved substantially towards 1. In Figure 2 their PMs are quite different, due to the neighbor structure. One region has 6 neighbors and a PM moved substantially towards 1 from the ML because the average of the ML estimates for its neighbors (1.69) is given considerable weight by the CAR prior. In contrast, the other region has only one neighbor and its ML = 1.41. The weight given by the CAR prior sufficiently reduces the variance to dampen the shrinkage produced by the exchangeable prior. The form of the prior has considerable influence on the posterior distribution and inferences computed from it.

5.0.4 GR and PM

Figures 3 and 4 compare GR, ML and PM estimates of relative risk for the exchangeable and CAR models respectively. The exchangeable and CAR priors produce very different GR estimates, but for each prior the spread

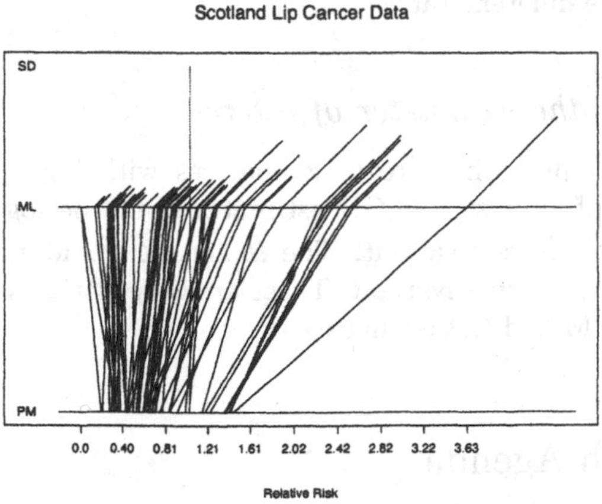

FIGURE 1. Maximum likelihood estimates (ML) of relative risk, their estimated standard errors (SD) and posterior mean (PM) relative risks computed from the exchangeable model. Note that when the MLE = 0, the estimated SD = 0.

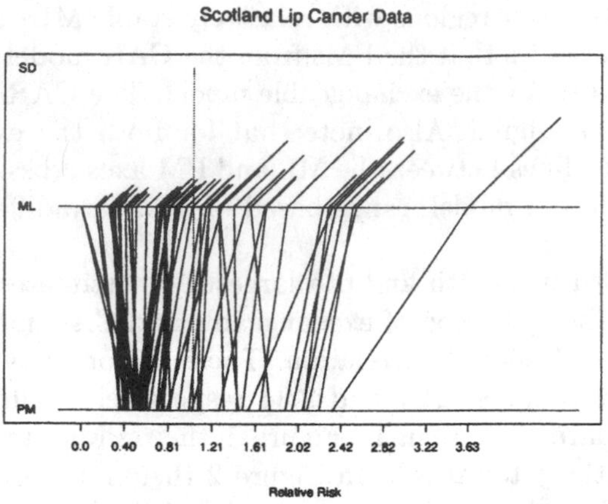

FIGURE 2. Maximum likelihood estimates (ML) of relative risk, their estimated standard errors (SD) and posterior mean (PM) relative risks computed from the CAR model. Note that when the MLE = 0, the estimated SD = 0.

of the GR estimates is between that for the ML and the PM. For each region, and in both the exchangeable and CAR models, the PM is closer to the prior mean than is the companion ML estimate. However, some of the GR estimates are further from the prior mean than is their companion ML estimate (the GR estimate is "stretched" away from the prior mean). Stretching is possible whenever region-specific sampling variances (controlled by the m_k in our application) have a large relative range. In this situation, a low-variance ML with rank below but not at K can have $\widehat{R} = K$ and GR estimate $\widehat{U}_K$. Finally, note that the lines on all graphs cross, indicating that, unlike the exchangeable case, the ML, PM, CB and GR can produce different ranks.

Transforming the parameter of interest

If we were to repeat the foregoing analyses with $\log(\rho)$ $(= \theta + \phi)$ as the parameter of interest, the GR estimates would be $\log(\widehat{\rho}^{\mathrm{GR}})$, they are monotone-transform equivariant. The $\overline{R}$ and $\widehat{R}$ would not change, they are monotone-transform invariant. These are very attractive properties not shared by the PM and CB estimates.

6 Research Agenda

The Bayesian formalism has many advantages in structuring an approach and in producing inferences with good Bayesian and frequentist properties. However, the approach must be used cautiously and models must be

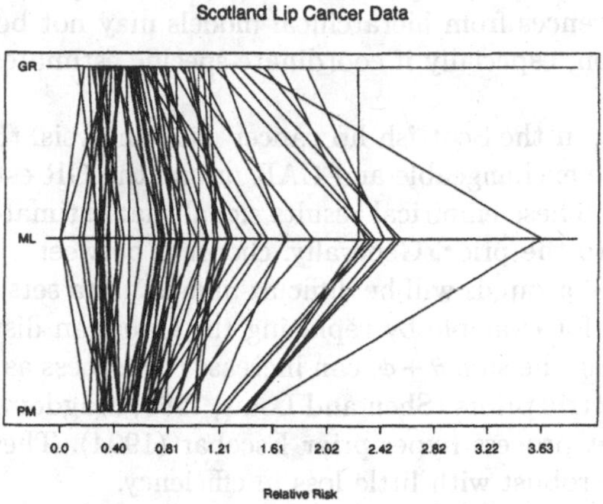

FIGURE 3. GR, ML, and PM estimates of relative risk for the exchangeable model.

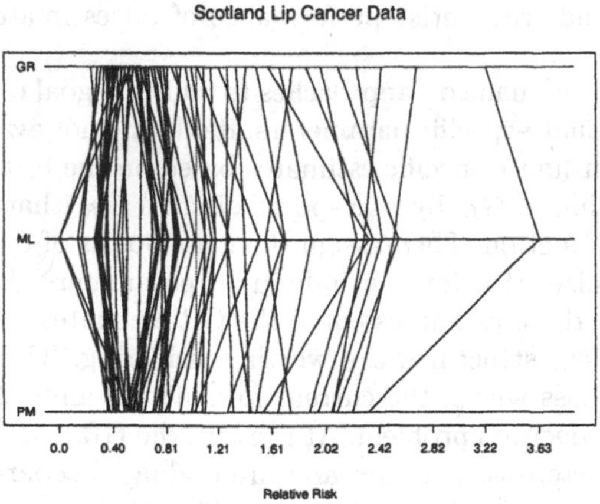

FIGURE 4. GR, ML, and PM estimates of relative risk for the CAR model.

sufficiently flexible to incorporate important stochastic features. Practical models must be robust. In addition to possible nonrobustness to misspecification of the sampling likelihood (a potential problem shared by all methods), inferences from hierarchical models may not be robust to prior mis-specification, especially if coordinate-specific estimates have relatively high variance.

For example, in the Scottish lip cancer data analysis, the PMs are very different for the exchangeable and CAR priors; the GR estimates are even more sensitive. These empirical results signal that estimates of G_K will be very sensitive to the prior. Generally, choosing between the two priors on purely empirical grounds will be difficult in most data sets. Broadening the class of priors, for example by replacing the Gaussian distribution by the t-family, or using the sum $\theta + \phi$, can increase robustness as can use of semi- and nonparametric priors (Shen and Louis, 1999; Magder and Zeger, 1996) or the Dirichlet process hyper-prior Escobar (1994). These last have the potential to be robust with little loss in efficiency.

Asymptotic analysis is standard, if information in the sampling distributions (the l_k) increases and K is constant or increases slowly, but "K-asymptotics" are the more interesting and relevant. Few results are available, especially for nonparametric priors.

S&L have made good progress in analyzing properties for two-stage, exchangeable models using SEL for the estimand of interest. They depend on large sample theory for order statistics for iid variables. Similar theory is not available for the noniid case and so nonexchangeable models require a mix of mathematical and simulation-based evaluations. Comparisons have not been made with estimates based on posterior modes or medians. Empirical Bayes and frequentist performance of all estimates needs further evaluation.

A start has been made on approaches to multiple-goal estimates for multivariate, coordinate-specific parameters. Ranks are not available in this situation, but coordinate-specific estimates based on the histogram are available. First, estimate G_K by a K-point discrete $\widehat{G}_K$ that optimizes some histogram loss function. Then assign the mass points of $\widehat{G}_K$ to coordinates so as to minimize SEL for the individual parameters. In the univariate parameter case these estimates equal the GR estimates.

Alternative loss structures are worth considering. These include absolute and "0/1" loss within the current formulation and reformulation as a multi-attribute decision problem. At present, the GR and CB estimates are developed in a step-wise fashion and are evaluated separately for each of the histogram, ranks and coordinate-specific parameters inferential goals. One reformulation uses a loss function which is a weighted sum of the losses in estimating G_K, the ranks and individual θ's. Preliminary analysis indicates that if the weight on SEL for estimating the individual θ's is 0, the GR estimates are optimal irrespective of the other two weights. Therefore, the GR estimates should emerge in the limit as the weight on the loss in estimating the individual θ's goes to 0, and they might be optimal for a range of weights.

Acknowledgments: Support provided by Grant 1R01-ES-07750 from the U.S. National Institute of Environmental Health Sciences. I thank the referees for their insightful comments and suggestions.

7 References

Besag, J., J.C York, and A. Molliè (1991). Bayesian image restoration, with two applications in spatial statistics (with discussion). *Ann. Inst. Statist. Math. 43*, 1–59.

Carlin, B.P. and T.A. Louis (2000). *Bayes and Empirical Bayes Methods for Data Analysis.* London: Chapman and Hall.

Conlon, E.M. and T.A Louis (1999). Addressing multiple goals in evaluating region-specific risk using Bayesian methods. In A. Lawson, A. Biggeri, D. Böhning, E. Lesaffre, J.-F. Viel, and R. Bertollini (Eds.), *Disease Mapping and Risk Assesment for Public Health*, Chapter 3, pp. 31–47. New York: Wiley.

Escobar, M.D. (1994). Estimating normal means with a Dirichlet process prior. *J. Amer. Statist. Assoc. 89*, 268–277.

Ghosh, M. (1992). Constrained Bayes estimates with applications. *J. Amer. Statist. Assoc. 87*, 533–540.

Gilks, W.R., S. Richardson, and D.J. Spiegelhalter (Eds.) (1996). *Markov Chain Monte Carlo in Practice.* London: Chapman and Hall.

International Agency for Research on Cancer (1985). *Scottish Cancer Atlas*, Volume 72 of *IARC Scientific Publications.* Lyon, France.

Laird, N.M. and T.A. Louis (1989). Bayes and empirical Bayes ranking methods. *J. Edu. Statist. 14*, 29–46.

Lindsay, B.G. (1983). The geometry of mixture likelihoods: a general theory. *Ann. Statist. 11*, 86–94.

Louis, T.A. (1984). Estimating a population of parameter values using Bayes and empirical Bayes methods. *J. Amer. Statist. Assoc. 79*, 393–398.

Magder, L.S. and S. Zeger (1996). A smooth nonparametric estimate of a mixing distribution using mixtures of Gaussians. *J. Amer. Statist. Assoc. 912*, 1141–1151.

Robbins, H (1983). Some thoughts on empirical Bayes estimation. *Ann. Statist. 1*, 713–723.

Shen, W. and T.A. Louis (1998). Triple-goal estimates in two-stage, hierarchical models. *J. Roy. Statist. Soc. Ser. B 60*, 455–471.

Shen, W. and T.A. Louis (1999). Empirical Bayes estimation via the smoothing by roughening approach. *J. Comput. Graph. 8*, 800–823.

Spiegelhalter, D.J., A. Thomas, N. Best, and W.R. Gilks (1995). BUGS: Bayesian inference using Gibbs sampling, version 0.50. Technical report, Medical Research Council Biostatistics Unit, Institute of Public Health, Cambridge University.

Teicher, H. (1961). Identifiability of mixtures. *Ann. Math. Statist. 32*, 244–248.

2

Empirical Bayes Estimators and EM Algorithms in One-Way Analysis of Variance Situations

I. Guttman

ABSTRACT The thesis of this paper is two-fold, namely that when there is a choice of working with a joint posterior or a marginal posterior, there may be an optimal choice of which posterior to use, so that care must be taken as to which posterior to work with, and, secondly, if using the EM algorithm for producing estimators, care must be taken with the choice of parameters to be declared "missing", for the wrong choice could lead to inconsistent estimators and/or estimators with poor mean square error behavior. These two propositions are exhibited for well defined hierarchical models in this paper. The indication that a choice of which posteriors to work with should be considered, was first made by O'Hagan (1976), and this is further discussed in Sun et al. (1996).

1 The Model

The hierarchical model of this section encompasses the one-way analysis of variance situation for the case that the error variance, denoted by σ_y^2, is assumed known. Since σ_y^2 is known, we do not need replications within any of the groups, and we work with the following hierarchical model (notation used is: "$\sim$" for "distributed as"; "IDN" for "independent normal...", and "IIDN" for "identically and independent normal...")

Stage 1: Given θ_j, σ_y^2, $y_j \sim \text{IDN}(\theta_j, \sigma_y^2)$, for $j = 1, \ldots, m$.

Stage 2: Given μ, σ_θ^2, $\theta_j \sim \text{IIDN}(\mu, \sigma_\theta^2)$, for $j = 1, \ldots, m$.

Stage 3: Given λ, ν, then

(i) $\sigma_\theta^2 \sim \nu\lambda \times (\chi_\nu^2)^{-1}$, independently of

(ii) μ, which has the uniform distribution over R^1.

2 Some Analysis Derived From the Hierarchical Model

Using Stages 1 and 2 of Section 1's hierarchical model, we have that

$$\begin{aligned} p(\mathbf{y}, \boldsymbol{\theta} \mid \mu, \sigma_\theta^2; \sigma_y^2) &= p(\mathbf{y} \mid \boldsymbol{\theta}; \mu, \sigma_\theta^2; \sigma_y^2) p(\boldsymbol{\theta} \mid \mu, \sigma_\theta^2; \sigma_y^2), \\ &= p(\mathbf{y} \mid \boldsymbol{\theta}; \sigma_y^2) p(\boldsymbol{\theta} \mid \mu, \sigma_\theta^2). \end{aligned} \tag{1}$$

Using Stages 1 and 2, feeding in the above factors and completing squares in θ_j's, it is easy to verify that

$$\begin{aligned} p(\mathbf{y}, \boldsymbol{\theta} \mid \mu, \sigma_\theta^2; \sigma_y^2) \propto (\sigma_\theta^2)^{-m/2} \exp -\frac{1}{2\tau^2} \sum_{j=1}^{m} (\theta_j - \theta_j^*)^2 \\ \times \exp -\frac{1}{2(\sigma_y^2 + \sigma_\theta^2)} \sum_{j=1}^{m} (y_j - \mu)^2, \end{aligned} \tag{2a}$$

where

$$\tau^2 = \left(\frac{1}{\sigma_y^2} + \frac{1}{\sigma_\theta^2}\right)^{-1} = \frac{\sigma_\theta^2 \sigma_y^2}{\sigma_\theta^2 + \sigma_y^2}, \tag{2b}$$

and

$$\theta_j^* = \left[(\sigma_y^2)^{-1} y_j + (\sigma_\theta^2)^{-1} \mu\right] / \left[(\sigma_y^2)^{-1} + (\sigma_\theta^2)^{-1}\right], \tag{2c}$$

$$= \mu + \frac{(\sigma_y^2)^{-1}}{(\sigma_y^2)^{-1} + (\sigma_\theta^2)^{-1}} (y_j - \mu). \tag{2d}$$

Both forms of θ_j^* are interesting—(2c) says "information" about θ_j supplied by θ_j^* is a weighted combination of sample (y_j) and prior (μ) information, with weights that are proportional to "precision"—specifically $(\sigma_y^2)^{-1}$ for y_j and $(\sigma_\theta^2)^{-1}$ for μ. The version in (2d) is also interesting in that it represents an updating of information about θ_j from the prior information, by adding to the prior mean μ, a shrinkage of the observed y_j to the prior information, with shrinkage coefficient which is exactly equal to the Kalman-filter matrix for this situation. It turns out that θ_j^* is the mean of the conditional posterior of θ_j, conditional on μ, σ_θ^2 and of course $\mathbf{y}$ and σ_y^2.

To see the latter, we first note that

$$p(\boldsymbol{\theta} \mid \mathbf{y}; \mu, \sigma_\theta^2; \sigma_y^2) \propto p(\mathbf{y}, \boldsymbol{\theta} \mid \mu, \sigma_\theta^2; \sigma_y^2), \tag{3}$$

so that on consulting (2), we immediately have, since $\mathbf{y}$, μ, σ_θ^2, σ_y^2 are given, that

$$p(\boldsymbol{\theta} \mid \mathbf{y}; \mu, \sigma_\theta^2; \sigma_y^2) = \prod_{j=1}^{m} N'_{\theta_j}(\theta_j^*, \tau^2), \tag{4}$$

where, in general, we have used the notation $N'_x(\delta, \xi)$ to stand for the probability density function (pdf) of a normal variable x, whose mean and variance are δ and ξ, respectively. For the record, then,

$$\theta_j \mid \mathbf{y}; \mu, \sigma_\theta^2; \sigma_y^2 \sim \text{IDN}(\theta_j^*, \tau^2), \tag{5}$$

so that

$$E(\boldsymbol{\theta} \mid \mathbf{y}; \mu, \sigma_\theta^2; \sigma_y^2) = \boldsymbol{\theta}^* = (\theta_1^*, \ldots, \theta_j^*, \ldots, \theta_m^*)', \tag{6}$$

and

$$V(\boldsymbol{\theta} \mid \mathbf{y}; \mu, \sigma_\theta^2; \sigma_y^2) = \tau^2 I_m,$$

with τ^2 given by (2a) and θ_j^* given by (2b)–(2c).

Now returning to (2), it is easy to see that on integrating out $\boldsymbol{\theta}$ we immediately have that the marginal pdf of the y_j's, given $(\mu, \sigma_\theta^2, \sigma_y^2)$ is

$$p(\mathbf{y} \mid \mu, \sigma_\theta^2, \sigma_y^2) = \prod_{j=1}^{m} N'_{y_j}(\mu, \sigma_y^2 + \sigma_\theta^2). \tag{7}$$

Note here that we have the IID structure underlying the discussion for "maximum likelihood procedures". This need not happen always but it does happen here, so that, in particular, estimates μ and σ_θ^2 produced by maximizing (7) over (μ, σ_θ^2) are consistent as $m \to \infty$. These estimates are provided by the theorem stated below, and are referred to as Empirical Bayes estimators (EB estimators).

Theorem 1. *The* EB *estimators of* (μ, σ_θ^2), *say* $(\widehat{\mu}, \widehat{\sigma}_\theta^2)$ *are given by*

$$\widehat{\mu} = \overline{y}, \quad \widehat{\sigma}_\theta^2 = \left[m^{-1} \sum_{1}^{m} (y_j - \overline{y})^2 - \sigma_y^2 \right]_+, \tag{8}$$

$$\{[a]_+ = a \quad \text{if} \quad a > 0 \text{ and zero otherwise}\}.$$

Further, if we introduce

$$\widehat{\theta}_j = E(\theta_j \mid \mathbf{y}, \widehat{\mu}, \widehat{\sigma}_\theta^2; \sigma_y^2) = \widehat{\tau}^2 \left[(\sigma_y^2)^{-1} y_j + (\widehat{\sigma}_\theta^2)^{-1} \widehat{\mu} \right], \tag{9}$$

where $\widehat{\tau}^2 = \left[(\sigma_y^2)^{-1} + (\widehat{\sigma}_\theta^2)^{-1} \right]^{-1}$, *then* $\widehat{\mu} = m^{-1} \sum_{j=1}^{m} \widehat{\theta}_j = \overline{y}$, *and if* $\widehat{\sigma}_\theta^2 > 0$, *we have that* $\widehat{\sigma}_\theta^2$ *satisfies*

$$\widehat{\sigma}_\theta^2 = m^{-1} \sum_{j=1}^{m} (\widehat{\theta}_j - \widehat{\mu})^2 + \widehat{\tau}^2. \tag{10}$$

Proof. See Appendix A1 of Guttman (1998). □

We remark that $\widehat{\theta}_j$ is referred to as the EB estimator of θ_j [see for example, Efron and Morris (1976)]. From (9), we can write $\widehat{\theta}_j$ in the form

$$\widehat{\theta}_j = \widehat{\mu} + \frac{\widehat{\sigma}_\theta^2}{\widehat{\sigma}_\theta^2 + \sigma_y^2} (y_j - \widehat{\mu}), \tag{11}$$

so that for the case $\widehat{\sigma}_\theta^2 > 0$, we find [see Appendix A1 of Guttman (1998)] that $\widehat{\sigma}_\theta^2$ satisfies

$$\widehat{\sigma}_\theta^2 = \left(\frac{\widehat{\sigma}_\theta^2}{\widehat{\sigma}_\theta^2 + \sigma_y^2}\right)^2 m^{-1}\sum_1^m (y_j - \widehat{\mu})^2 + \widehat{\tau}^2. \tag{12}$$

We will return in short order to 2, but now turn to the 3-stage model of Section 1 to determine various posteriors based on all the information available from the 3 stages. Indeed, we first note that we may write

$$p(\mathbf{y}, \boldsymbol{\theta}, \mu, \sigma_\theta^2 \mid \sigma_y^2) \propto (\sigma_\theta^2)^{-(m+v/2+1)} \times \exp\left[-\frac{1}{2\sigma_y^2}\sum_{j=1}^m (y_j - \theta_j)^2 - \frac{1}{2\sigma_\theta^2}\sum_{j=1}^m (\theta_j - \mu)^2 - \frac{\nu\lambda}{2\sigma_\theta^2}\right]. \tag{13}$$

Since

$$p(\boldsymbol{\theta}, \mu, \sigma_\theta^2 \mid \mathbf{y}; \sigma_y^2) \propto p(\mathbf{y}, \boldsymbol{\theta}, \mu, \sigma_\theta^2 \mid \sigma_y^2),$$

we immediately have that the expression "to the right" of the sign of proportionality in (13) is the form of the posterior of $(\boldsymbol{\theta}, \mu, \sigma_\theta^2)$ and we call this, of course, the joint posterior of $(\boldsymbol{\theta}, \mu, \sigma_\theta^2)$, given the data $\mathbf{y}$ and σ_y^2. We have the following theorem.

Theorem 2. *The mode of the joint posterior $p(\boldsymbol{\theta}, \mu, \sigma_\theta^2 \mid \mathbf{y}; \sigma_y^2)$ has coordinates $(\widetilde{\boldsymbol{\theta}}, \widetilde{\mu}, \widetilde{\sigma}_\theta^2)$ which are such that*

$$\widetilde{\theta}_j = \widetilde{\tau}^2\left[(\sigma_y^2)^{-1} y_j + (\widetilde{\sigma}_\theta^2)^{-1}\widetilde{\mu}\right], \quad \widetilde{\tau}^2 = \left[(\sigma_y^2)^{-1} + (\widetilde{\sigma}_\theta^2)^{-1}\right]^{-1}, \tag{14a}$$

$$\widetilde{\mu} = m^{-1}\sum_{j=1}^m \widetilde{\theta}_j = \overline{y}, \tag{14b}$$

$$\widetilde{\sigma}_\theta^2 = (m + \nu + 2)^{-1}\left[\sum_{j=1}^m (\widetilde{\theta}_j - \widetilde{\mu})^2 + \nu\lambda\right]. \tag{14c}$$

Proof. See Appendix A2 of Sun et al. (1996). Note that in view of (14b), $\widetilde{\theta}_j$ is a weighted combination of y_j and $\overline{y}$. □

We remark that the estimators of Theorem 2.2 given by (14a), (14b) and (14c) are referred to in the literature as Lindley-Smith estimators—see Sun et al. (1996).

Now if concern lies solely in the hierarchical parameters (μ, σ_θ^2), then the marginal posterior of (μ, σ_θ^2), given $\mathbf{y}$ and σ_y^2 is readily obtained by integrating out $\boldsymbol{\theta}$ in (13), and the question of its modes and comparison with the mode of the joint posterior supplied by (14a)–(14c) is of interest. We first have the following theorem.

Theorem 3. *The mode of the marginal posterior density* $p(\mu, \sigma_\theta^2 \mid \mathbf{y}; \sigma_y^2)$ *is given by the coordinates* $(\tilde{\tilde{\mu}}, \tilde{\tilde{\sigma}}_\theta^2)$, *which are such that*

$$\tilde{\tilde{\mu}} = \bar{y}, \quad \tilde{\tilde{\sigma}}_\theta^2 = (m + \nu + 2)^{-1}\left[\sum_{j=1}^{m} (\tilde{\tilde{\theta}}_j - \tilde{\tilde{\mu}})^2 + m\tilde{\tilde{\tau}}^2 + \nu\lambda\right], \tag{15}$$

where we have introduced

$$\tilde{\tilde{\theta}}_j = \tilde{\tilde{\tau}}^2 \left[(\sigma_y^2)^{-1} y_j + (\tilde{\tilde{\sigma}}_\theta^2)^{-1}\tilde{\tilde{\mu}}\right], \quad \tilde{\tilde{\tau}}^2 = \left[(\sigma_y^2)^{-1} + (\tilde{\tilde{\sigma}}_\theta^2)^{-1}\right]^{-1}. \tag{16}$$

Proof. [1] See Appendix A3 of Guttman (1998). □

We summarize as follows: after a bit of algebra we may write $\tilde{\sigma}_\theta^2$ of (14c) and $\tilde{\tilde{\sigma}}_\theta^2$ of (15) as, respectively

$$\tilde{\sigma}_\theta^2 = (m + \nu + 2)^{-1}\left\{\left[\frac{\tilde{\sigma}_\theta^2}{\tilde{\sigma}_\theta^2 + \sigma_y^2}\right]^2 \sum_{j=1}^{m} (y_j - \bar{y})^2 + \nu\lambda\right\}, \tag{17a}$$

and

$$\tilde{\tilde{\sigma}}_\theta^2 = (m + \nu + 2)^{-1}\left\{\left[\frac{\tilde{\tilde{\sigma}}_\theta^2}{\tilde{\tilde{\sigma}}_\theta^2 + \sigma_y^2}\right]^2 \sum_{j=1}^{m} (y_j - \bar{y})^2 + \nu\lambda + m\tilde{\tilde{\tau}}^2\right\}. \tag{17b}$$

Because of the form of $\tilde{\tilde{\sigma}}_\theta^2$ that calls for an addition of the term $m\tilde{\tilde{\tau}}^2$ in $\tilde{\tilde{\sigma}}_\theta^2$ when compared to the form of $\tilde{\sigma}_\theta^2$, a suspicion may arise at this point as to whether or not $\tilde{\sigma}_\theta^2 < \tilde{\tilde{\sigma}}_\theta^2$. We turn now to this question and its implications in the following section.

3 Large Sample Properties of the Estimates: The Collapsing Effect

In this section, we will inquire into "large sample" properties of the estimators derived in Section 2. Accordingly, the approximations we discuss in this section hold "almost surely" with respect to the measure over the sample space of the y_j's, given at (7).

Indeed, letting

$$T^2 = \sum_{j=1}^{m} (y_j - \bar{y})^2, \tag{18a}$$

---

[1]Parathentically, the notation we use may be remembered by the mnemonic that alphabetically "j" < "m" so that one wiggle means an estimator derived from the j=joint posterior and two wiggles (1 < 2) means an estimator derived from the marginal posterior.

we will inquire about the estimators of Section 2 when $m \to \infty$ in such a way that ν, λ, and $m^{-1}T^2$ are fixed. (We note that the requirement that $m^{-1}T^2$ is fixed as $m \to \infty$ merely says that the average information about the "between populations" effect remains comparable as $m \to \infty$.)

We first note that the above conditions for m large imply that we may write (17a) and (17b) in approximate form as

$$\widetilde{\sigma}_\theta^2 \simeq \left(\frac{\widetilde{\sigma}_\theta^2}{\widetilde{\sigma}_\theta^2 + \sigma_y^2}\right)^2 \frac{T^2}{m}, \tag{18b}$$

and

$$\widetilde{\widetilde{\sigma}}_\theta^2 \simeq \left(\frac{\widetilde{\widetilde{\sigma}}_\theta^2}{\widetilde{\widetilde{\sigma}}_\theta^2 + \sigma_y^2}\right)^2 \frac{T^2}{m} + \widetilde{\widetilde{\tau}}^2, \tag{18c}$$

with

$$\widetilde{\widetilde{\tau}}^2 = \left((\widetilde{\widetilde{\sigma}}_\theta^2)^{-1} + (\sigma_y^2)^{-1}\right)^{-1} = \sigma_y^2 \widetilde{\widetilde{\sigma}}_\theta^2 / [\sigma_y^2 + \widetilde{\widetilde{\sigma}}_\theta^2]. \tag{18d}$$

We note here that since $\widetilde{\mu} = \widetilde{\widetilde{\mu}} = \overline{y}$, $\widetilde{\theta}_j$ and $\widetilde{\widetilde{\theta}}_j$ are both linear functions of y_j and $\overline{y}$. We can now state the following important theorem.

Theorem 4. *Under the conditions cited above,*

(*a*) $\widetilde{\widetilde{\sigma}}_\theta^2 \simeq (T^2/m - \sigma_y^2)_+ = \widehat{\sigma}_\theta^2$ *that is, the* marginal *model estimate* $\widetilde{\widetilde{\sigma}}_\theta^2$ *of* (15) *and/or* (17b) *is asymptotically equivalent to* $\widehat{\sigma}_\theta^2$, *the* EB *estimate of* σ_θ^2, *so that* (*see remark below* (7)) $\widetilde{\widetilde{\sigma}}_\theta^2$ *is consistent for* σ_θ^2.

(*b*) *The joint model estimate of* σ_θ^2 *given in* (14c) *and/or* (17a) *satisfies*

$$\widetilde{\sigma}_\theta^2 \simeq \frac{T^2}{2m} - \sigma_y^2 + \sqrt{\left(\frac{T^2}{2m}\right)^2 - \frac{T^2}{m}\sigma_y^2} \quad \text{if} \quad T^2/m > 4\sigma_y^2, \tag{19}$$

$\widetilde{\sigma}_\theta^2 = 0$ *otherwise.*

Proof. See Appendix B1 of Guttman (1998). □

We may use Theorem 4 directly, for since $\sqrt{(T^2/2m)^2 - \sigma_y^2 T^2/m} < T^2/2m$, then, *interestingly*, for the case $T^2/m > 4\sigma_y^2$, we now have

$$\widetilde{\sigma}_\theta^2 < \frac{T^2}{2m} - \sigma_y^2 + \frac{T^2}{2m} = \frac{T^2}{m} - \sigma_y^2 = \widehat{\sigma}_\theta^2 \simeq \widetilde{\widetilde{\sigma}}_\theta^2, \tag{20}$$

which is to say (see (19)) that $\widetilde{\sigma}_\theta^2$ is underestimating σ_θ^2 (at least for large m), since $\widehat{\sigma}_\theta^2$ (and its asymptotic equivalent $\widetilde{\widetilde{\sigma}}_\theta^2$) is consistent for σ_θ^2. This important result is called the *collapsing phenomenon*, a term first coined by Tom Leonard, and incorporated in Sun et al. (1996). Note then, that when modes are of interest (and they often are, if only because then integrations may be avoided), then the choice of which posterior to work with is of crucial importance.

In turn, the above raises the following question: Recall from Section 2 that $\boldsymbol{\theta} \mid \mathbf{y}; \mu, \sigma_\theta^2; \sigma_y^2 \sim N(\boldsymbol{\theta}^*, \tau^2 I_m)$, with τ^2 given by (2a) and $\boldsymbol{\theta}^*$ of (6) is such that its jth component is given by (2c)–(2d). The interesting question that now arises is, which of

$$\widetilde{\theta}_j = E(\theta_j \mid \mathbf{y}; \widetilde{\mu}, \widetilde{\sigma}_\theta^2; \sigma_y^2), \tag{21a}$$

$$\widetilde{\widetilde{\theta}}_j = E(\theta_j \mid \mathbf{y}; \widetilde{\widetilde{\mu}}, \widetilde{\widetilde{\sigma}}_\theta^2; \sigma_y^2), \tag{21b}$$

is preferred? The answer is that using a mean square criterion, $\widetilde{\widetilde{\boldsymbol{\theta}}}$ is preferred to $\widetilde{\boldsymbol{\theta}}$, and we now discuss this in the next section.

4 Mean Square Considerations

In this section, we will inquire into the large sample properties of the estimators of the θ_j's developed in the previous sections. In particular, then, we will assume that the distribution of the y_j's are as given by Stage 1 of the model of Section 1, that is, $\mathbf{y} \mid \boldsymbol{\theta}; \sigma_y^2 \sim N(\boldsymbol{\theta}, \sigma_y^2 I_m)$.

We will make use of the following

Definition 1. Let $Q_m = m^{-1} \sum_1^m (\theta_j - \bar{\theta})^2$, $\bar{\theta} = m^{-1} \sum_1^m \theta_j$, and assume that $\theta_1, \theta_2, \ldots,$ are given, and such that

$$Q = \lim_{m\to\infty} \frac{1}{m} \sum_{j=1}^{m} (\theta_j - \bar{\theta})^2, \tag{22}$$

exists.

Now using the assumption that $\mathbf{y} \mid \boldsymbol{\theta}; \sigma_y^2 \sim N(\boldsymbol{\theta}, \sigma_y^2 I_m)$, we may state the following theorem:

Theorem 5. *Under first stage assumptions (see Section 3),*

$$m^{-1}T^2 = m^{-1} \sum_{J=1}^{m} (y_j - \bar{y})^2 \to Q + \sigma_y^2, \tag{23}$$

as $m \to \infty$.

Proof. See Appendix C1 of Guttman (1998). □

The above result (23) is not unexpected, for after all, viewed in the one-way analysis of variance context, T^2 is a between sum of squares, well known to reflect, apart from "error", information about the variation of the θ_j's.

We also need for this section the following definition.

Definition 2. The mean square error (MSE) for estimators g_j of θ_j, say $\mathbf{g} = (g_1, \ldots, g_j, \ldots, g_m)'$ for $\boldsymbol{\theta} = (\theta_1, \ldots, \theta_j, \ldots, \theta_m)'$, is given by

$$\text{MSE}(\mathbf{g}) = m^{-1} E\{(\mathbf{g} - \boldsymbol{\theta})'(\mathbf{g} - \boldsymbol{\theta})\}, \tag{24}$$

where the expectation is taken with respect to the (sampling) distribution of $\mathbf{g}$, given $\boldsymbol{\theta}$.

Now as already noted, $\tilde{\theta}_j$ and $\tilde{\tilde{\theta}}_j$ are linear functions of y_j and $\bar{y}$. So consider the estimator of θ_j, say g_j, which is such that

$$g_j(a) = y_j - a(y_j - \bar{y}), \tag{25a}$$

or, equivalently,

$$g_j(a) = (1 - a)y_j + a\bar{y}. \tag{25b}$$

The form (25a) is that of a "shrinkage to the mean" estimator of θ_j, while the form (25b) is a weighted combination of y_j and $\bar{y}$. It is straightforward to determine that

$$E[g_j(a)] = \theta_j + b_j, \tag{26a}$$

where the bias of $g_j(a)$ is given by

$$b_j = -a(\theta_j - \bar{\theta}), \tag{26b}$$

and, further, that the variance of $g_j(a)$ is

$$\text{Var}[g_j(a)] = \sigma_y^2[(1 - a + a/m)^2 + a^2(m - 1)/m^2], \tag{27a}$$

so that for large m,

$$\text{Var}[g_j(a)] \cong \sigma_y^2(a - 1)^2. \tag{27b}$$

Now using Definition 2, it is easy to verify that

$$\begin{aligned} \text{MSE}(\mathbf{g}) &= m^{-1} \sum_1^m E(g_j - \theta_j)^2, \\ &= a^2 Q_m + \sigma_y^2[(1 - a + a/m)^2 + a^2(m - 1)/m^2], \end{aligned} \tag{28}$$

so that as $m \to \infty$, we have that

$$\text{MSE}(\mathbf{g}) \to a^2 Q + (a - 1)^2 \sigma_y^2, \tag{29}$$

with $a^2 Q$ representing the bias contributions and $(a - 1)^2 \sigma_y^2$ representing the variance contributions to the limiting MSE($\mathbf{g}$). It turns out, as is easily verified that

$$\min_a [a^2 Q + (a - 1)^2 \sigma_y^2], \tag{30}$$

occurs at a^*, where

$$a^* = \sigma_y^2 / (Q + \sigma_y^2), \tag{31}$$

and indeed the minimum value of the limiting MSE(**g**) of (29) is

$$(a^*)^2 Q + (a^* - 1)^2 \sigma_y^2 = \left[Q^{-1} + (\sigma_y^2)^{-1}\right]^{-1} = \tau^{2*}, \tag{32}$$

that is,

$$\min_a \left[\text{limiting MSE}(\boldsymbol{\theta})\right] = \tau^{2*}. \tag{33}$$

At this point, we make the following remarks:

(1) If $a = 0$, then $g_j(0) = y_j$ or $\mathbf{g} = \mathbf{y}$ and $\mathbf{y}$ is unbiased for $\boldsymbol{\theta}$ so that

$$\text{MSE}(\mathbf{g}) = \text{MSE}(\mathbf{y}) = \sigma_y^2.$$

(2) If we assign "a" the value

$$\sigma_y^2/(\widehat{\sigma}_\theta^2 + \sigma_y^2) = \widehat{a}, \tag{34}$$

then for the case $\widehat{\sigma}_\theta^2 > 0$, we have

$$\begin{aligned} g_j(\widehat{a}) &= y_j - \widehat{a}(y_j - \overline{y}), \\ &= (1 - \widehat{a})y_j + \widehat{a}\overline{y}, \end{aligned} \tag{35}$$

and after some algebra involving (34), we find

$$g_j(\widehat{a}) = \widehat{\theta}_j = \widehat{\tau}^2 \left[(\sigma_y^2)^{-1} y_j + (\widehat{\sigma}_\theta^2)^{-1}\overline{y}\right], \tag{36}$$

with

$$\widehat{\tau}^2 = \left[(\sigma_y^2)^{-1} + (\widehat{\sigma}_\theta^2)^{-1}\right]^{-1}.$$

Now for large m, $\widetilde{\widetilde{\sigma}}_\theta^2 \simeq \widehat{\sigma}_\theta^2$ so that

$$g_j(\widehat{a}) \simeq \widehat{\theta}_j \simeq \widetilde{\widetilde{\theta}}_j, \tag{37}$$

and of course,

$$\widehat{a} \simeq \widetilde{\widetilde{a}} = \sigma_y^2(\widehat{\sigma}_\theta^2 + \sigma_y^2) \simeq \sigma_y^2(\widetilde{\widetilde{\sigma}}_\theta^2 + \sigma_y^2). \tag{38}$$

But from (8), we have that when $\widehat{\sigma}_\theta^2 > 0$,

$$\widetilde{\widetilde{\sigma}}_\theta^2 + \sigma_y^2 \simeq \widehat{\sigma}_\theta^2 + \sigma_y^2 = \frac{T^2}{m} \to Q + \sigma_y^2, \tag{39a}$$

from Theorem 5. Hence, we have that, in probability, $\widehat{a}$ (and $\widetilde{\widetilde{a}}$) is such that

$$\widehat{a} = \sigma_y^2(\widehat{\sigma}_\theta^2 + \sigma_y^2) \to \sigma_y^2(Q + \sigma_y^2) = a^*, \tag{39b}$$

as $m \to \infty$. Thus

$$\text{MSE}(\widehat{\boldsymbol{\theta}}) \cong \text{MSE}(\widetilde{\widetilde{\boldsymbol{\theta}}}) \to \text{MSE}\big(\mathbf{g}(a^*)\big) = \tau^{2*},$$

that is, in words, $\widetilde{\widetilde{\boldsymbol{\theta}}}$ has the minimum limiting MSE, namely τ^{2*}.

Now when we assign the value for "a" equal to

$$\widetilde{a} = \sigma_y^2(\widetilde{\sigma}_\theta^2 + \sigma_y^2), \tag{40}$$

we immediately find that

$$g_j(\widetilde{a}) = \widetilde{\theta}_j, \tag{41}$$

the Lindley–Smith estimator of θ_j given in Theorem 2. But recall that Theorem 4 says that $\widetilde{\sigma}_\theta^2$ may be written as

$$\widetilde{\sigma}_\theta^2 = \begin{cases} T^2/2m - \sigma_y^2 + \sqrt{(T^2/2m)^2 - T^2\sigma^2 y/m} & \text{if } T^2/m > 4\sigma_y^2, \\ 0 & \text{otherwise,} \end{cases} \tag{42}$$

so that since $T^2/m \to Q + \sigma_y^2$ as $m \to \infty$, we have, as is easily seen, that

$$\widetilde{\sigma}_\theta^2 \to \widetilde{\sigma}_\theta^{2^{(0)}} = \begin{cases} (Q - \sigma_y^2)/2 + \frac{1}{2}\sqrt{(Q + \sigma_y^2)(Q - 3\sigma_y^2)} & \text{if } Q > 3\sigma_y^2, \\ 0 & \text{otherwise.} \end{cases} \tag{43}$$

For large m, using (40) we have that

$$\widetilde{a} \simeq \sigma_y^2(\widetilde{\sigma}^{2^{(0)}} + \sigma_y^2), \tag{44}$$

so that (see (29)) we have that

$$\text{MSE}(\widetilde{\boldsymbol{\theta}}) \simeq \left(\frac{\sigma_y^2}{\widetilde{\sigma}_\theta^{2^{(0)}} + \sigma_y^2}\right)^2 Q + \left(\frac{\widetilde{\sigma}_\theta^{2^{(0)}}}{\widetilde{\sigma}_\theta^{2^{(0)}} + \sigma_y^2}\right)^2 \sigma_y^2, \tag{45}$$

for $Q > 3\sigma_y^2$. Hence, using the definition of minimum, we have that

$$\text{MSE}(\widetilde{\boldsymbol{\theta}}) > \text{MSE}(\widetilde{\widetilde{\boldsymbol{\theta}}}), \tag{46}$$

if $Q > 3\sigma_y^2$. Further, if $Q < 3\sigma_y^2$, then $\sigma_\theta^{2^{(0)}} = 0$, so that $\widetilde{a}$ (see (44)) is equal to one and we have that, using (45),

$$\text{MSE}(\widetilde{\boldsymbol{\theta}}) = Q > \text{MSE}(\widetilde{\widetilde{\boldsymbol{\theta}}}) = Q\frac{\sigma_y^2}{\sigma_y^2 + Q}. \tag{47}$$

In summary then, $\text{MSE}(\widetilde{\boldsymbol{\theta}})$ is larger than the $\text{MSE}(\widetilde{\widetilde{\boldsymbol{\theta}}})$, so that $\widetilde{\widetilde{\boldsymbol{\theta}}}$, is preferred to $\widetilde{\boldsymbol{\theta}}$ no matter what the relation of Q and σ_y^2 may be. This together with the result that $\widetilde{\sigma}_\theta^2 < \widetilde{\widetilde{\sigma}}_\theta^2$, with $\widetilde{\widetilde{\sigma}}_\theta^2$ consistent for σ_θ^2, shows that working with the posterior of (μ, σ_θ^2) is a must, for the joint posterior of $\boldsymbol{\theta}$, μ, σ_θ^2 yields estimators of $\boldsymbol{\theta}$ and σ_θ^2 which do not perform well. Indeed, $\widetilde{\sigma}_\theta^2$ is not consistent for σ_θ^2, and in fact, if the true "state of nature" is such that $Q < 3\sigma_y^2$, $\widetilde{\sigma}_\theta^2$ is consistent for 0, a startling result.

5 The EM Algorithm in the One-Way Situation: Case σ_y^2 Unknown

The usual framework for one-way Analysis of Variance may, at the "Stage 1" level, be stated as

$$y_{ij} \sim N(\theta_j, \sigma_y^2), \quad i = 1, \dots, n_j; \ j = 1, \dots, m, \tag{48}$$

with all y_{ij} independent, so that in particular

$$\overline{y}_j \sim N(\theta_j, \sigma_y^2/n_j), \quad j = 1, \dots, m. \tag{49}$$

If indeed σ_y^2 known as in previous sections, and $n_j = n_0$, then

$$\text{Var}(g_j) = \sigma_y^2/n_0 = \text{Var}(\overline{y}_j), \tag{50}$$

and the previous sections in effect re-labelled $\overline{y}_j$ as y_j, $\text{Var}(\overline{y}_j)$ as σ_y^2, etc.

But here, if σ_y^2 is unknown, we do need a source of sample information about σ_y^2, and in this section we consider the following more usual hierarchical model when $(\boldsymbol{\theta}, \mu, \sigma_\theta^2, \sigma_y^2)$ are all unknown, which of course is a slightly extended hierarchical model to that given in Section 2.

Stage 1: Given θ_j, σ_y^2, $y_{ij} \sim \text{IDN}(\theta_j, \sigma_y^2)$, $i = 1, \dots, n_j$; $j = 1, \dots, m$.

Stage 2: Given μ, σ_θ^2, $\theta_j \sim \text{IIDN}(\mu, \sigma_\theta^2)$.

Stage 3: μ, σ_y^2, σ_θ^2 independent and such that

(i) $p(\mu)d\mu \propto d\mu$.
(ii) $\sigma_y^2 \sim \nu_0\lambda_0 \times (\chi_{\nu_0}^2)^{-1}$.
(iii) $\sigma_\theta^2 \sim \nu\lambda \times (\chi_\nu^2)^{-1}$.

Needless to say, there are a great number of similarities for results obtained under the hierarchical model of this section to results obtained in Sections 2–4. These are sketched in Appendix D1 of Guttman (1998), and see also Sun et al. (1996). Suffice it to say at this point that we are able to show that a collapsing effect exists here. It turns out, firstly, that for large m,

$$\begin{aligned} \tilde{\tilde{\sigma}}_\theta^2 &\simeq \hat{\sigma}_\theta^2, \quad \hat{\sigma}_\theta^2 \text{ consistent for } \sigma_\theta^2, \\ \tilde{\tilde{\sigma}}_y^2 &\simeq \hat{\sigma}_y^2, \quad \hat{\sigma}_y^2 \text{ consistent for } \sigma_y^2, \end{aligned} \tag{51}$$

and that if $n_j = n_0$, then $\tilde{\tilde{\boldsymbol{\theta}}}$ is preferred to $\tilde{\boldsymbol{\theta}}$ (see (52) below) in the mean square sense. For the definitions of $\tilde{\theta}_j$ and $\tilde{\tilde{\theta}}_j$, we have for this case ($n_j = n_0$), that $\tilde{\mu} = \tilde{\tilde{\mu}} = \overline{y}$, so that

$$\begin{aligned} \tilde{\theta}_j &= \tilde{\gamma}^2 \left[n_0(\tilde{\sigma}_y^2)^{-1}\overline{y}_j + (\sigma_\theta^2)^{-1}\overline{y}\right], \quad \tilde{\gamma}^2 = \left[n_0(\tilde{\sigma}_y^2)^{-1} + (\tilde{\sigma}_\theta^2)^{-1}\right]^{-1}, \\ \tilde{\tilde{\theta}}_j &= \tilde{\tilde{\gamma}}^2 \left[n_0(\tilde{\tilde{\sigma}}_y^2)^{-1}\overline{y}_j + (\sigma_\theta^2)^{-1}\overline{y}\right], \quad \tilde{\tilde{\gamma}}^2 = \left[n_0(\tilde{\tilde{\sigma}}_y^2)^{-1} + (\tilde{\tilde{\sigma}}_\theta^2)^{-1}\right]^{-1}. \end{aligned} \tag{52}$$

And further, it turns out that if interest is in $\text{Var}(\bar{y}_j \mid n_j \equiv n_0) = \delta^2$, where

$$\delta^2 = \sigma_\theta^2 + (n_0)^{-1}\sigma_y^2, \tag{53a}$$

that we can show, see Guttman (1998, Appendix D1), that

$$\tilde{\delta}^2 = \tilde{\sigma}_\theta^2 + \tilde{\sigma}_y^2/n_0 < \tilde{\tilde{\delta}}^2 = \tilde{\tilde{\sigma}}_\theta^2 + \tilde{\tilde{\sigma}}_y^2/n_0, \tag{53b}$$

with $\tilde{\tilde{\delta}}^2$ consistent for δ^2. Here, the expectations are taken with respect to $y_j \mid \mu, \sigma_\theta^2, \sigma_y^2$—see Stages 1 and 2 of the model of this section, and Appendix D1 of Guttman (1998), where the needed details are given.

With this put in the background, we will now discuss the implementation of the EM algorithm for obtaining estimates of parameters of the model of this section. We first note that this hierarchical model of this section is such that

$$\begin{aligned} p(\mathbf{y}, \boldsymbol{\theta}, \mu, \sigma_y^2, \sigma_\theta^2) &\propto (\sigma_y^2)^{-[(N+\nu_0)/2+1]}(\sigma_\theta^2)^{-[(m+\nu)/2+1]} \\ &\quad \times \exp -\frac{1}{2}\Bigg\{(\sigma_y^2)^{-1}\left[S_w + \sum_{j=1}^m n_j(\bar{y}_j - \theta_j)^2 + \nu_0\lambda_0\right] \\ &\qquad + (\sigma_\theta^2)^{-1}\left[\sum_{j=1}^m (\theta_j - \mu)^2 + \nu\lambda\right]\Bigg\}, \end{aligned} \tag{54a}$$

with

$$\mathbf{y} = (\mathbf{y}_1', \ldots, \mathbf{y}_m')', \quad \bar{y}_j = \mathbf{1}_{n_j}'\mathbf{y}_j/n_j \tag{54b}$$

and

$$S_w = \sum_{j=1}^m \sum_{i=1}^{n_j} (y_{ij} - \bar{y}_j)^2, \quad N = \sum_{j=1}^m n_j. \tag{54c}$$

With this as the setting, and identifying, in the language of Dempster et al. (1977), the complete data as $(\mathbf{y}, \boldsymbol{\theta}, \mu, \sigma_y^2, \sigma_\theta^2)$, we now declare that $\boldsymbol{\theta}$ is missing and perform the EM algorithm for this case. We are able to state the following theorem.

Theorem 6. *The marginal modal estimators $\tilde{\tilde{\mu}}$, $\tilde{\tilde{\sigma}}_y^2$ and $\tilde{\tilde{\sigma}}_\theta^2$ obtained from the marginal posterior density $p(\mu, \sigma_y^2, \sigma_\theta^2 \mid \mathbf{y})$ are identical to the estimators produced by the* EM *algorithm when $\boldsymbol{\theta}$ is regarded as missing.*

Some discussion is appropriate here. To begin with, (54) may be used to find the posterior of $\boldsymbol{\theta}$, μ, $\sigma^2 y$, σ_θ^2, given $\mathbf{y}$ easily, and from that integration with respect to $\boldsymbol{\theta}$ yields the marginal posterior of the parameters mentioned in Stage 3. The mode is then easily found and the equations are supplied in Appendix D2 of Guttman (1998).

Now to apply the EM algorithm, since θ is missing, the "expectation" step (the E-step) of the algorithm says that expectations are to be taken with respect to the conditional posterior distribution of $\boldsymbol{\theta}$, given $\mathbf{y}$, μ, σ_y^2, σ_θ^2. But in direct analogy to the work in Section 2, we easily find that

$$p(\theta_j \mid \mathbf{y}; \mu, \sigma_\theta^2, \sigma_y^2) = N'_{\theta_j}(\theta_j^*, \gamma_j^2), \tag{55a}$$

where

$$\theta_j^* = E(\theta_j \mid \mathbf{y}; \mu, \sigma_y^2, \sigma_\theta^2) = \gamma_j^2 \left[n_j(\sigma_y^2)^{-1} y_j + (\sigma_\theta^2)^{-1}\mu\right], \tag{55b}$$

$$\gamma_j^2 = \mathrm{Var}(\theta_j \mid \mathbf{y}; \mu, \sigma_y^2, \sigma_\theta^2) = \left[n_j(\sigma_y^2)^{-1} + (\sigma_\theta^2)^{-1}\right]^{-1}. \tag{55c}$$

Suppose then, we have concluded the pth stage of the EM algorithm, obtaining the estimates $\mu^{(p)}$, $\sigma_y^{2(p)}$ and $\sigma_\theta^{2(p)}$. We proceed to the $(p+1)$st stage by first performing the E-step. We will need the notation

$$E(\theta_j \mid \mathbf{y}; \mu^{(p)}, \sigma_y^{2(p)}, \sigma_\theta^{2(p)}) \\ = \gamma_j^{2(p)} \left[n_j(\sigma_y^{2(p)})^{-1} y_j + (\sigma_\theta^{2(p)})^{-1}\mu^{(p)}\right] = \theta_j^{(p)}, \tag{56}$$

for the E-step says calculate

$$H(\mu, \sigma_y^2, \sigma_\theta^2; \mu^{(p)}, \sigma_y^{2(p)}, \sigma_\theta^{2(p)}) = E\left[\log p(\mathbf{y}, \boldsymbol{\theta}, \mu, \sigma_\theta^2, \sigma_y^2)\right], \tag{57}$$

where, as stated, but for emphasis, the expectation is taken with respect to $\boldsymbol{\theta} \mid \mathbf{y};\ \mu^{(p)}$, $\sigma_y^{2(p)}$, $\sigma_\theta^{2(p)}$. In (56), $\gamma_j^{2(p)}$ denotes (55c) evaluated at $\sigma_y^{2(p)}$ and $\sigma_\theta^{2(p)}$, so that

$$\gamma_j^{2(p)} = \left[n_j(\sigma_y^{2(p)})^{-1} + (\sigma_\theta^{2(p)})^{-1}\right]^{-1}. \tag{58}$$

For this expectation of (57), we simply have to apply the identity

$$E\left[(\theta_j - b)^2 \mid \mathbf{y}, \mu^{(p)}, \sigma_y^{2(p)}, \sigma_\theta^{2(p)}\right] = \gamma_j^{2(p)} + (\theta_j^{(p)} - b)^2, \tag{59}$$

where b is successively, $\overline{y}_j$ and $\mu^{(p)}$. Consulting (54), and using (59), and after some algebra it can be shown [Appendix D2, Guttman (1998)] that

$$H(\mu, \sigma_y^2, \sigma_\theta^2; \mu^{(p)}, \sigma_y^{2(p)}, \sigma_\theta^{2(p)}) \\ = \text{const} + \log p(\mathbf{y}, \boldsymbol{\theta}^{(p)}, \mu, \sigma_y^2, \sigma_\theta^2) - \frac{1}{2}\{\Sigma n_j \gamma_j^{2(p)} \sigma_y^2 + \Sigma \gamma_j^{2(p)}/\sigma_\theta^2\}. \tag{60}$$

Having now completed the "E-step", we now proceed to the "M-step" and maximize H with respect to $(\mu, \sigma_y^2, \sigma_\theta^2)$, denoting the "place" of the maximum by $\mu^{(p+1)}$, $\sigma_y^{2(p+1)}$, $\sigma_\theta^{2(p+1)}$. We have [see Appendix D2 of Guttman

(1998)] the resulting equations:

$$\mu^{(p+1)} = \bar{\theta}^{(p)} = m^{-1} \sum_{j=1}^{m} \theta_j^{(p)},$$

$$\sigma_y^{2(p+1)} = \left\{ S_w + \sum_{j=1}^{m} n_j \left[(\theta_j^{(p)} - \overline{y}_j)^2 + \gamma_j^{2(p)} \right] + \nu_0 \lambda_0 \right\} \Big/ (N + \nu_0 + 2), \tag{61}$$

$$\sigma_\theta^{2(p+1)} = \left\{ \sum_{j=1}^{m} \left[(\theta_j^{(p)} - \bar{\theta}^{(p)})^2 + \gamma_j^{2(p)} + \nu\lambda \right] \right\} / (m + \nu + 2),$$

and where we introduce, in accordance with (56),

$$\theta_j^{(p+1)} = \gamma_j^{2(p+1)} \left[n_j (\sigma_y^{2(p+1)})^{-1} \overline{y}_j + (\sigma_\theta^{2(p+1)})^{-1} \mu^{(p+1)} \right],$$

with $\gamma_j^{2(p+1)}$ defined in obvious fashion. Now these equations are interesting from the following viewpoint. If we determine the marginal $p(\mu, \sigma_\theta^2, \sigma_y^2 \mid \mathbf{y})$ and its mode, $\tilde{\tilde{\mu}}$, $\tilde{\tilde{\sigma}}_\theta^2$, $\tilde{\tilde{\sigma}}_y^2$ and introduce

$$\tilde{\tilde{\theta}}_j = \tilde{\tilde{\gamma}}^2 \left[n_j (\tilde{\tilde{\sigma}}_y^2)^{-1} \overline{y}_j + (\tilde{\tilde{\sigma}}_\theta^2)^{-1} \tilde{\tilde{\mu}} \right], \tag{62}$$

we find that we have to solve equations of the same form as equations (61)—here we are thinking of the EM algorithm being terminated at desired level of accuracy at Stage $p' + 1$, so that for example, $\mu^{(p'+1)} = \mu^{(p')} = \mu^*$, say, etc., to that level of accuracy, and μ^*, σ_θ^{2*}, σ_y^{2*}, θ^* are inserted in the equations (61). But from previous discussion we know that, in particular, $\tilde{\tilde{\sigma}}_\theta^2$ and $\tilde{\tilde{\sigma}}_y^2$ are consistent for σ_θ^2 and σ_y^2 respectively, and $\tilde{\tilde{\theta}}$ has minimum mean square error. [Details and proofs given in Appendix D1 of Guttman (1998), but see Sections 2–4 of this paper]. In addition, $\tilde{\tilde{\delta}}^2$ of (53b) is consistent.

In summary, then, so far, the application of the EM algorithm when $\boldsymbol{\theta}$ is declared missing from $(\mathbf{y}, \boldsymbol{\theta}, \mu, \sigma_\theta^2, \sigma_y^2)$ leads to estimators with good properties. But as indicated in the introduction, care must be exercised in "what is declared missing" when applying the EM algorithm, and indeed, we have the following theorem.

Theorem 7. *The joint model estimators* $(\tilde{\boldsymbol{\theta}}, \tilde{\mu}, \tilde{\sigma}_y^2, \tilde{\sigma}_\theta^2)$ *obtained from the posterior* $p(\boldsymbol{\theta}, \mu, \sigma_y^2, \sigma_\theta^2 \mid \mathbf{y})$ *are, for large* N, m, *equivalent to the estimators found using the EM algorithm when* $(\mu, \sigma_y^2, \sigma_\theta^2)$ *is regarded as missing.*

The same manipulations as in Theorem 6 are used here, but of course, expectations are taken with respect to the conditional posterior of μ, σ_y^2, σ_θ^2, given $\boldsymbol{\theta}$, $\mathbf{y}$, which is to say, that if we have completed the pth stage of the EM algorithm, having obtained the estimate $\boldsymbol{\theta}^{(p)}$, we proceed to the $(p+1)$st stage by first performing the E-step, calculating

$$E\{\log p(\mathbf{y}, \boldsymbol{\theta}, \mu, \sigma_\theta^2, \sigma^2 y)\} = K(\boldsymbol{\theta}, \boldsymbol{\theta}^{(p)}). \tag{63}$$

Here expectation, is conducted with respect to $(\mu, \sigma_\theta^2, \sigma_y^2)$, given $\boldsymbol{\theta}^{(p)}$, $\mathbf{y}$. We then maximize K of (63) with respect to $\boldsymbol{\theta}$, and denote the "place" of the maximum by $\boldsymbol{\theta}^{(p+1)}$.

Now introduce the notation (and definitions)

$$\begin{aligned}
E\left[\mu \mid \mathbf{y}, \boldsymbol{\theta}^{(p)}, \sigma_\theta^{2(p)}\right] &= \mu^{(p)}, \\
E\left[(\sigma_\theta^2)^{-1} \mid \mathbf{y}, \boldsymbol{\theta}^{(p)}\right] &= \frac{1}{\sigma_\theta^{2(p)}} = \left[\frac{\sum(\theta_j^{(p)} - \bar{\theta}^{(p)})^2 + \nu\lambda}{m+\nu-1}\right]^{-1}, \\
E[(\sigma_y^2)^{-1} \mid \bar{y}, \theta^{(p)}] &= \frac{1}{\sigma_y^{2(p)}} \\
&= \left[\frac{S_w + \sum_{j=1}^m n_j(\bar{y}_j - \theta_j^{(p)})^2 + \nu_0\lambda_0}{N+\nu_0}\right]^{-1}.
\end{aligned} \tag{64}$$

Then, it turns out that $\boldsymbol{\theta}^{(p+1)}$, the place of the maximum of K is at $\boldsymbol{\theta}^{(p+1)}$, where the jth component of this vector is given by

$$\theta_j^{(p+1)} = \gamma_j^{2(p)}\left[n_j(\sigma_y^{2(p)})^{-1}\bar{y}_j + (\sigma_\theta^{2(p)})^{-1}\mu^{(p)}\right], \tag{65}$$

with

$$\gamma_j^{2(p)} = \left[n_j(\sigma_y^{2(p)})^{-1} + \sigma_\theta^{2(p)}\right]^{-1}.$$

The results (65) and (64), which are proved in Appendix D2 of Guttman (1998), are again interesting from the following point of view—if the joint posterior $p(\boldsymbol{\theta}, \mu, \sigma_\theta^2, \sigma_y^2 \mid \mathbf{y})$ is determined (see (54)), and its mode found, say $(\widetilde{\boldsymbol{\theta}}, \widetilde{\mu}, \widetilde{\sigma}_\theta^2, \widetilde{\sigma}_y^2)$, then the equations have the solution (64)–(65), except that the divisor for $\widetilde{\sigma}_\theta^2$, $N + \nu_0 + 2$ would have to be replaced by $N + \nu_0$, and the divisor for $\widetilde{\sigma}_y^2$, $m + \nu + 2$ would have to be replaced by $m + \nu - 1$, which for large m and N are asymptotically equivalent.

But we know that the collapsing effect for δ^2 reported in (53b) occurs, in that $\widetilde{\delta}^2 = \widetilde{\sigma}_\theta^2 + \widetilde{\sigma}_y^2/n_0$ is not consistent for δ^2, and indeed $\widetilde{\delta}^2 < \widetilde{\widetilde{\delta}}^2$ with $\widetilde{\widetilde{\delta}}^2$ consistent for δ^2. Further, $\widetilde{\widetilde{\boldsymbol{\theta}}}$ has larger MSE than $\widetilde{\widetilde{\boldsymbol{\theta}}}$, so we have the striking result that says—beware when applying the EM algorithm in hierarchical situations, for the declaration of what is missing can lead to (misleading) estimators which do not have desired properties.

6 References

Dempster, A., N. Laird, and D. Rubin (1977). Likelihood from incomplete data via the EM algorithm (with discussion). *J. Roy. Statist. Soc. Ser. B 39*, 1–38.

Efron, B. and C. Morris (1976). Multivariate empirical Bayes and estimation of covariance matrices. *Ann. Statist. 4*, 141–150.

Guttman, I. (1998). Empirical Bayes estimators and EM algorithms in one-way analysis of variance situations. Technical report, Department of Statistics, SUNY at Buffalo.

O'Hagan, A. (1976). On posterior joint and marginal modes. *Biometrika 63*, 329–333.

Sun, L., J. Hsu, I. Guttman, and T. Leonard (1996). Bayesian methods for variance components. *J. Amer. Statist. Assoc. 91*, 743–752.

3

EB and EBLUP in Small Area Estimation

J.N.K. Rao

ABSTRACT Sample survey data can be used to derive reliable direct estimates for large areas or domains, but sample sizes in small areas or domains are seldom large enough for direct estimators to provide adequate precision for small areas. This makes it necessary to employ indirect estimators that borrow strength from related areas. We provide empirical Bayes (EB) or empirical best linear unbiased prediction (EBLUP) estimators under two basic models: unit level and area level. Methods for measuring the variability of the EB estimator are compared. Simple modifications to the estimator of mean squared error, proposed by Prasad and Rao (1990), are given. These estimators are area-specific, unlike the Prasad-Rao estimator, in the sense of depending on area-specific data.

1 Introduction

Reliable direct estimates for large areas or subpopulations (domains) of interest can be obtained from sample surveys using only area-specific data. But sample sizes are rarely large enough for direct estimates to provide adequate precision for small geographical areas such as counties or small domains such as age-sex-race groups within a geographical area. This makes it necessary to employ indirect estimates that borrow strength from related areas. We refer the reader to Ghosh and Rao (1994) for a recent review of methods used for small area estimation. We use the term "small area" to denote either a small geographical area or a small domain.

In this article, we focus on model-based indirect estimators obtained from small area models that relate a characteristic of interest, y, to auxiliary variables, $\mathbf{x}$, with known population characteristics often ascertained from external sources such as a previous census or administrative records. In particular, we consider empirical Bayes (EB) and empirical best linear unbiased prediction (EBLUP) estimation of small area means under two types of models. In the first type, only area-specific auxiliary data $\mathbf{x}_i = (x_{i1}, \ldots, x_{ip})'$ are available, and a specified function, $\theta_i = g(\overline{Y}_i)$, of the ith small area mean $\overline{Y}_i$ is assumed to be related to $\mathbf{x}_i$. In particular, we assume the following model that links the θ_i's:

$$\theta_i = \mathbf{x}_i'\beta + v_i, \quad i = 1, \ldots, m, \tag{1}$$

where m ($> p$) is the number of small areas of interest, $\mathbf{x}_i$'s are known vectors of constants, and the v_i's are random small area effects assumed to be independent identically distributed (iid) with mean 0 and variance σ_v^2. Fay and Herriot (1979) used (1) with $\theta_i = \log \overline{Y}_i$ in the context of estimating per capita income (PCI) for small places using the associated county values of log (PCI) ascertained from the previous census as x_i. For a binary variable, $\theta_i = \log\{P_i/(1-P_i)\}$ is commonly used, where $\overline{Y}_i = P_i$ is the population proportion in the ith area. We assume that direct estimators $\widehat{\theta}_i = g(\widehat{\overline{Y}}_i)$ of θ_i can be obtained from a sample survey and that

$$\widehat{\theta}_i = \theta_i + e_i, \quad i = 1, \ldots, m, \tag{2}$$

where the e_i's are sampling errors with $E(e_i \mid \theta_i) = 0$ and $\mathrm{Var}(e_i \mid \theta_i) = \psi_i$, and the sampling variances, ψ_i, are assumed to be known. In practice, specification of ψ_i may not be easy, although the method of generalized variance function (Wolter, 1995, Chapter 5), based on sampling error models, is often used to ascertain ψ_i. Also, the assumption $E(e_i \mid \theta_i) = 0$ may not hold in the case of nonlinear functions $\theta_i = g(\overline{Y}_i)$ if the sample size, n_i, is unduly small, even if the direct estimator $\widehat{\overline{Y}}_i$ is unbiased, i.e., $E(\widehat{\overline{Y}}_i \mid \overline{Y}_i) = \overline{Y}_i$. The case of unknown ψ_i is difficult to handle in the context of small areas without additional assumptions.

Combining the sampling model (2) with the linking model (1) we get a type 1 model:

$$\widehat{\theta}_i = \mathbf{x}_i'\beta + v_i + e_i. \tag{3}$$

Various extensions of the basic type 1 model (3) have been studied in the literature see Ghosh and Rao (1994).

In the second type of models, element-specific auxiliary data $\mathbf{x}_{ij} = (x_{ij1}, \ldots, x_{ijp})'$ are used to model the element-values y_{ij} in terms of $\mathbf{x}_{ij}$ and random small effects v_i. For example, a nested error regression model is often used (Battese et al., 1988; Prasad and Rao, 1990):

$$y_{ij} = \mathbf{x}_{ij}'\beta + v_i + e_{ij}, \quad j = 1, \ldots, N_i; \; i = 1, \ldots, m, \tag{4}$$

where $v_i \overset{\text{iid}}{\sim} N(0, \sigma_v^2)$ are independent of element errors $e_{ij} \overset{\text{iid}}{\sim} N(0, \sigma_e^2)$ and N_i is the number of population elements in the ith area. In the case of a binary variable y, it is often assumed that y_{ij} are independent Bernoulli random variables with $\mathrm{pr}(y_{ij} = 1 \mid p_{ij}) = p_{ij}$ and that $\log\{p_{ij}/(1 - p_{ij})\} = x_{ij}'\beta + v_i$ see, e.g., Farrell et al. (1997). More general models on $\log\{p_{ij}/(1 - p_{ij})\}$ can be formulated; for example, models with random regression coefficients (Malec et al., 1997). In this paper, we focus on (4) and assume that the sample data $\{y_{ij}, \mathbf{x}_{ij}, j = 1, \ldots, n_i;\ i = 1, \ldots, m\}$ also obey the model (4), i.e., selection bias is absent. The basic type 2 model (4) is appropriate for single-stage sampling within areas (e.g., simple random sampling). For two-stage sampling within areas, a two-fold nested error regression model may be used (Stukel and Rao, 1999).

Section 2 gives the EB estimator, $\widehat{\theta}_i^{\mathrm{EB}}$, of θ_i under the area-level model (3) and compares different estimators of the mean squared error (MSE) of $\widehat{\theta}_i^{\mathrm{EB}}$ proposed in the literature; the MSE estimators measure the variability associated with $\widehat{\theta}_i^{\mathrm{EB}}$. Section 3 gives the EB estimator, $\overline{y}_i^{\mathrm{EB}}$, of the ith area mean $\overline{Y}_i$ under the element-level model (4) and compares different estimators of the MSE of $\overline{y}_i^{\mathrm{EB}}$.

2 Type 1 Model

If we assume that the errors v_i and e_i are independent and normally distributed, then the "optimal" estimator of realized θ_i is given by the conditional expectation of θ_i given $\widehat{\theta}_i$:

$$E(\theta_i \mid \widehat{\theta}_i, \boldsymbol{\beta}, \sigma_v^2) = \widehat{\theta}_i^{\mathrm{B}} = \gamma_i \widehat{\theta}_i + (1 - \gamma_i)\mathbf{x}_i'\boldsymbol{\beta}, \tag{5}$$

where $\gamma_i = \sigma_v^2/(\sigma_v^2 + \psi_i)$. Note that (5) follows from the well-known result that the posterior distribution of θ_i is $N(\hat{\theta}_i^{\mathrm{B}}, \gamma_i\psi_i)$ if $\hat{\theta}_i \mid \theta_i \sim N(\theta_i, \psi_i)$, $\theta_i \sim N(\mu_i = \mathbf{x}_i'\boldsymbol{\beta}, \sigma_v^2)$ and $\boldsymbol{\beta}$, ψ_i and σ_v^2 are known (see e.g., Casella and Berger (1990, p. 299)). The estimator $\widehat{\theta}_i^{\mathrm{B}} = \widehat{\theta}_i^{\mathrm{B}}(\boldsymbol{\beta}, \sigma_v^2)$ is the Bayes estimator under squared error loss and it is optimal in the sense that its mean squared error, $\mathrm{MSE}(\widehat{\theta}_i^{\mathrm{B}}) = E(\widehat{\theta}_i^{\mathrm{B}} - \theta_i)^2$, is smaller than the MSE of any other estimator of θ_i, not necessarily linear in the $\widehat{\theta}_i$'s. The estimator $\widehat{\theta}_i^{\mathrm{B}}$ depends on β and σ_v^2 which are unknown in practice. For a given σ_v^2, the weighted least squares (WLS) estimator of $\boldsymbol{\beta}$ is given by

$$\widetilde{\boldsymbol{\beta}} = \widetilde{\boldsymbol{\beta}}(\sigma_v^2) = \left[\Sigma_i \mathbf{x}_i\mathbf{x}_i'/(\sigma_v^2 + \psi_i)\right]^{-1}\left[\Sigma_i \mathbf{x}_i\widehat{\theta}_i/(\sigma_v^2 + \psi_i)\right]. \tag{6}$$

The estimator (6) is also the maximum likelihood estimator under normality because $\hat{\theta}_i \sim N(\mathbf{x}_i'\boldsymbol{\beta}, \sigma_v^2 + \psi_i)$ using (3), or $\widehat{\boldsymbol{\theta}} = (\hat{\theta}_1, \ldots, \hat{\theta}_m)' \sim N_m\left[\mathbf{X}\boldsymbol{\beta}, \mathbf{V} = \mathrm{diag}_i(\sigma_v^2 + \psi_i)\right]$ with $\mathbf{X} = \mathrm{col}_i(\mathbf{x}_i')$. Substituting $\widetilde{\boldsymbol{\beta}}$ for $\boldsymbol{\beta}$ in $\widehat{\theta}_i^{\mathrm{B}}$, we get an EB estimator $\widetilde{\theta}_i^{\mathrm{EB}} = \widehat{\theta}_i^{\mathrm{B}}(\widetilde{\boldsymbol{\beta}}, \sigma_v^2)$ which is also the best linear unbiased prediction (BLUP) estimator (Prasad and Rao, 1990) or a linear EB estimator (Ghosh and Lahiri, 1987) without the normality assumption.

The EB estimator $\widetilde{\theta}_i^{\mathrm{EB}}$ depends on σ_v^2 which may be estimated either by the method of moments or by the restricted maximum likelihood (REML) method. A simple moment estimator of σ_v^2 is given by $\widehat{\sigma}_v^2 = \max(\widetilde{\sigma}_v^2, 0)$, where

$$\widetilde{\sigma}_v^2 = (m - p)^{-1}\left[\Sigma_i(y_i - \mathbf{x}_i'\boldsymbol{\beta}^*)^2 - \Sigma_i(1 - h_{ii})\psi_i\right], \tag{7}$$

where $\boldsymbol{\beta}^* = (\Sigma_i \mathbf{x}_i \mathbf{x}_i')^{-1}(\Sigma_i \mathbf{x}_i \widehat{\theta}_i)$ is the ordinary least squares (OLS) estimator of $\boldsymbol{\beta}$, and $h_{ii} = \mathbf{x}_i'(\Sigma_i \mathbf{x}_i \mathbf{x}_i')^{-1}\mathbf{x}_i$. An alternative method of moments estimator can be obtained by solving iteratively $\Sigma_i(y_i - \mathbf{x}_i'\widetilde{\boldsymbol{\beta}})^2/(\sigma_v^2 + \psi_i) = m - p$ in conjunction with (6) and letting $\widehat{\sigma}_v^2 = 0$ whenever a positive solution does not exist (Fay and Herriot, 1979). Now substituting $\widehat{\sigma}_v^2$ for σ_v^2 in $\widetilde{\theta}_i^{\mathrm{EB}}$ we get the final EB estimator $\widehat{\theta}_i^{\mathrm{EB}} = \theta_i^{\mathrm{B}}\left[\widetilde{\boldsymbol{\beta}}(\widehat{\sigma}_v^2), \widehat{\sigma}_v^2\right]$. Note that $\widehat{\theta}_i^{\mathrm{EB}}$ does not require the normality assumption if a moment estimator of σ_v^2 is used. The estimator $\widehat{\theta}_i^{\mathrm{EB}}$ is also the empirical BLUP (EBLUP) estimator. The EB estimator of $\overline{Y}_i$ is taken as $g^{-1}(\widehat{\theta}_i^{\mathrm{EB}})$. Note that $\widehat{\theta}_i^{\mathrm{EB}}$ is a weighted average of the direct estimator $\widehat{\theta}_i$ and the regression synthetic estimator $\mathbf{x}_i'\widehat{\boldsymbol{\beta}}$, where the weight $\widehat{\gamma}_i = \widehat{\sigma}_v^2/(\widehat{\sigma}_v^2 + \psi_i)$ reflects the model variance σ_v^2 relative to the total variance $\sigma_v^2 + \psi_i$. Thus the EB estimator takes proper account of the between area variation relative to the precision of the direct estimator.

A measure of variability associated with $\widehat{\theta}_i^{\mathrm{EB}}$ is given by its MSE. Under normality of the errors, we have

$$\mathrm{MSE}(\widehat{\theta}_i^{\mathrm{EB}}) = E(\widehat{\theta}_i^{\mathrm{EB}} - \theta_i)^2 = E(\widetilde{\theta}_i^{\mathrm{EB}} - \theta_i)^2 + E(\widehat{\theta}_i^{\mathrm{EB}} - \widetilde{\theta}_i^{\mathrm{EB}})^2, \qquad (8)$$

see Kackar and Harville (1984); in general, the covariance term $E(\widehat{\theta}_i^{\mathrm{EB}} - \widetilde{\theta}_i^{\mathrm{EB}})(\widetilde{\theta}_i^{\mathrm{EB}} - \theta_i)$ is not zero (Lahiri and Rao, 1995). It now follows from (8) that $\mathrm{MSE}(\widehat{\theta}_i^{\mathrm{EB}})$ is always larger than the MSE of the BLUP estimator $\widetilde{\theta}_i^{\mathrm{EB}}$. Also, the last term of (8) is not tractable analytically, unlike $\mathrm{MSE}(\widetilde{\theta}_i^{\mathrm{EB}}) = E(\widetilde{\theta}_i^{\mathrm{EB}} - \theta_i)^2$. We have

$$\mathrm{MSE}(\widetilde{\theta}_i^{\mathrm{EB}}) = g_{1i}(\sigma_v^2) + g_{2i}(\sigma_v^2) = M_{1i}(\sigma_v^2), \qquad (9)$$

where

$$g_{1i}(\sigma_v^2) = \gamma_i \psi_i,$$

and

$$g_{2i}(\sigma_v^2) = (1 - \gamma_i)^2 \mathbf{x}_i' \left[\Sigma_i \mathbf{x}_i \mathbf{x}_i'(\sigma_v^2 + \psi_i)\right]^{-1} \mathbf{x}_i,$$

without the normality assumption. The leading term $g_{1i}(\sigma_v^2)$ is of order $O(1)$ whereas the last term, $g_{2i}(\sigma_v^2)$, is of order $O(m^{-1})$ for large m, under regularity conditions: $\max_i\{\mathbf{x}_i'(\mathbf{X}'\mathbf{X})^{-1}\mathbf{x}_i\} = O(m^{-1})$ and $0 < \psi_L \le \psi_i \le \psi_U < \infty$ for all i; see Prasad and Rao (1990). Also, the leading term shows that the MSE of $\widetilde{\theta}_i^{\mathrm{EB}}$ can be substantially smaller than $\mathrm{MSE}(\widehat{\theta}_i) = \psi_i$, the MSE of the direct estimator $\widehat{\theta}_i$ under the model, when $\gamma_i = \sigma_v^2/(\sigma_v^2 + \psi_i)$ is small or the model variance, σ_v^2, is small relative to the sampling variance ψ_i. Note that $\widetilde{\theta}_i^{\mathrm{EB}}$ gives more weight to the "synthetic" component, $\mathbf{x}_i'\widetilde{\boldsymbol{\beta}}$, than to $\widehat{\theta}_i$ when γ_i is small.

The last term of (8) may be approximated, for large m, as

$$E\left(\widehat{\theta}_i^{\mathrm{EB}} - \widetilde{\theta}_i^{\mathrm{EB}}\right)^2 \approx g_{3i}(\sigma_v^2), \qquad (10)$$

where

$$\begin{aligned} g_{3i}(\sigma_v^2) &= [\psi_i^2/(\sigma_v^2+\psi_i)^4]E(\hat{\theta}_i - \mathbf{x}_i'\boldsymbol{\beta})^2\overline{V}(\tilde{\sigma}_v^2), \\ &= [\psi_i^2/(\sigma_v^2+\psi_i)^3]\overline{V}(\tilde{\sigma}_v^2), \end{aligned} \tag{11}$$

and $\overline{V}(\tilde{\sigma}_v^2)$ is the asymptotic variance of $\tilde{\sigma}_v^2$ (Prasad and Rao, 1990). The neglected terms in the approximation (9) are of lower order than $O(m^{-1})$; see Prasad and Rao (1990). For the simple moment estimator $\tilde{\sigma}_v^2$ given by (7), we have

$$\overline{V}(\tilde{\sigma}_v^2) = 2m^{-2}\Sigma_i(\sigma_v^2+\psi_i)^2 = h_i(\sigma_v^2), \text{ say.}$$

Combining (8) and (9) we get

$$\text{MSE}(\boldsymbol{\theta}_i^{\text{EB}}) \approx M_{2i}(\sigma_v^2) = M_{1i}(\sigma_v^2) + g_{3i}(\sigma_v^2). \tag{12}$$

The approximation $M_{2i}(\sigma_v^2)$ is correct to terms of order $O(m^{-1})$. Datta and Lahiri (1997) have shown that the approximation (12) also holds for the REML estimator of σ_v^2.

We now turn to the estimation of $\text{MSE}(\hat{\theta}_i^{\text{EB}})$. It is customary to ignore the variability associated with $\hat{\sigma}_v^2$, i.e., neglect the g_{3i}-term in (11), and use $M_{1i}(\hat{\sigma}_v^2)$ as an estimator of $\text{MSE}(\hat{\theta}_i^{\text{EB}})$. This can lead to significant underestimation because $g_{3i}(\sigma_v^2)$ is of the same order as $g_{2i}(\sigma_v^2)$ in $M_{1i}(\hat{\sigma}_v^2)$. A correct, approximately unbiased estimator of $\text{MSE}(\hat{\theta}_i^{\text{EB}})$ is given by

$$\text{MSE}(\hat{\theta}_i^{\text{EB}}) = M_{1i}(\hat{\sigma}_v^2) + 2g_{3i}(\hat{\sigma}_v^2), \tag{13}$$

(Prasad and Rao, 1990). The bias of (13) is of lower order than m^{-1}. Lahiri and Rao (1995) have shown that (13) is robust to nonnormality of the small area effects, v_i, in the sense it remains approximately unbiased. Note that the normality of the sampling errors, e_i, is not as restrictive as the normality of v_i, due to the central limit theorem's effect on the direct estimators $\hat{\theta}_i$.

Butar and Lahiri (1997) note that (13) is not area-specific in the sense that it does not depend on the area-specific direct estimator $\hat{\theta}_i$, although $\mathbf{x}_i$ is involved. But it is easy to find other choices, using the form (11) for $g_{3i}(\sigma_v^2)$, that yield area-specific estimators of $\text{MSE}(\hat{\theta}_i^{\text{EB}})$. For example, noting that $\hat{\boldsymbol{\beta}} = \tilde{\boldsymbol{\beta}}(\hat{\sigma}_v^2)$ is a consistent estimator of $\boldsymbol{\beta}$, we obtain two different area-specific estimators:

$$\text{MSE}_1(\hat{\theta}_i^{\text{EB}}) = M_{1i}(\hat{\sigma}_v^2) + 2[\psi_i^2/(\hat{\sigma}_v^2+\psi_i)^4](\hat{\theta}_i - \mathbf{x}_i'\hat{\boldsymbol{\beta}})^2 h_i(\hat{\sigma}_v^2), \tag{14}$$

$$\begin{aligned} \text{MSE}_2(\hat{\theta}_i^{\text{B}}) &= M_{1i}(\hat{\sigma}_v^2) + g_{3i}(\hat{\sigma}_v^2) \\ &\quad + [\psi_i^2/(\hat{\sigma}_v^2+\psi_i)^4](\hat{\theta}_i - \mathbf{x}_i'\hat{\boldsymbol{\beta}})^2 h_i(\hat{\sigma}_v^2). \end{aligned} \tag{15}$$

Both these estimators are also approximately unbiased. Note that the use of area-specific squared residual $(\hat{\theta}_i - \mathbf{x}_i'\hat{\boldsymbol{\beta}})^2$ induces instability in the estimators (14) and (15) but its effect should be small because only the $O(m^{-1})$ term, $2g_{3i}(\hat{\sigma}_v^2)$, is changed.

Alternative measures of variability associated with $\widehat{\theta}_i^{\mathrm{EB}}$ have also been proposed. If $\boldsymbol{\beta}$ and σ_v^2 are given, the posterior (conditional) variance $V(\theta_i \mid \widehat{\theta}_i, \boldsymbol{\beta}, \sigma_v^2) = g_{1i}(\sigma_v^2)$ measures the variability associated with $\widehat{\theta}_i^{\mathrm{B}}$. Replacing σ_v^2 by $\widehat{\sigma}_v^2$, a naive EB approach uses the estimated posterior variance, $g_{1i}(\widehat{\sigma}_v^2)$, as a measure of variability associated with $\widehat{\theta}_i^{\mathrm{EB}}$. This leads to severe underestimation of MSE($\widehat{\theta}_i^{\mathrm{EB}}$) as seen from (12). Note that we are treating $\boldsymbol{\beta}$ and σ_v^2 as fixed, unknown parameters, so no prior distributions are involved. However, if we adopt a hierarchical Bayes (HB) approach by assuming the hyper-parameters $\boldsymbol{\beta}$ and σ_v^2 as random with a specified prior distribution, then the posterior mean $\widehat{\theta}_i^{\mathrm{HB}} = E(\theta_i \mid \widehat{\boldsymbol{\theta}})$ is used as the estimator of θ_i and the posterior variance $V(\theta_i \mid \widehat{\boldsymbol{\theta}})$ as a measure of variability associated with $\widehat{\theta}_i^{\mathrm{HB}}$, where $\widehat{\boldsymbol{\theta}} = (\widehat{\theta}_1, \ldots, \widehat{\theta}_m)'$. We can express $E(\theta_i \mid \widehat{\boldsymbol{\theta}})$ and $V(\theta_i \mid \widehat{\boldsymbol{\theta}})$ as

$$E(\theta_i \mid \widehat{\boldsymbol{\theta}}) = E_{\boldsymbol{\beta}, \sigma_v^2}\big[\widehat{\theta}_i^{\mathrm{B}}(\boldsymbol{\beta}, \sigma_v^2)\big], \tag{16}$$

and

$$V(\theta_i \mid \widehat{\theta}) = E_{\boldsymbol{\beta}, \sigma_v^2}\big[g_{1i}(\sigma_v^2)\big] + V_{\boldsymbol{\beta}, \sigma_v^2}\big[\widehat{\theta}_i^{\mathrm{B}}(\boldsymbol{\beta}, \sigma_v^2)\big], \tag{17}$$

where $E_{\boldsymbol{\beta}, \sigma_v^2}$ and $V_{\boldsymbol{\beta}, \sigma_v^2}$ respectively denote the expectation and variance with respect to the posterior distribution of $\boldsymbol{\beta}$ and σ_v^2 given the data $\widehat{\boldsymbol{\theta}}$. It follows from (16) and (17) that $\widehat{\theta}_i^{\mathrm{EB}} = \widehat{\theta}_i^{\mathrm{B}}(\widehat{\boldsymbol{\beta}}, \widehat{\sigma}_v^2)$ tracks $E(\theta_i \mid \widehat{\boldsymbol{\theta}})$ well, but $g_{1i}(\widehat{\sigma}_v^2)$ is a good approximation only to the first variance term on the right hand side of (17). Consequently, the naive estimator $g_{1i}(\widehat{\sigma}_v^2)$ can lead to severe underestimation of the true posterior variance, $V(\theta_i \mid \widehat{\boldsymbol{\theta}})$.

Laird and Louis (1989) proposed a bootstrap method to imitate the form (17) and thus account for the variability in $\widehat{\boldsymbol{\beta}}$ and $\widehat{\sigma}_v^2$. Repeated bootstrap samples, $\theta_i^* \overset{\text{inds}}{\sim} N(\mathbf{x}_i'\widehat{\boldsymbol{\beta}}, \widehat{\sigma}_v^2)$, $i = 1, \ldots, m$ are drawn and the estimates $\boldsymbol{\beta}^*$ and σ_v^{*2} are then calculated from the bootstrap data $\{\theta_i^*, x_i, i = 1, \ldots, m\}$. This leads to the following measure of variability associated with $\widehat{\theta}_i^{\mathrm{EB}}$:

$$\mathrm{MSE_{LL}}(\widehat{\theta}_i^{\mathrm{EB}}) = E_*\big[g_{1i}(\sigma_v^{*2})\big] + V_*\big[\widehat{\theta}_i^{\mathrm{B}}(\boldsymbol{\beta}^*, \sigma_v^{*2})\big], \tag{18}$$

which has a form similar to (16), where E_* and V_* respectively denote the bootstrap expectation and the bootstrap variance. Butar and Lahiri (1997) obtained an analytical approximation to (18), for large m, as

$$\mathrm{MSE_{ILL}}(\widehat{\theta}_i^{\mathrm{EB}}) = M_{1i}(\widehat{\sigma}_v^2) + \big[\psi_i^2/(\widehat{\sigma}_v^2 + \psi_i)^4\big] h_i(\widehat{\sigma}_v^2)(\widehat{\theta}_i - \mathbf{x}_i'\widehat{\boldsymbol{\beta}})^2. \tag{19}$$

Comparing (19) with (15) it follows that the bootstrap measure of variability is not second-order correct, i.e., its bias involves terms of order m^{-1}. In particular, $E\big[\mathrm{MSE_{ILL}}(\widehat{\theta}_i^{\mathrm{EB}})\big] \approx \mathrm{MSE}(\widehat{\theta}_i^{\mathrm{EB}}) - g_{3i}(\sigma_v^2)$. To remove this bias, Butar and Lahiri (1997) proposed the estimator

$$\mathrm{MSE_{BL}}(\widehat{\theta}_i^{\mathrm{EB}}) = \mathrm{MSE_{ILL}}(\widehat{\theta}_i^{\mathrm{EB}}) + g_{3i}(\widehat{\sigma}_v^2), \tag{20}$$

which is second-order correct and also area-specific. Note that (20) is identical to our area-specific estimator (15).

Morris (1983) used $\left[(t-p-2)/(t-p)\right]\widehat{\sigma}_v^2/(\widehat{\sigma}_v^2+\psi_i)$ to estimate γ_i in (5) and obtained an estimator of MSE of the resulting EB estimator. In the special case of equal sampling variances, i.e., $\psi_i = \psi$, his MSE estimator is an approximation to the HB posterior variance, $V(\theta_i \mid \widehat{\boldsymbol{\theta}})$, under a flat improper prior distribution on the hyper-parameters $\boldsymbol{\beta}$ and σ_v^2 (Ghosh, 1992).

Under the design-based approach, the small area parameters θ_i are regarded as fixed, i.e., only the sampling model (2) is used. It is natural, therefore, to use the conditional MSE, $E\left[(\widehat{\theta}_i^{\text{EB}}-\theta_i)^2 \mid \boldsymbol{\theta}\right]$ as the measure of variability associated with $\widehat{\theta}_i^{\text{EB}}$. By expressing $\widehat{\theta}_i^{\text{EB}}$ as $\widehat{\theta}_i+k_i(\widehat{\theta}_1,\ldots,\widehat{\theta}_m)$ and using Stein's (1981) lemma, an unbiased estimator of the conditional MSE of $\widehat{\theta}_i^{\text{EB}}$ can be obtained, under normality of the sampling errors e_i. Hwang and Rao87 in work as yet unpublished obtained this estimator explicitly for the special case of no auxiliary data $\mathbf{x}_i$, using the moment estimators (7) of σ_v^2. They also conducted a limited simulation study by first generating a realization of $\boldsymbol{\theta}$ from the linking model and then creating a large number of simulation runs $\widehat{\boldsymbol{\theta}}$ from the sampling model (2) for specified ψ_i. Their results suggest that the model-based MSE estimators track the conditional MSE quite well even under moderate violations of the assumed linking model. Moreover, the model-based estimators are much more stable than the unbiased estimator. Only in extreme cases, such as a large outlier θ_i, the model-based estimator might perform poorly compared to the unbiased estimator. Rivest (1996) made a systematic study of unbiased estimators of conditional MSE of shrinkage estimators, including EB estimators and bench-marked EB estimators that agree with a director estimator at an aggregate level.

3 Type 2 Model

Turning to the nested error regression model (4), we assume that N_i is large for simplicity so that $\overline{Y}_i \approx \overline{\mathbf{X}}_i'\boldsymbol{\beta} + v_i$, where $\overline{\mathbf{X}}_i = \Sigma_j \mathbf{x}_{ij}/N_i$ and $\overline{Y}_i = \Sigma_j y_{ij}/N_i$ are the ith area means. For fixed σ_v^2 and σ_e^2 the EB estimator of $\overline{Y}_i$ is given by,

$$\overline{y}_i^{\text{EB}}(\sigma_v^2,\sigma^2) = \gamma_i\left[\overline{y}_i + (\overline{\mathbf{X}}_i - \overline{\mathbf{x}}_i)'\widetilde{\boldsymbol{\beta}}\right] + (1-\gamma_i)\overline{\mathbf{X}}_i'\widetilde{\boldsymbol{\beta}}, \tag{21}$$

where $\widetilde{\boldsymbol{\beta}} = \widetilde{\boldsymbol{\beta}}(\sigma_v^2,\sigma_e^2)$ is the WLS estimator of $\boldsymbol{\beta}$, $\gamma_i = \sigma_v^2(\sigma_v^2+\sigma_e^2/n_i)^{-1}$ and $\overline{y}_i$, $\overline{\mathbf{x}}_i$ are the sample means. Now replacing σ_v^2 and σ_e^2 by either moment estimators or REML estimators $\widehat{\sigma}_v^2$ and $\widehat{\sigma}_e^2$, we get the EB estimator $\overline{y}_i^{\text{EB}} = \overline{y}_i^{\text{EB}}(\widehat{\sigma}_v^2,\widehat{\sigma}_e^2)$ which is a weighted average of the "survey regression" estimator $\overline{y}_i + (\overline{\mathbf{X}}_i - \overline{\mathbf{x}}_i)'\widehat{\boldsymbol{\beta}}$ and the regression synthetic estimator $\overline{\mathbf{X}}_i'\widehat{\boldsymbol{\beta}}$. The weight

$\widehat{\gamma}_i = \widehat{\sigma}_v^2(\widehat{\sigma}_v^2 + \widehat{\sigma}_e^2/n_i)^{-1}$ reflects the model variance σ_v^2 relative to the total variance.

For large m, an approximation to $\text{MSE}(\overline{y}_i^{\text{EB}})$ is given by

$$\text{MSE}(\overline{y}_i^{\text{EB}}) \approx M_{2i}(\sigma_v^2, \sigma^2) = M_{1i}(\sigma_v^2, \sigma_e^2) + g_{3i}(\sigma_v^2, \sigma_e^2), \tag{22}$$

where

$$M_{1i}(\sigma_v^2, \sigma_e^2) = g_{1i}(\sigma_v^2, \sigma_e^2) + g_{2i}(\sigma_v^2, \sigma_e^2),$$

with

$$g_{1i}(\sigma_v^2, \sigma_e^2) = (1 - \gamma_i)\sigma_v^2,$$

and

$$g_{2i}(\sigma_v^2, \sigma_e^2) = (\overline{\mathbf{X}}_i - \gamma_i\overline{\mathbf{x}}_i)' A^{-1}(\overline{\mathbf{X}}_i - \gamma_i\overline{\mathbf{x}}_i),$$

where

$$\mathbf{A} = \sum_{i=1}^{m}\left(\sum_{j=1}^{n_i} \mathbf{x}_{ij}\mathbf{x}_{ij}' - \gamma_i n_i \overline{\mathbf{x}}_i \mathbf{x}_i'\right).$$

Further,

$$\begin{aligned} g_{3i}(\sigma_v^2, \sigma_e^2) &= n_i^{-2}\left(\sigma_v^2 + \frac{\sigma_e^2}{n_i}\right)^{-4} E(\overline{y}_i - \overline{\mathbf{x}}_i\beta)^2 h_i(\sigma_v^2, \sigma_e^2), \\ &= n_i^{-2}\left(\sigma_v^2 + \frac{\sigma_e^2}{n_i}\right)^{-3} h_i(\sigma_v^2, \sigma_e^2), \end{aligned} \tag{23}$$

where

$$h_i(\sigma_v^2, \sigma_e^2) = \text{var}(\sigma_v^2\widehat{\sigma}_e^2 - \sigma_e^2\widehat{\sigma}_v^2).$$

The approximation is valid for both moment and REML estimators and it is correct to terms of order $O(m^{-1})$ under regularity conditions (Prasad and Rao, 1990; Datta and Lahiri, 1997).

Prasad and Rao (1990) gave an approximately unbiased estimator of $\text{MSE}(\overline{y}_i^{\text{EB}})$ as

$$\text{MSE}(\overline{y}_i^{\text{EB}}) = M_{1i}(\widehat{\sigma}_v^2, \widehat{\sigma}_e^2) + 2g_{3i}(\widehat{\sigma}_v^2, \widehat{\sigma}_e^2). \tag{24}$$

The bias of (24) is of lower order than m^{-1}. As noted by Butar and Lahiri (1997), the estimator is not area-specific. But using the form (23), we can obtain area-specific estimators as in Section 2. We have two different area-specific estimators:

$$\text{MSE}_1(\overline{y}_i^{\text{EB}}) = M_{1i}(\widehat{\sigma}_v^2, \widehat{\sigma}_e^2) + 2n_i^{-2}\left(\widehat{\sigma}_v^2 + \frac{\widehat{\sigma}_e^2}{n_i}\right)^{-4} h_i(\widehat{\sigma}_v^2, \widehat{\sigma}_e^2)(\overline{y}_i - \overline{x}_i'\widehat{\beta})^2, \tag{25}$$

$$\begin{aligned} \text{MSE}_2(\overline{y}_i^{\text{EB}}) = {} & M_{1i}(\widehat{\sigma}_v^2, \widehat{\sigma}_e^2) + g_{3i}(\widehat{\sigma}_v^2, \widehat{\sigma}_e^2) \\ & + n_i^{-2}\left(\widehat{\sigma}_v^2 + \frac{\widehat{\sigma}_e^2}{n_i}\right)^{-4} h_i(\widehat{\sigma}_v^2, \widehat{\sigma}_e^2)(\overline{y}_i - \overline{\mathbf{x}}_i'\widehat{\beta})^2, \end{aligned} \tag{26}$$

where $\widetilde{\beta} = \widetilde{\beta}(\widehat{\sigma}_v^2, \widehat{\sigma}_e^2)$. Both (25) and (26) are approximately unbiased.

A bootstrap estimator of $\text{MSE}(\overline{y}_i^{\text{EB}})$ is obtained along the lines of Section 2. Butar and Lahiri (1997) obtained an analytical approximation to the bootstrap estimator, $\text{MSE}_{\text{LL}}(\overline{y}_i^{\text{EB}})$, and showed that its bias is $-g_{3i}(\widehat{\sigma}_v^2, \widehat{\sigma}_e^2)$ to terms of order m^{-1}. By correcting for this bias, they arrived at a second-order correct, area specific estimator which is identical to our area-specific estimator (26).

Fuller (1989) obtained an approximation to the conditional MSE of the EB estimator of a small area mean under a general framework, where the conditioning is on the data in the area of interest. For the type 2 model, this conditional MSE is given by $E[(\overline{y}_i^{\text{EB}} - \overline{Y}_i)^2 \mid \mathbf{y}_i]$, where $\mathbf{y}_i = (y_{i1}, \ldots, y_{in_i})'$. He also obtained an estimator of the conditional MSE. This estimator is area-specific and it would be interesting to spell it out for the type 2 model and compare it with our area-specific estimators (25) and (26). Singh et al. (1998) obtained a Monte Carlo alternative to the Prasad-Rao estimator (24), but it is not area-specific.

The EB estimator $\overline{y}_i^{\text{EB}}$ is not design-consistent, unlike $\widehat{\theta}_i^{\text{EB}}$, if the survey weights are unequal. Prasad and Rao (1995) proposed a pseudo-EBLUP estimator which depends on the survey weights and is design-consistent. They have also obtained a model-based estimator of its MSE.

For the case of a binary variable y, Farrell et al. (1997) used the bootstrap method to obtain an estimator of MSE of the EB estimator of P_i, the population proportion in the ith area.

Acknowledgments: This research was supported by a grant from the Natural Sciences and Engineering Research Council of Canada. The author is thankful to the referees for useful comments.

4 References

Battese, G.E., R.M. Harter, and W.A. Fuller (1988). An error-component model for prediction of county crop areas using survey and satellite data. *J. Amer. Statist. Assoc. 83*, 28–36.

Butar, F.B. and P. Lahiri (1997). On the measures of uncertainty of empirical Bayes small-area estimators. Technical report, Department of Mathematics and Statistics, University of Nebraska-Lincoln.

Casella, G. and R.L. Berger (1990). *Statistical Inference.* Belmont, California: Wadsworth.

Datta, G. and P. Lahiri (1997). A unified measure of uncertainty of estimated best linear unbiased predictor in small-area estimation problems.

Technical report, Department of Mathematics and Statistics, University of Nebraska-Lincoln.

Farrell, P., B. MacGibbon, and T.J. Tomberlin (1997). Empirical Bayes estimators of small area proportions in multistage designs. *Statistica Sinica 7*, 1065–1083.

Fay, R.E. and R.A. Herriot (1979). Estimates of income for small places: an application of James-Stein procedures to census data. *J. Amer. Statist. Assoc. 74*, 269–277.

Fuller, W.A. (1989). Prediction of true values for the measurement error model. In *Conference on Statistical Analysis of Measurement Error Models and Applications*. Humboldt State University.

Ghosh, M. (1992). Constrained Bayes estimates with applications. *J. Amer. Statist. Assoc. 87*, 533–540.

Ghosh, M. and P. Lahiri (1987). Robust empirical Bayes estimation of means from stratified samples. *J. Amer. Statist. Assoc. 87*, 1153–1162.

Ghosh, M. and J.N.K. Rao (1994). Small area estimation: An appraisal. *Statist. Sci. 9*, 55–93.

Kackar, R.N. and D.A. Harville (1984). Approximation for standard errors of estimators of fixed and random effect in mixed linear models. *J. Amer. Statist. Assoc. 79*, 853–862.

Lahiri, P. and J.N.K. Rao (1995). Robust estimation of mean squared error of small area estimators. *J. Amer. Statist. Assoc. 90*, 758–766.

Laird, N.M. and T.A. Louis (1989). Bayes and empirical Bayes ranking methods. *J. Edu. Statist. 14*, 29–46.

Malec, D., J. Sedransk, C.L. Moriarity, and F.B. LeClere (1997). Small area inference for binary variables in the national health interview survey. *J. Amer. Statist. Assoc. 92*, 815–826.

Morris, C.N. (1983). Parametric empirical Bayes inference: Theory and applications. *J. Amer. Statist. Assoc. 78*, 47–65.

Prasad, N.G.N. and J.N.K. Rao (1990). The estimation of the mean squared error of small-area estimators. *J. Amer. Statist. Assoc. 85*, 163–171.

Prasad, N.G.N. and J.N.K. Rao (1995). On robust small area estimation using a simple random effects model. Technical report, Laboratory for Research in Statistics and Probability, Carleton University.

Rivest, L.P. (1996). An estimator for the mean squared error of small area estimates. Technical report, Department of Mathematics and Statistics, University of Laval.

Singh, A.C., D.M. Stukel, and D. Pfefferman (1998). Bayes versus frequentist measures of error in small area estimation. *J. Roy. Statist. Soc. Ser. B 60*, 377–396.

Stein, C. (1981). Estimation of the mean of a multivariate normal distribution. *Ann. Statist. 9*, 1135–1151.

Stukel, D.M. and J.N.K. Rao (1999). On small-area estimation under two-fold nested error regression models. *J. Statist. Planning and Inf. 78*, 131–147.

Wolter, K.M. (1995). *Introduction to Variance Estimation.* New York: Springer-Verlag.

4

Semiparametric Empirical Bayes Estimation in Linear Models

P.K. Sen

ABSTRACT The empirical Bayes methodology, in a parametric setup, incorporates a conjugate prior in the estimation of the shrinkage factor that characterizes the estimators. The situation becomes more complex in semiparametric and (even more in) nonparametric models. Using first-order asymptotic representations for semiparametric and nonparametric estimators it is shown that the Zellner (Gaussian) *g*-prior on the regression parameters can be readily adopted to formulate suitable empirical Bayes (point) estimators that are essentially related to the robust Stein-rule estimators. Adaptive versions are also considered in the same vein.

1 Introduction

Bayes, empirical Bayes (EB) and related procedures, mostly having a parametric flavor, have received considerable attention from theoretical as well as application perspectives. At the same time, it has also been recognised that in interdisciplinary scientific investigations, standard parametric statistical models may generally have a limited scope (due to plausible departures from model based regularity assumptions), and semiparametrics (and nonparametrics) fare better in some respects. Though Dirichlet priors have been used in some nonparametric approaches, basic robustness issue surfaces in this context too. In various fields of application, semiparametric models that compromise robustness and efficiency are being used more and more. In the current study we confine ourselves to regression (and mostly) linear models. We are confronted with two basic queries: *To what extent can a simple parametric* (*such as a normal theory*) *linear model be adopted in practice*? *Moreover, to what extent does the* EB *methodology work out beyond the conventional parametric paradigms*?

In regression analysis, linear models occupy a focal place. Often, Box–Cox type transformations are incorporated to induce linearity of regression relationships to a greater extent, though in that process, the normality of the error components may be distorted. Thus, even if a linear regression model is contemplated, a semiparametric model that allows the error distribution to be rather arbitrary may have greater practical appeal. Nonparametric estimators in linear models are based on rank statistics and enjoy global distributional robustness properties (Puri and Sen, 1971, 1985), while semiparametric estimators have greater affinity to M-and L-statistics which put more emphasis on local robustness aspects (Huber, 1973; Jurečková and Sen, 1996) around a stipulated model. An appraisal of this robustness picture from ***estimating functions*** perspective has recently been made by Sen (1997).

In a parametric EB setup, a conjugate prior is generally incorporated in the formulation of the Bayes shrinkage factor with an EB interpretation and estimation protocol; their finite sample justifications may no longer be strictly tenable in a semiparametric or nonparametric setup. Even in ***generalized linear models*** (GLM) that pertain mostly to densities belonging to the so called ***exponential family*** (McCullagh and Nelder, 1989), such interpretations may not be universally true. As such, there may be a need for some ***asymptotics*** to eliminate the impasse. In this perspective, we examine the adaptability of the classical EB approach in linear models when robust and/or nonparametric estimators are used instead of a classical ***least squares*** estimators (LSE) or normal theory ***maximum likelihood estimators*** (MLE). In this study, we shall mainly concentrate on point estimators, and examine the interplay of semiparametrics and EB methodology. The EB methodology to deal with the confidence set estimation problems in a comparable level of generality will be presented in a subsequent study.

Section 2 deals with a very brief review of the normal theory MLE and their Stein-rule versions, and in Section 3, following Ghosh et al. (1989), an EB formulation based on Zellner's (1986) g-prior is recapitulated. The findings are extended in Section 4 to BAN (best asymptotically normal) point estimators. The main results on empirical Bayes estimators for semiparametric linear models are then discussed in Section 5. Semiparametric adaptive empirical Bayes estimators are also considered there. Some concluding remarks are presented in the concluding section.

2 Preliminary Notions

Let us consider the classical univariate linear model:

$$\mathbf{Y} = (Y_1, \ldots, Y_n)' = \mathbf{X}\boldsymbol{\beta} + \mathbf{e}; \quad \mathbf{e} = (e_1, \ldots, e_n)', \tag{1}$$

where the error variables e_i are independent and identically distributed (iid) random variables (rv), $\boldsymbol{\beta}$ is an unknown parameter (p-vector), $\mathbf{X}$ is a known (design) matrix of order $n \times p$ (whose row vectors are denoted by $\mathbf{x}_1, \ldots, \mathbf{x}_n$ respectively), and without any loss of generality, we assume that $\mathbf{X}$ is of rank p $(1 \leq p \leq n)$. We are primarily intertested in the estimation of the parameter $\boldsymbol{\beta}$ when there may be some ***uncertainty about the constraints.***

In a classical parametric (normal theory) setup, it is assumed that

$$\mathbf{e} \sim \mathcal{N}_n(\mathbf{0}, \sigma^2 \mathbf{I}_n), \quad 0 < \sigma^2 < \infty. \tag{2}$$

In this setup, the LSE and MLE of $\boldsymbol{\beta}$ are the same, and are given by

$$\widehat{\boldsymbol{\beta}}_{\text{LS}} = (\mathbf{X}'\mathbf{X})^{-1}\mathbf{X}'\mathbf{Y}. \tag{3}$$

It is easy to verify that

$$E(\widehat{\boldsymbol{\beta}}_{\text{LS}}) = \boldsymbol{\beta}; \quad \mathbf{D}(\widehat{\boldsymbol{\beta}}_{\text{LS}}) = \sigma^2(\mathbf{X}'\mathbf{X})^{-1}. \tag{4}$$

For the above model, the log-likelihood function is given by

$$l_n(\boldsymbol{\beta}) = -\frac{n}{2}\log(2\pi) - n\log\sigma - \frac{1}{2\sigma^2}\|\mathbf{Y} - \mathbf{X}\boldsymbol{\beta}\|^2, \tag{5}$$

so the information matrix (on $\boldsymbol{\beta}$) is given by

$$\mathcal{I}_n(\boldsymbol{\beta}) = \sigma^2(\mathbf{X}'\mathbf{X}). \tag{6}$$

Thus, $\widehat{\boldsymbol{\beta}}_{LS}$ is unbiased, efficient and sufficient for $\boldsymbol{\beta}$. However, led by the remarkable observation of Stein (1956), we may conclude that for $p \geq 3$, with respect to a *quadratic risk* function, $\widehat{\boldsymbol{\beta}}_{\text{LS}}$ is not *admissible*; it can be dominated by a class of *shrinkage* estimators, now referred to as the Stein-rule estimators. Similarly, with respect to a *generalized Pitman closeness criterion* (GPCC), $\widehat{\boldsymbol{\beta}}_{\text{LS}}$ is not admissible, for $p \geq 2$. In this case also, the Stein-rule estimators dominate the scene. Motivated by this, we outline some Stein-rule estimators for the linear model in (1), rewritten as

$$\mathbf{Y} = \mathbf{X}_1\boldsymbol{\beta}_1 + \mathbf{X}_2\boldsymbol{\beta}_2 + \mathbf{e}, \tag{7}$$

where $\mathbf{X}$ is partitioned into $\mathbf{X}_1$ and $\mathbf{X}_2$ of order $n \times p_1$ and $n \times p_2$ respectively (so that $p = p_1 + p_2$), and similarly $\boldsymbol{\beta}$ is partitioned into $\boldsymbol{\beta}_1$ and $\boldsymbol{\beta}_2$. We are primarily interested in estimating $\boldsymbol{\beta}_1$ when it is plausible that $\boldsymbol{\beta}_2$ is *close to* a specified $\boldsymbol{\beta}_2^o$ (that we can take without any loss of generality as $\mathbf{0}$). For example, in a factorial design, $\boldsymbol{\beta}_1$ may refer to the main-effects of the factors and $\boldsymbol{\beta}_2$ the interaction effects of various orders. A similar situation is encountered in a polynomial regression model. Actually, in a canonical reduction of the parameter such a partition can always be posed as an alternative to setting a null hypothesis as H_0: $\mathbf{C}\boldsymbol{\beta} = \mathbf{0}$ where $\mathbf{C}$ is a $p_2 \times p$ matrix of known constants (and of rank p_2). Here, in particular, we have $\mathbf{C} = (\mathbf{0}, \mathbf{I}_{p_2})$. Then under the restraint that $\boldsymbol{\beta}_2 = \mathbf{0}$, the restricted MLE/LSE of $\boldsymbol{\beta}_1$ is given by

$$\widehat{\boldsymbol{\beta}}_{\text{RLS},1} = (\mathbf{X}_1'\mathbf{X}_1)^{-1}\mathbf{X}_1'\mathbf{Y}, \tag{8}$$

and we partition the unrestricted estimator as

$$\widehat{\boldsymbol{\beta}}_{\text{LS}} = (\widehat{\boldsymbol{\beta}}_{\text{LS},1}', \boldsymbol{\beta}_{\text{LS},2}')'. \tag{9}$$

Consider then the hypothesis testing problem for

$$H_0: \boldsymbol{\beta}_2 = \mathbf{0} \text{ against } H_1: \boldsymbol{\beta}_2 \neq \mathbf{0}. \tag{10}$$

Let $\mathcal{L}_n$ be the conventional ANOVA (analysis of variance) test statistic, standardized in such a way that asymptotically under the null hypothesis it has the central chi-square distribution with p_2 degrees of freedom (DF). Then, typically a Stein-rule estimator of $\boldsymbol{\beta}_1$ can be posed as

$$\widehat{\boldsymbol{\beta}}^{S}_{\mathrm{LS},1} = \widehat{\boldsymbol{\beta}}_{\mathrm{RLS},1} + (1 - k/\mathcal{L}_n)(\widehat{\boldsymbol{\beta}}_{\mathrm{LS},1} - \widehat{\boldsymbol{\beta}}_{\mathrm{RLS},1}), \tag{11}$$

where $k(\geq 0)$ is a suitable shrinkage factor, often taken equal to $p_2 - 2$ when $p_2 \geq 2$. Side by side, we may also list the so called ***positive-rule shrinkage*** estimator:

$$\widehat{\boldsymbol{\beta}}^{S+}_{\mathrm{LS},1} = \widehat{\boldsymbol{\beta}}_{\mathrm{RLS},1} + (1 - k/\mathcal{L}_n)^{+}(\widehat{\boldsymbol{\beta}}_{\mathrm{LS},1} - \widehat{\boldsymbol{\beta}}_{\mathrm{RLS},1}), \tag{12}$$

where $a^+ = \max(a, 0)$. From a related perspective, GPCC, which will be considered later on, we arrive at the following version of a Stein-rule estimator of $\boldsymbol{\beta}_1$; this corresponds to a special case of $k = m_{p_2}$, where m_q stands for the median of the central chi-square d.f. with q DF (and it is known that $q - 1 < m_q < q - 1/2$, $\forall q \geq 1$, and for large q, $m_q \sim q - 2/3$).

$$\tilde{\boldsymbol{\beta}}^{S}_{\mathrm{LS},1} = \widehat{\boldsymbol{\beta}}_{\mathrm{RLS},1} + (1 - m_{p_2}/\mathcal{L}_n)(\widehat{\boldsymbol{\beta}}_{\mathrm{LS},1} - \widehat{\boldsymbol{\beta}}_{\mathrm{RLS},1}). \tag{13}$$

Thus, it suffices to consider here $p_2 \geq 1$. It is also possible to define a positive-rule version $\tilde{\boldsymbol{\beta}}^{S+}_{\mathrm{LS},1}$ by replacing in (13) the shrinkage factor $(1 - m_{p_2}/\mathcal{L}_n)$ by $(1 - m_{p_2}/\mathcal{L}_n)^+$. There are many other variants of such Stein-rule estimators. Among these, the ***preliminary test estimator*** (PTE) deserves mention. This can be posed as

$$\widehat{\boldsymbol{\beta}}^{\mathrm{PT}}_{\mathrm{LS},1} = \widehat{\boldsymbol{\beta}}_{\mathrm{RLS},1} I(\mathcal{L}_n < \mathcal{L}_{n,\alpha}) + \widehat{\boldsymbol{\beta}}_{\mathrm{LS},1} I(\mathcal{L}_N \geq \mathcal{L}_{n,\alpha}), \tag{14}$$

where $\mathcal{L}_{n,\alpha}$ stands for the critical value of $\mathcal{L}_{n,\alpha}$ at the level of significance $\alpha (0 < \alpha < 1)$. Technically, the Stein-rule estimator does not involve the level of significance (α), and has some theoretical advantages over the PTE, though operationally a PTE may be more intuitive and convenient.

3 Empirical Bayes Interpretations

We mainly follow the line of attack of Ghosh et al. (1989). In the usual Bayesian setup, we assume that

$$\mathbf{Y} \mid \boldsymbol{\beta} \sim \mathcal{N}_n(\mathbf{X}\boldsymbol{\beta}, \sigma^2 \mathbf{I}_n), \quad 0 < \sigma^2 < \infty, \tag{15}$$

and the *prior* Π on $\boldsymbol{\beta}$ is given by

$$\boldsymbol{\beta} \sim \mathcal{N}_p(\boldsymbol{\nu}, \tau^2 \mathbf{V}), \tag{16}$$

where $\mathbf{V}$ is positive definite (p.d.) and $0 < \tau^2 < \infty$. Then the *posterior* distribution of $\boldsymbol{\beta}$, given $\mathbf{Y}$, is

$$\mathcal{N}_p\left[\boldsymbol{\nu} + \left(\mathbf{X}'\mathbf{X} + \frac{\sigma^2}{\tau^2}\mathbf{V}^{-1}\right)^{-1} \mathbf{X}'(\mathbf{Y} - \mathbf{X}\boldsymbol{\nu}), \boldsymbol{\Gamma}\right], \tag{17}$$

where

$$\boldsymbol{\Gamma} = \sigma^2 \left(\mathbf{X}'\mathbf{X} + \frac{\sigma^2}{\tau^2}\mathbf{V}^{-1}\right)^{-1} \tag{18}$$

For a p.d. $\mathbf{Q}$, consider the (nonnegative) quadratic loss function

$$L(\mathbf{b}, \boldsymbol{\beta}) = (\mathbf{b} - \boldsymbol{\beta})'\mathbf{Q}(\mathbf{b} - \boldsymbol{\beta}). \tag{19}$$

Then the Bayes estimator of $\boldsymbol{\beta}$ is given by the posterior mean, i.e.,

$$\begin{aligned} \widehat{\boldsymbol{\beta}}_B &= \boldsymbol{\nu} + \left(\mathbf{X}'\mathbf{X} + \frac{\sigma^2}{\tau^2}\mathbf{V}^{-1}\right)^{-1} \mathbf{X}'(\mathbf{Y} - \mathbf{X}\boldsymbol{\nu}) \\ &= \boldsymbol{\nu} + \left(\mathbf{I} + \frac{\sigma^2}{\tau^2}\mathbf{V}^{-1}(\mathbf{X}'\mathbf{X})^{-1}\right)^{-1} (\widehat{\boldsymbol{\beta}}_{\text{LS}} - \boldsymbol{\nu}). \end{aligned} \tag{20}$$

Under the so called Zellner (1986) *g-prior* (Arnold, 1981),

$$\mathbf{V} = (\mathbf{X}'\mathbf{X})^{-1}, \tag{21}$$

we have

$$\mathbf{I} + \frac{\sigma^2}{\tau^2}\mathbf{V}^{-1}(\mathbf{X}'\mathbf{X})^{-1} = \left(1 + \frac{\sigma^2}{\tau^2}\right)\mathbf{I}, \tag{22}$$

so that the Bayes estimator simplifies to

$$\widehat{\boldsymbol{\beta}}_B = \boldsymbol{\nu} + (1 - B)(\widehat{\boldsymbol{\beta}}_{\text{LS}} - \boldsymbol{\nu}), \tag{23}$$

where the Bayes (shrinkage) factor is given by

$$B = \sigma^2(\sigma^2 + \tau^2)^{-1} (\geq 0). \tag{24}$$

Recall that as in Section 2, our primary interest lies in the estimation of the component parameter $\boldsymbol{\beta}_1$. With this end in mind, we write $\boldsymbol{\nu} = (\boldsymbol{\nu}_1', \boldsymbol{\nu}_2')'$, and obtain the posterior distribution of $\boldsymbol{\beta}_1$, given $\mathbf{Y}$, as

$$\mathcal{N}_{p_1}\left[\nu_1 + (1 - B)(\widehat{\boldsymbol{\beta}}_{\text{LS},1} - \nu_1), \sigma^2(1 - B)\mathbf{C}_{11.2}^{-1}\right], \tag{25}$$

where we let

$$\begin{aligned} \mathbf{X}'\mathbf{X} &= \mathbf{C} = ((\mathbf{C}_{ij}))_{i,j=1,2}; \\ \mathbf{C}_{ii.j} &= \mathbf{C}_{ii} - \mathbf{C}_{ij}\mathbf{C}_{jj}^{-1}\mathbf{C}_{ji}, \quad \text{for} \quad i, j = 1, 2. \end{aligned} \tag{26}$$

Note that ν_1, σ^2, τ^2, and hence, B are all unknown quantities. In an EB approach, in order to deemphasize the uncertainty of the prior Π, we substitute their estimates derived from the marginal distribution of $\mathbf{Y}$, (given $\mathbf{X}$). In the current context, in line with Section 2, we set

$$\boldsymbol{\nu}' = (\boldsymbol{\nu}_1', \mathbf{0}'), \text{ i.e., } \quad \boldsymbol{\nu}_2 = \mathbf{0}, \quad \boldsymbol{\nu}_1 \text{ arbitrary.} \tag{27}$$

Also note that the RLSE of $\boldsymbol{\beta}_1$ is $\widehat{\boldsymbol{\beta}}_{\text{RLS},1} = \mathbf{C}_{11}^{-1}(\mathbf{X}_1'\mathbf{Y})$, while the marginal distribution of $\mathbf{Y}$ is

$$\mathcal{N}_n(\mathbf{X}_1\boldsymbol{\nu}_1, \sigma^2\mathbf{I}_n + \tau^2\mathbf{P}_\mathbf{X}), \tag{28}$$

where the *projection matrix* is defined as

$$\mathbf{P}_\mathbf{X} = \mathbf{X}(\mathbf{X}'\mathbf{X})^{-1}\mathbf{X}'. \tag{29}$$

Further, noting that $\mathbf{X}_1'\mathbf{Y} = \mathbf{C}_{11}\widehat{\boldsymbol{\beta}}_{\text{RLS},1} = \mathbf{C}_{11}\widehat{\boldsymbol{\beta}}_{\text{LS},1} + \mathbf{C}_{12}\widehat{\boldsymbol{\beta}}_{\text{LS},2}$, we have

$$\widehat{\boldsymbol{\beta}}_{\text{LS},1} = \widehat{\boldsymbol{\beta}}_{\text{RLS},1} - \mathbf{C}_{11}^{-1}\mathbf{C}_{12}\widehat{\boldsymbol{\beta}}_{\text{LS},2}. \tag{30}$$

Ghosh et al. (1989) invoked the completeness and sufficiency properties of $(\widehat{\boldsymbol{\beta}}_{\text{RLS},1}, \widehat{\boldsymbol{\beta}}_{\text{LS},2}'\mathbf{C}_{22.1}\widehat{\boldsymbol{\beta}}_{\text{LS},2}, \|\mathbf{Y} - \mathbf{X}\widehat{\boldsymbol{\beta}}_{\text{LS}}\|^2)$, noticed that

$$\begin{gathered} E(\widehat{\boldsymbol{\beta}}_{\text{RLS},1}) = \boldsymbol{\nu}_1, \quad \|\mathbf{Y} - \mathbf{X}\widehat{\boldsymbol{\beta}}_{\text{LS}}\|^2/\sigma^2 \sim \chi^2_{n-p}, \\ (\tau^2 + \sigma^2)^{-1}(\widehat{\boldsymbol{\beta}}_{\text{LS},2}\mathbf{C}_{22.1}\widehat{\boldsymbol{\beta}}_{\text{LS},2}) \sim \chi^2_{p_2}, \end{gathered} \tag{31}$$

and thereby considered the estimator $\widehat{\boldsymbol{\beta}}_{\text{RLS},1}$ for $\boldsymbol{\nu}_1$, the uniformly minimum variance (UMV) estimator $(p_2-2)/(\widehat{\boldsymbol{\beta}}_{\text{LS},2}'\mathbf{C}_{22.1}\widehat{\boldsymbol{\beta}}_{\text{LS},2})$ for $(\sigma^2+\tau^2)^{-1}$, and the best scale invariant estimator $S_e^2 = \|\mathbf{Y} - \mathbf{X}\widehat{\boldsymbol{\beta}}_{\text{LS}}\|^2/(n-p+2)$ for σ^2. These led them to the following EB estimator of $\boldsymbol{\beta}_1$:

$$\widehat{\boldsymbol{\beta}}_{\text{EB},1} = \widehat{\boldsymbol{\beta}}_{\text{RLS},1} + \left[1 - \frac{(p_2-2)S_e^2}{\widehat{\boldsymbol{\beta}}_{\text{LS},2}'\mathbf{C}_{22.1}\widehat{\boldsymbol{\beta}}_{\text{LS},2}}\right](\widehat{\boldsymbol{\beta}}_{\text{LS},1} - \widehat{\boldsymbol{\beta}}_{\text{RLS},1}). \tag{32}$$

A positive-rule EB estimator can then be easily obtained from (32) by replacing the shrinkage factor by its nonnegative part, that is

$$\widehat{\boldsymbol{\beta}}^+_{\text{EB},1} = \widehat{\boldsymbol{\beta}}_{\text{RLS},1} + \left[1 - \frac{(p_2-2)S_e^2}{\widehat{\boldsymbol{\beta}}_{\text{LS},2}'\mathbf{C}_{22.1}\widehat{\boldsymbol{\beta}}_{\text{LS},2}}\right]^+(\widehat{\boldsymbol{\beta}}_{\text{LS},1} - \widehat{\boldsymbol{\beta}}_{\text{RLS},1}). \tag{33}$$

Similarly, replacing $p_2 - 2$ by a constant k, a general shrinkage estimator can be interpreted in an EB fashion. For example, if instead of the UMV estimator of $(\sigma^2+\tau^2)^{-1}$, we consider its median-unbiased (MU) estimator $m_{p_2}/(\widehat{\boldsymbol{\beta}}_{LS,2}'\mathbf{C}_{22.1}\boldsymbol{\beta}_{LS,2})$ (also a sufficient statistic), we would arrive at the EB estimator proposed in (13). However, it follows from Ghosh et al. (1989) that under the quadratic loss mentioned before, the risk of such an empirical

Bayes estimator is minimized at $k = p_2 - 2$. Hence, we prefer to use the estimator in (32), though we may remark that there could be an optimal choice of k other than $p_2 - 2$ if we use some other loss function.

With the quadratic loss defined in (19), we consider the risk (i.e., the expected loss) $r(\cdot)$ of these estimators. It follows from Ghosh et al. (1989) that

$$r(\widehat{\boldsymbol{\beta}}_{\mathrm{LS},1}) - r(\widehat{\boldsymbol{\beta}}_{\mathrm{EB},1}) \\ = \mathbf{B}\frac{(n-p)(p_2-2)}{p_2(n-p+2)} \text{ trace } (\mathbf{C}_{21}\mathbf{C}_{11}^{-1}\mathbf{Q}\mathbf{C}_{11}^{-1}\mathbf{C}_{12}\mathbf{C}_{22.1}^{-1}), \quad (34)$$

and this is nonnegative for all $p_2 > 2$. It also suggests that whenever $\mathbf{C}_{12}$ is a null matrix, there is no reduction in risk even if p_2 is greater than 2. Also, we have

$$r(\widehat{\boldsymbol{\beta}}_{\mathrm{RLS},1}) - r(\widehat{\boldsymbol{\beta}}_{\mathrm{EB},1}) \geq 0 \text{ iff } \left(\frac{1-B}{B}\right)^2 \geq \frac{2(n-p+p_2)}{p_2(n-p+2)}. \quad (35)$$

Whenever τ^2 is small compared to σ^2, $B \sim 1$, and hence, $\widehat{\boldsymbol{\beta}}_{\mathrm{RLS},1}$ performs better than the EB estimator. Since the right hand side of (35) converges to $2/p_2$ as n becomes large (and $p_2 > 2$), the EB estimator dominates the RLSE for a range of τ^2/σ^2. In passing, we may also remark that the positive-rule EB estimator dominates the usual EB estimator under quadratic loss, and we will have a similar property under some other loss functions as well.

Let us look into this relative picture in the light of the generalized GPCC. For two rival estimators, say, $\hat{\boldsymbol{\theta}}_1$ and $\hat{\boldsymbol{\theta}}_2$, of a common parameter $\boldsymbol{\theta}$, and a loss function $L(\mathbf{a}, \mathbf{b})$, as defined in (19), the GPCC of $\hat{\boldsymbol{\theta}}_1$ with respect to $\hat{\boldsymbol{\theta}}_2$ is defined as

$$P(\hat{\boldsymbol{\theta}}_1, \hat{\boldsymbol{\theta}}_2 \mid \boldsymbol{\theta}) = P\{L(\hat{\boldsymbol{\theta}}_1, \boldsymbol{\theta}) < L(\hat{\boldsymbol{\theta}}_2, \boldsymbol{\theta}) \mid \boldsymbol{\theta}\} \\ + \frac{1}{2}P\{L(\hat{\boldsymbol{\theta}}_1, \boldsymbol{\theta}) = L(\hat{\boldsymbol{\theta}}_2, \boldsymbol{\theta}) \mid \boldsymbol{\theta}\}. \quad (36)$$

Then $\hat{\boldsymbol{\theta}}_1$ is regarded as closer to $\boldsymbol{\theta}$ than $\hat{\boldsymbol{\theta}}_2$ if the right hand side of (36) is $\geq 1/2$ for all $\boldsymbol{\theta}$, with the strict inequality sign holding for some $\boldsymbol{\theta}$. We refer to Keating et al. (1993) for some discussion of the GPCC.

Sen et al. (1989) have examined the Stein paradox in the light of the GPCC. Some further results in this parametric setup are due to Sen and Sengupta (1991), and Saleh and Sen (1991). We may summarize these findings as follows. In the light of GPCC, as well as in quadratic risk,

$$\widehat{\boldsymbol{\beta}}^{+}_{\mathrm{EB},1} \succ \widehat{\boldsymbol{\beta}}_{\mathrm{EB},1} \succ \widehat{\boldsymbol{\beta}}_{\mathrm{LS},1}. \quad (37)$$

On the other hand, the PTE fails to dominate any of the three estimators considered above, though in a close neighborhood of $\boldsymbol{\beta}_2 = \mathbf{0}$, it performs better than them. Unfortunately, under GPCC, there may not be a unique best shrinkage estimator even within the class of Stein-rule estimators.

Ghosh and Sen (1991) extended the GPCC concept to a Bayesian setup and introduced the *posterior Pitman closeness* (PCC) measure. Let $\Pi(\boldsymbol{\theta})$ be a prior distribution defined on $\boldsymbol{\Theta}$, and let $\boldsymbol{\delta}_1, \boldsymbol{\delta}_2$ be two Bayes estimators of $\boldsymbol{\theta}$ under the prior $\Pi(\boldsymbol{\theta})$. Then

$$\mathrm{PPC}_{\Pi}(\boldsymbol{\delta}_1, \boldsymbol{\delta}_2 \mid \mathbf{Y}) = P_{\Pi}\{L(\boldsymbol{\delta}_1, \boldsymbol{\theta}) < L(\boldsymbol{\delta}_2, \boldsymbol{\theta})\} + \frac{1}{2} P_{\Pi}\{L(\boldsymbol{\delta}_1, \boldsymbol{\theta}) = L(\boldsymbol{\delta}_2, \boldsymbol{\theta})\}. \tag{38}$$

Thus, $\boldsymbol{\delta}_1$ is said to be posterior Pitman closer to $\boldsymbol{\theta}$ than $\boldsymbol{\delta}_2$, under the prior $\Pi(\boldsymbol{\theta})$, provided

$$\mathrm{PPC}_{\Pi}(\boldsymbol{\delta}_1, \boldsymbol{\delta}_2 \mid \mathbf{Y}) \geq \frac{1}{2}, \quad \forall \mathbf{Y}, \text{ a.e.}, \tag{39}$$

with strict inequality holding for some $\mathbf{Y}$.

For real valued θ, $\mathcal{M}(\theta \mid \mathbf{Y})$, the *posterior median* of θ, given $\mathbf{Y}$, has the PPC_{Π} property. For vector $\boldsymbol{\theta}$, characterizations of multivariate posterior medians have been considered by a host of researchers in the past few years. In particular, if the posterior distribution of $\boldsymbol{\theta}$, given $\mathbf{Y}$, is diagonally symmetric about (a location) $\boldsymbol{\delta}(\mathbf{Y})$, then $\boldsymbol{\delta}(\mathbf{Y})$ is a posterior median of $\boldsymbol{\theta}$, and it has the PPC_{Π}-property (Sen, 1991). It follows from the above results that the dominance of the empirical Bayes estimator over the LSE holds under the PPC criterion as well.

4 BAN Estimators: Empirical Bayes Versions

We consider the same linear model as in Section 2, but deemphasize the normality of the errors. Thus, we let as in (1), $\mathbf{Y} = \mathbf{X}\boldsymbol{\beta} + \mathbf{e}$ where $\mathbf{e} = (e_1, \ldots, e_n)'$, and the e_i are iid r.v's with a probability density function $f(e)$, where the form of f is free from $\boldsymbol{\beta}$; typically a location-scale family of density is contemplated here, and the normal density assumed in Section 2 is an important member of this class. Under suitable regularity assumption (on f, $\mathbf{X}$ and $\boldsymbol{\beta}$), $\widehat{\boldsymbol{\beta}}_{\mathrm{ML}}$, the MLE of $\boldsymbol{\beta}$, is BAN in the sense that asymptotically (as $n \to \infty$)

$$\sqrt{n}(\widehat{\boldsymbol{\beta}}_{\mathrm{ML}} - \boldsymbol{\beta}) \sim \mathcal{N}_p\{\mathbf{0}, n[\mathcal{I}(f)]^{-1}\mathbf{C}^{-1}\}, \tag{40}$$

where $\mathbf{C}$ is defined in (26), and it is tacitly assumed that $\lim_{n\to\infty} n^{-1}\mathbf{X}'\mathbf{X}$ exists, and $\mathcal{I}(f)$, the Fisher information for the p.d.f. f, is defined as

$$\mathcal{I}(f) = E\{[-f'(e)/f(e)]^2\}; \tag{41}$$

note that by assumption, $\mathcal{I}(f)$ does not depend on $\boldsymbol{\beta}$. The *estimating equation* (EE) for the MLE is given by

$$\sum_{i=1}^{n} \mathbf{x}_i[-f'(Y_i - \mathbf{x}_i\widehat{\boldsymbol{\beta}})/f(Y_i - \mathbf{x}_i\widehat{\boldsymbol{\beta}})] = \mathbf{0}'. \tag{42}$$

This asymptotic normality of the MLE, and attainment of the information limit for its asymptotic covariance matrix are shared by a large class of estimators that are known as the BAN or first-order efficient estimators. Following the line of attack of Jurečková and Sen (1996), we may assume that for such a BAN estimator of $\boldsymbol{\beta}$, denoted by $\widehat{\boldsymbol{\beta}}_n$, under suitable regularity assumptions, the following *first-order asymptotic distributional representation* (FOADR) result holds:

$$\widehat{\boldsymbol{\beta}}_n - \boldsymbol{\beta} = \sum_{i=1}^{n} \mathbf{c}_{ni}\phi(e_i) + \mathbf{R}_n, \tag{43}$$

where the score function $\phi(\cdot)$ is given by

$$\phi(x) = -f'(x)/f(x), \quad (-\infty < x < \infty), \tag{44}$$

the $\mathbf{c}_{ni}(= \mathbf{C}^{-1}\mathbf{x}_i')$ depend on the matrix $\mathbf{X}$, and

$$n^{1/2}\|\mathbf{R}_n\| \to 0, \text{ in a suitable mode;} \tag{45}$$

for the EB approach, we assume that this convergence holds in the second mean, while for the GPCC approach, convergence in probability suffices.

In the context of such BAN estimators, we conceive of the same Zellner g-prior, namely that

$$\boldsymbol{\beta} \sim \mathcal{N}_p(\boldsymbol{\nu}, \tau^2\mathbf{V}), \tag{46}$$

where as in (21), we choose $\mathbf{V} = (\mathbf{X}'\mathbf{X})^{-1} = \mathbf{C}^{-1}$. Then writing $\widehat{\boldsymbol{\beta}}_n - \boldsymbol{\nu} = \widehat{\boldsymbol{\beta}}_n - \boldsymbol{\beta} + \boldsymbol{\beta} - \boldsymbol{\nu}$, we obtain by convolution a FOADR for $\widehat{\boldsymbol{\beta}}_n - \boldsymbol{\nu}$ that yields the following (asymptotic) marginal law:

$$\mathcal{N}_p(\mathbf{0}, \{[\mathcal{I}(f)]^{-1} + \tau^2\}\mathbf{V}). \tag{47}$$

As a result, in the FOADR for the posterior distribution of $\boldsymbol{\beta}$, given $\widehat{\boldsymbol{\beta}}_n$, the principal component has the following law:

$$\mathcal{N}_p\left\{\boldsymbol{\nu} + \left[1 + \frac{1}{\mathcal{I}(f)\tau^2}\right]^{-1}(\widehat{\boldsymbol{\beta}}_n - \boldsymbol{\nu}), \boldsymbol{\Gamma}\right\}, \tag{48}$$

where

$$\boldsymbol{\Gamma} = \frac{1}{\mathcal{I}(f)}\left[1 + \frac{1}{\mathcal{I}(f)\tau^2}\right]^{-1}\mathbf{V} = \{\tau^2/[1 + \mathcal{I}(f)\tau^2]\}\mathbf{V}. \tag{49}$$

As such, we may proceed as in Section 2, and by some standard arguments, conclude that in a FOADR representation for the posterior mean (as well as generalized median) of $\boldsymbol{\beta}$, the principal term is

$$\widehat{\boldsymbol{\beta}}_{nB} = \boldsymbol{\nu} + (1 - B)(\widehat{\boldsymbol{\beta}}_n - \boldsymbol{\nu}), \tag{50}$$

where the Bayes (shrinkage) factor B is given by

$$B = \left[1 + \mathcal{I}(f)\tau^2\right]^{-1}, \tag{51}$$

and the other terms in this FOADR are all negligible in a suitable norm.

Given this asymptotic representation for the Bayes BAN-estimator, noting the structural analogy with the LSE, we may proceed as in Section 3, and formulate the following EB estimators. Let $\breve{\boldsymbol{\beta}}_{n,1}$ be the BAN estimator of $\boldsymbol{\beta}_1$ under the reduced model: $\mathbf{Y} = \mathbf{X}_1\boldsymbol{\beta}_1 + \mathbf{e}$. Further, partition the BAN estimator $\widehat{\boldsymbol{\beta}}_n$ as $(\widehat{\boldsymbol{\beta}}'_{n,1}, \widehat{\boldsymbol{\beta}}'_{n,2})'$. Then note by the FOADR for $\widehat{\boldsymbol{\beta}}_n - \boldsymbol{\nu}$ considered in (47),

$$\frac{\mathcal{I}(f)}{1 + \mathcal{I}(f)\tau^2}(\widehat{\boldsymbol{\beta}}'_{n,2}\mathbf{C}_{22.1}\widehat{\boldsymbol{\beta}}_{n,2}) \sim \chi^2_{p_2}. \tag{52}$$

As such, by analogy with Section 3, we estimate

$$\frac{\mathcal{I}(f)}{1 + \tau^2\mathcal{I}(f)} \text{ by } \frac{p_2 - 2}{\widehat{\boldsymbol{\beta}}'_{n,2}\mathbf{C}_{22.1}\widehat{\boldsymbol{\beta}}_{n,2}}. \tag{53}$$

Moreover, by the FOADR for $\widehat{\boldsymbol{\beta}}_n - \boldsymbol{\nu}$, considered in (47),

$$V_n^* = \frac{-1}{n}\sum_{i=1}^{n} \frac{\partial^2}{\partial x^2}\log f(x)\big|_{x = Y_i - \mathbf{x}_i\widehat{\boldsymbol{\beta}}_n} \xrightarrow{\mathcal{P}} \mathcal{I}(f), \quad \text{as } n \to \infty. \tag{54}$$

As a result, we estimate

$$\left[1 + \mathcal{I}(f)\tau^2\right]^{-1} \text{ by } \frac{(p_2 - 2)}{V_n^*(\widehat{\boldsymbol{\beta}}'_{n,2}\mathbf{C}_{22.1}\widehat{\boldsymbol{\beta}}_{n,2})}. \tag{55}$$

At this stage, we denote the conventional log-likelihood ratio type test statistic (or its asymptotically equivalent form based on the Wald or the Rao score statistics), for testing $H_0: \boldsymbol{\beta}_2 = \mathbf{0}$ vs. $H_1: \boldsymbol{\beta}_2 \neq \mathbf{0}$, by $\mathcal{L}_{n2}$. Then, using the FOADR results stated above, it can be shown that under the null as well as contiguous alternatives,

$$\left|V_n^*(\widehat{\boldsymbol{\beta}}'_{n,2}\mathbf{C}_{22.1}\widehat{\boldsymbol{\beta}}_{n,2}) - \mathcal{L}_{n2}\right| \xrightarrow{\mathcal{P}} 0, \quad \text{as} \quad n \to \infty. \tag{56}$$

Consequently, using the above approximation, we arrive at the following EB version of a BAN estimator:

$$\widehat{\boldsymbol{\beta}}_{n\,\mathrm{EB},1} = \breve{\boldsymbol{\beta}}_{n,1} + \left(1 - \frac{p_2 - 2}{\mathcal{L}_{n2}}\right)(\widehat{\boldsymbol{\beta}}_{n,1} - \breve{\boldsymbol{\beta}}_{n,1}). \tag{57}$$

In this form, the EB version agrees with the shrinkage MLE version considered in detail in Sen (1986), though the EB interpretation was not explored there. A positive-rule version of (57) is also from the above discussion. Moreover, if in (52) we use the median m_{p_2} of $\chi^2_{p_2}$, and plug in (54), then instead of (55), we estimate

$$[1+\mathcal{I}(f)\tau^2]^{-1} \quad \text{by} \quad \frac{m_{p_2}}{V_n^*(\widehat{\beta}'_{n,2}\mathbf{C}_{22.1}\widehat{\beta}_{n,2})}, \tag{58}$$

and this leads to an alternative estimator where in (57) we are to replace $(p_2 - 2)$ by m_{p_2}; a positive-rule version also can be formulated by obvious modification of the shrinkage factor.

In this context, if we want to justify the asymptotic risk computations in a conventional sense (i.e., as the limit of the actual risk when n is made to increase indefinitely), then we need to impose the regularity assumptions for the Hájek–LeCam–Inagaki *regular estimators*, as have been displayed in detail in LeCam (1986). Some of these stringent regularity assumptions can be avoided to a certain extent by adopting the measure *asymptotic distributional risk* (ADR) that is based directly on the FOADR itself (wherein the remainder term is neglected). We refer to Sen (1986) where the ADR concept has been elaborated and incorporated in the study of asymptotic properties of shrinkage MLE's. In that setup, in the FOADR, it suffices to show that the remainder term is $o_p(n^{-1/2})$ (while the others need the same order in quadratic mean). In this respect, the situation is much more handy with the GPCC, as formulated in the preceding section; there the limits involve only the probability distributions, and hence, $o_p(n^{-1/2})$ characterization for the remainder term in the FOADR suffices. For certain general isomorphism of GPCC and quadratic risk dominance results, we may refer to Sen (1994), and these findings based on suitable FOADR results, pertain to such Bayes and EB estimators as well. Further, the GPCC dominance results hold for a broader range of p_2 values.

5 Semiparametric Empirical Bayes Estimators

The BAN estimation methodology provides the access to semiparametrics in a very natural way, and EB versions can be worked out in an analogous manner. The main concern on the unrestricted use of parametric EB estimators is their vulnerability or nonrobustness to possible model departures that can either arise due to nonlinearity of the model and/or heteroscedaticity of the errors, or due to a plausible departure from the assumed form of the error density function. In a semiparametric setup, the error distribution may be taken as arbitrary, although the linearity of the model is assumed. In this way, we retain the primary emphasis on the linearity of the model, and want to draw conclusions on the regression parameter without making necessarily stringent distributional assumption on the error component. In that scheme, we may allow for some *local* or *global* departures from an assumed form of the error density; in the former case, there is a stronger emphasis on robustness to local departures (or infinitesimal robustness) without much compromise on the (asymptotic) efficiency. M-estimators and M-statistics (Huber, 1973; Hampel et al., 1986) are particularly useful in this context. L-estimators have also found their way to robust estimation in linear models, though they have generally computational complications in other than location and scale models viz. (Jurečková and Sen, 1996). In the case of global robustness, however, rank based procedures have greater appeal. R-estimators, and their siblings: *regression rank scores* estimators (Gutenbrunner and Jurečková, 1992), and *regression quantiles* (Koenker and Bassett, 1978) are more popular in this context. On the other hand, if the basic linearity of the model is questionable, then nonparametric regression function formulation may be more appealing, wherein the form of the regression function is allowed to be smooth but rather arbitrary. Thus, a nonparametric regression formulation may be comparatively more robust. However, there is a price that we may need to pay for this option. The finite dimensional parameter vector (i.e., $\boldsymbol{\beta}$), considered in earlier sections, has to be replaced by a functional parameter, i.e., the regression function, and its estimation in a nonparametric fashion entails slower rates of convergence, as well as, a possible lack of optimality properties even in an asymptotic setup. Development of EB estimators of such functional parameters is not contemplated in the current study, and we shall confine ourselves only to semiparametric linear models. The procedures to be considered here are generally more robust than standard parametric ones (referred to in earlier sections) as long as the postulated linearity of the model holds, although they may not be robust to possible departures from the assumed linearity of the model. There are other semiparametric models, such as the multiplicative intensity process models that are basically related to the Cox (1972) *proportional hazards model* (PHM). Such PHM's generally relate to the *hazard regression* problem wherein the baseline hazard function is treated as arbitrary (that is nonparametric) while the regression on the covariates is treated as parametric, and the statistical methodology evolved with the development of the *partial likelihood* principle that has been extended to a more general context wherein *matrix valued counting processes* are incorporated to facilitate statistical analysis that has a predominant asymptotic flavor. Here also the hazard regression need not be a finite-dimensional parametric function, and in a more general setup, the regression parameters may be *time-dependent* resulting in a functional parameter space (Murphy and Sen, 1991). For the finite dimensional parametric hazard function formulations, we may refer to the monograph of Anderson et al. (1993) for a unified, up-to-date treatment, and Pedroso de Lima and Sen (1997) for generalization to some multivariate cases. Development of empirical Bayes procedures for such matrix valued counting process needs a somewhat different (and presumably more complex) approach; we shall not enter into these discussions in the current study. Keeping this scenario in mind, we consider specifically the following semiparametric linear model (wherein we use the same notations as introduced in Section 2):

$$\begin{aligned} Y_i &= \mathbf{x}_i\boldsymbol{\beta} + e_i, \quad i = 1, \ldots, n, \\ e_i &\sim f(e), \quad e \in \mathcal{R}, \end{aligned} \tag{59}$$

where the form of f, though unknown, is assumed to be free from $\boldsymbol{\beta}$. It is also assumed that f has a finite Fisher information (with respect to location) $\mathcal{I}(f)$. As in earlier sections, we assume that $\boldsymbol{\beta}$ has the Zellner g-prior, namely

$$\begin{aligned} &\boldsymbol{\beta} \sim \mathcal{N}_p(\boldsymbol{\nu}, \tau^2\mathbf{V}), \\ \mathbf{V} = (\mathbf{X}'\mathbf{X})^{-1}; \quad &\tau^2 = c\big[\mathcal{I}(f)\big]^{-1}, \ c > 0. \end{aligned} \tag{60}$$

Note that here we have a parametric regression model, a parametric prior on $\boldsymbol{\beta}$, a nonparametric p.d.f. $f(e)$. That and why is we term it a semiparametric linear model. We show that EB estimators exist for such semiparametric models, and they correspond to the so called Stein-rule or shrinkage estimators (based on robust statistics) that have been studied extensively in the literature.

In a frequentist setup, let $\widehat{\boldsymbol{\beta}}_n$ be a suitable estimator of $\boldsymbol{\beta}$. Such an estimator can be chosen from a much wider class of ***robust*** estimators: M, $L-$ and R-estimators of regression (and location) parameters, regression quantile estimators, regression rank scores estimators, and many other robust estimators belong to this class. We refer to Jurečková and Sen (1996, Chapters 3 to 7), for a detailed coverage, where a basic FOADR result has been exploited in a systematic manner. Based on this exploitation, we assume that the following FOADR result holds for $\widehat{\boldsymbol{\beta}}_n$:

$$\widehat{\boldsymbol{\beta}}_n - \boldsymbol{\beta} = \sum_{i=1}^{n} \mathbf{c}_{ni}\phi(e_i) + \mathbf{R}_n, \tag{61}$$

where $\phi(\cdot)$ is a suitable score function, which may generally depend on the unknown density $f(\cdot)$, normalized so that

$$\int \phi(e)f(e)\mathrm{de} = 0, \quad \int \phi^2(e)f(e)\mathrm{de} = \sigma_\phi^2(< \infty). \tag{62}$$

The regression vectors $\mathbf{c}_{ni} = (\mathbf{X}'\mathbf{X})^{-1}\mathbf{x}_i'$ depend on $\mathbf{X}$ and are assumed to satisfy the generalized Noether condition:

$$\max_{\{1\le i\le n\}}\mathbf{c}_{ni}'\left(\sum_{j=1}^{n} \mathbf{c}_{nj}\mathbf{c}_{nj}'\right)^{-1} \mathbf{c}_{ni} \to 0. \tag{63}$$

The remainder term $\mathbf{R}_n$ is assumed to satisfy the condition

$$n^{1/2}\|\mathbf{R}_n\| \xrightarrow{\mathcal{P}} 0, \quad \text{as} \quad n \to \infty. \tag{64}$$

We may need to strengthen the mode of convergence to the mean-square norm if we are to deal with the conventional quadratic risk criterion, while for the GPCC, convergence in probability suffices. Note that whenever the density $f(\cdot)$ has a finite Fisher information $\mathcal{I}(f)$, by the Cramér–Rao information inequality,

$$\sigma_\phi^2 \geq [\mathcal{I}(f)]^{-1}, \tag{65}$$

where the equality sign holds when $\widehat{\boldsymbol{\beta}}_n$ is a BAN estimator of $\boldsymbol{\beta}$. For the class of estimators referred to above, there is a subclass that comprise the BAN estimators, and this facilitates the incorporation of the methodology presented in the preceding section.

By virtue of (59), (60) and the above asymptotic representation, we may proceed as in the case of BAN estimators, treated in the preceding section, and by (asymptotic) convolution (for $\widehat{\boldsymbol{\beta}}_n - \boldsymbol{\beta}$ and $\boldsymbol{\beta} - \boldsymbol{\nu}$) obtain the marginal (asymptotic) distribution of $\widehat{\boldsymbol{\beta}}_n - \boldsymbol{\nu}$ (as multinormal with null mean vector and dispersion matrix as the sum of the two dispersion matrices that appear in (60) and in the FOADR in (61):

$$\mathcal{N}_p\big[\mathbf{0}, (\tau^2 + \sigma_\phi^2)(\mathbf{X}'\mathbf{X})^{-1}\big]. \tag{66}$$

As a result (Sen, 1998), the asymptotic posterior distribution of $\boldsymbol{\beta}$, given $\widehat{\boldsymbol{\beta}}_n$, is multinormal with the following form:

$$\mathcal{N}_p\left[\boldsymbol{\nu} + \frac{\tau^2}{\tau^2 + \sigma_\phi^2}(\widehat{\boldsymbol{\beta}}_N - \boldsymbol{\nu}), \frac{\tau^2\sigma_\phi^2}{\tau^2 + \sigma_\phi^2}(\mathbf{X}'\mathbf{X})^{-1}\right]. \tag{67}$$

To pose the EB versions, we consider the reduced model:

$$Y_i = \mathbf{x}_{i(1)}\boldsymbol{\beta}_1 + e_i, \quad i = 1, \ldots, n, \tag{68}$$

where we partition $\boldsymbol{\beta}$ and $\mathbf{x}_i$ as in before, and the p.d.f. $f(\cdot)$ of the e_i is defined as in the case of the full model. Let $\breve{\boldsymbol{\beta}}_{n,1}$ be the corresponding estimator for $\boldsymbol{\beta}_1$. For this estimator, under the reduced model, we have a similar FOADR result where the $\mathbf{c}_{ni}$ are to be replaced by

$$\mathbf{c}_{ni}^{*(1)} = (\mathbf{X}_1'\mathbf{X}_1)^{-1}\mathbf{x}_{i(1)}', \quad i = 1, \ldots, n. \tag{69}$$

Further, for testing $H_0\colon \boldsymbol{\beta}_2 = \mathbf{0}$ against $H_1\colon \boldsymbol{\beta}_2 \neq \mathbf{0}$, let $\mathcal{L}_{n2}^*$ be a suitable test statistic that has asymptotically under the null hypothesis a central chi-square distribution with p_2 DF. Actually, in some cases (such as dealing with L-estimators,) we deal with a quadratic form in $\widehat{\boldsymbol{\beta}}_{n,2}$ while in some other cases (such as R-and M-estimators), we deal with suitable aligned rank or M-statistics as test statistics that are asymptotically equivalent to a quadratic form in the estimators $\widehat{\boldsymbol{\beta}}_{n,2}$.

As in Section 3, we may write $B = \sigma_\phi^2(\sigma_\phi^2 + \tau^2)^{-1}$ so that the (asymptotic) Bayes version can be written as

$$\widehat{\boldsymbol{\beta}}_{B,n} = \boldsymbol{\nu} + (1 - B)(\widehat{\boldsymbol{\beta}}_n - \boldsymbol{\nu}), \tag{70}$$

where on $\boldsymbol{\nu}$ we adopt the same simplifications as outlined in Section 3. As such, we need to restrict ourselves to EB estimates of $\boldsymbol{\nu}_1$ (treating $\boldsymbol{\nu}_2 = \mathbf{0}$), and the Bayes (shrinkage) factor B. We may analogously estimate $\boldsymbol{\nu}_1$ by the reduced model based estimator $\breve{\boldsymbol{\beta}}_{n,1}$. The estimation of B can be carried out in several ways: either estimating $(\sigma_\phi^2 + \tau^2)^{-1}$ from the marginal distribution of $\widehat{\boldsymbol{\beta}}_{n,2}$ (using a similar quadratic form as in Section 3) and σ_ϕ^2 from the distribution of $\widehat{\boldsymbol{\beta}}$, or using the test statistic $\mathcal{L}_{n2}^*$ to estimate B directly. While these alternative approaches are asymptotically equivalent, from a computational point of view, the second approach seems to be more useful. Hence, as in the case of BAN estimators, we consider the following EB version of the estimator of $\boldsymbol{\beta}$:

$$\widehat{\boldsymbol{\beta}}_{\mathrm{EB},n,1} = \breve{\boldsymbol{\beta}}_{n,1} + \left(1 - \frac{p_2 - 2}{\mathcal{L}_{n2}}\right)(\widehat{\boldsymbol{\beta}}_{n,1} - \breve{\boldsymbol{\beta}}_{n,1}). \tag{71}$$

This corresponds to the usual Stein-rule or shrinkage versions of such robust estimators that have been extensively considered in the literature during the past fifteen years, and reported in a systematic manner in Jurečková and Sen (1996); detailed references to the relevant literature are also cited there. Based on median unbiasedness, we may modify the shrinkage factor in (71) along the lines in (58), i.e., we replace $(p_2 - 2)$ by m_{p_2}. Positive-rule versions of such EB estimators are also available; we need to replace the shrinkage factor $(\cdot)$ by $(\cdot)^+$. Moreover, PTE versions can also be posed along the same line, though their EB interpretations may not be so apparent.

The EB interpretation of such semiparametric shrinkage estimators, as outlined above, enables us to incorporate the vast literature of shrinkage robust estimators (Jurečková and Sen, 1996) in the study of the properties of such estimators. In particular, the relative picture of (asymptotic) risks of various empirical Bayes estimators (based on diverse robust statistics) as well as their GPCC comparisons remains comparable to that portrayed in Sections 3 and 4; the basic difference comes in the related noncentrality parameters, and these are related to each other by the usual *Pitman asymptotic relative efficiency* (PARE) measure. For this reason, we skip the details here, but mention the following main results:

(i) If the influence function in the FOADR of $\widehat{\boldsymbol{\beta}}_n - \boldsymbol{\beta}$, i.e., the score $\phi(e)$ in (61), agrees with the conventional Fisher score function, i.e., $\phi_f(e) \equiv -f'(e)/f(e)$, then the corresponding empirical version (as considered above) is an EB BAN estimator, and hence, shares the properties ascribed to such estimators.

(ii) By virtue of (i), it is possible to choose a robust EB estimator such that it becomes stochastically equivalent to an EB BAN estimator (considered in Section 4) for a specific type of underlying density. This property may particularly be important in the context of local robustness aspects when an anticipated density $f(.)$ is in the picture.

(iii) In particular, if we use a rank based (R- or regression rank scores) estimator of $\boldsymbol{\beta}$ that incorporates the so called the *normal score* generating function in its formulation, such an estimator is (asymptotically) at least as efficient as the LSE (considered in Section 2) for a large class of underlying densities, so that the corresponding empirical Bayes version would perform better than the ones considered in Section 3 when the underlying density is not normal. At the same time, it retains robustness to a certain extent.

(iv) Along the same line as in (iii), an adaptive EB version of robust estimators of $\boldsymbol{\beta}$ can be formulated (using the results of Hušková and Sen (1985)) that combines the robustness and asymptotic efficiency properties to a greater extent. For an adaptive estimator $\widehat{\boldsymbol{\beta}}_{\mathrm{AEB},n}$ we have a FOADR as in (61) wherein, in the right hand side, the score function $\phi(e_i)$ is to be replaced by $\phi_f(e_i)$ through an adaptive $\widehat{\phi}(e_i)$ that is chosen from the data-set in such a way that

$$\|\widehat{\phi} - \phi_f\|^2 = \int_{\mathbf{R}} \{\widehat{\phi}(e) - \phi_f(e)\}^2 f(e) \mathrm{d}e \xrightarrow{p} 0, \quad \text{as} \quad n \to \infty, \tag{72}$$

where the Fisher-score function $\phi_f(e)$ is defined as

$$\phi_f(e) = \left[\mathcal{I}(f)\right]^{-1}\left[-f'(e)/f(e)\right], \quad e \in \mathbf{R}, \tag{73}$$

and where we assume that the density f has a finite Fisher information $\mathcal{I}(f)$.

We may remark that the adaptive score function $\widehat{\phi}(\cdot)$ is actually used in the construction of $\widehat{\boldsymbol{\beta}}_{\mathrm{AEB},n}$, though the above asymptotic equivalence result enables us to make use of the FOADR result with $\widehat{\phi}(e_i)$ replaced by $\phi_f(e_i)$. Further, based on this, we may as well claim that $\widehat{\boldsymbol{\beta}}_{\mathrm{AEB},n}$ is a BAN estimator of $\boldsymbol{\beta}$. Therefore, noting that

$$\begin{aligned} &\int_{\mathbf{R}} \left[-f'(e)/f(e)\right] f(e) \mathrm{d}e = 0, \\ &\int_{\mathbf{R}} \left[-f'(e)/f(e)\right]^2 f(e) \mathrm{d}e = \mathcal{I}(f) < \infty, \end{aligned} \tag{74}$$

we can use an orthonormal system $\left\{\left[P_k(u), u \in (0,1)\right]; k \geq 0\right\}$ in a Fourier series representation:

$$\phi_f(u) \sim \sum_{k \geq 0} \gamma_k P_k(u), \quad u \in (0,1), \tag{75}$$

where $\int_0^1 P_k(u)\,du = 0$, $\int_0^1 P_k(u)P_q(u)\,du = \delta_{kq} = 0, 1$ according as $k = q$ or not, for $k, q \geq 0$, and the γ_k are the Fourier coefficients. As a notable case of the $[P_k(\cdot)]$, the Legendre polynomial system was advocated by Hušková and Sen (1985), though there are other possibilities based on suitable trigonometric systems.

Two basic steps are used in the adaptive estimation. First, instead of the infinite series, one considers a finite one, namely,

$$\sum_{k \leq K} \gamma_k P_k(u), \text{ for some positive integer } K, \tag{76}$$

where K is generally adaptive (i.e., data-dependent). Secondly, one estimates the coefficients γ_k, $k = 0, \ldots, K$ based on suitable robust procedures, so that plugging in these estimates in the above finite series representation, one gets the adaptive $\widehat{\phi}(\,)$. For details, we refer to Hušková and Sen (1985) where other pertinent references are also cited. Once this adaptive estimator of $\boldsymbol{\beta}$ is obtained, one can convolute the same with the g-prior on $\boldsymbol{\beta}$ to obtain an asymptotic adaptive Bayes estimator. That in turn leads to the adaptive EB estimator in the same way as in (70) and (71). This adaptive EB estimator is robust and asymptotically efficient too. However, this generally entails a slightly slower rate of convergence. We also refer to Jurečková and Sen (1996, Section 8.5), where asymptotically efficient adaptive robust estimation in the location model has been reviewed in a systematic manner. Our dissemination here extends the findings to a broader class of shrinkage adaptive robust estimators in linear models as well.

Modern computational facilities indeed make it possible to explore such alternative robust empirical Bayes estimators with a view to their implementation in practice too.

6 Some Concluding Remarks

The foundation of empirical Bayes estimators based on robust statistics and their asymptotic isomorphism to shrinkage or Stein-rule versions rest on two basic factors:

(i) FOADR results that permit asymptotic Gaussian laws under suitable regularity assumptions, and

(ii) the incorporation of the Zellner (1986) g-prior that permits the convolution result in a manner compatible with the standard parametric cases.

The g-prior that incorporates the reciprocal of the design matrix $(\mathbf{X}'\mathbf{X})$ in its dispersion matrix clearly conveys the following message.

Whenever with increasing n, $n^{-1}\mathbf{X}'\mathbf{X}$ converges to a p.d. matrix, say $(\mathbf{C}_o)$, as has been assumed in this study (as well as in Ghosh et al. (1989)

and elsewhere in the literature), the prior has increasing (with n) concentration around $\boldsymbol{\nu}$. By having the rate of convergence of this prior comparable with that of $\widehat{\boldsymbol{\beta}}_n - \boldsymbol{\beta}$, the convolution result comes up in a handy nondegenerate (Gaussian) form, and this facilitates computation and simplification of the posterior distributions. This is also comparable to the general asymptotics for shrinkage robust estimators of regression parameters, where the dominance results relate to a small neighborhood of the assumed pivot that has a diameter of the order $n^{-1/2}$, and beyond this shrinking neighborhood the Stein-rule estimator becomes equivalent in risk to the original version; we refer to Jurečková and Sen (1996) for some discussion of such dominance results for shrinkage robust estimators. Thus, in reality, we adopt a sequence of Zellner priors that matches the rate of convergence of the frequentist estimators, and in that sense, our findings relate to a *local* empirical Bayes setup.

The FOADR approach advocated here has the main advantage of identifying the influence curves (IC) of the estimators in a visible manner. It is, of course, not necessary to force the use of such FOADR results. For example, for various types of rank based statistics, L-, M-, and R-estimators, U-statistics, von Mises functionals, and Hadamard differentiable statistical functionals, asymptotic normality results have been derived by alternative approaches by a host of researchers; we refer to Sen (1981) where, in particular, a unified *martingale* approach has been advocated. Such results also permit the convolution result for the marginal law of $\widehat{\boldsymbol{\beta}}_n - \boldsymbol{\nu}$, though in a less visible manner. The main advantage of using a FOADR approach is that whenever a second-order asymptotic distributional representation (SOADR) holds, a precise order for the remainder term can be studied, and this can be incorporated in the study of the rate of approach to the desired asymptotic results. For such SOADR results for some important members of the family of robust estimators, we may again refer to Jurečková and Sen (1996). More work along this line is under way, and it is anticipated that they would be of considerable importance in the study of the asymptotic properties of robust empirical Bayes estimators.

Our findings also pertain to general models (not necessarily linear ones) provided we have a finite dimensional parameter and we incorporate an appropriate Zellner-type prior to obtain the convolution law in a simple way. In particular, if we work with BAN estimators then such a prior can be essentially related to the information matrix related to the parameter as may be assumed to exist. However, the simplification we have in the linear model that estimates of $\boldsymbol{\beta}$ have dispersion matrices that are scalar multiples of the matrix $(\mathbf{X}'\mathbf{X})^{-1}$ may not be generally true in such a case, and that may call for some further adjustments and additional regularity assumptions to obtain the convolution distribution in a natural form that permits a Gaussian posterior distribution. These need to be worked out on a case by case basis.

We have not made use of the conventional Dirichlet priors (on the d.f. F) that are generally used in nonparametric Bayes estimation are based on iid observations. But, our findings may also be linked to such priors under appropriate differentiability conditions on the parameters that are regarded as functionals of the d.f. F. The details are to be provided in a separate communication.

Finally, we would like to mention that though the proposed approach works well for point estimation theory, there are some roadblocks for confidence set estimation. In a linear model, at least in an asymptotic setup, the estimator, or the estimating function that yields the estimator, is assumed to be (multi-)normally distributed, so that well known distribution theory relating to quadratic norms for multinormal laws can be used to construct suitable confidence sets. In the multiple parameter case, as encountered here, the Scheffé and Tukey methods of construction of confidence sets in MANOCOVA models, as well as conventional likelihood ratio statistic based confidence sets are the most popular. Confidence sets based on Roy's (1953) ***union-intersection principle*** (UIP) and the Step-down procedures all share a similar property; in addition, they are usable in more complex models where the earlier approaches might have considerable operational difficulties. Besides, in MANOCOVA models, the UI-confidence sets enjoy an optimality property that may not be shared by the others. On the other hand, if we look into the Stein-rule estimators, even in the classical parametric cases, we would observe that ***they are not normally distributed even in an asymptotic setup.*** Although the distribution theory of Stein-rule estimators, or their positive-rule versions, have been extensively studied in the literature (at least in the normal theory and asymptotic cases), the distribution theory of such quadratic norms may not generally come in a handy form so as to facilitate the construction of confidence sets having some optimal or desirable properties. As such, identifying the isomorphism of the EB point estimators with such robust Stein-rule estimators may not provide us with enough to construct suitable confidence sets that dominate the usual confidence sets in some sense. We need to attack this problem from a somewhat different angle, and relegate this study to a future communication.

Acknowledgments: The author is grateful to both the reviewers for their helpful comments on the original manuscript.

7 References

Anderson, P.K., O. Borgan, R.D. Gill, and N. Keiding (1993). ***Models Based on Counting Processes.*** New York: Springer-Verlag.

Arnold, S.F. (1981). *The Theory of Linear Models and Multivariate Analysis.* New York: Wiley.

Cox, D.R. (1972). Regression models and life tables (with discussion). *J. Roy. Statist. Soc. B 34*, 187–220.

Ghosh, M., A.M.E. Saleh, and P.K. Sen (1989). Empirical Bayes subset estimation in regression models. *Statist. Decisions 7*, 15–35.

Ghosh, M. and P.K. Sen (1991). Bayesian Pitman closeness. *Comm. Statist. A–Theory and Methods 14*, 1511–1529.

Gutenbrunner, C. and J. Jurečková (1992). Regression rank scores and regression quantiles. *Ann. Statist. 20*, 305–330.

Hampel, F.R., P.J. Rousseeuw, E. Ronchetti, and W. Stahel (1986). *Robust Statistics—The Approach Based on Influence Functions.* New York: Wiley.

Huber, P.J. (1973). Robust regression: Asymptotics, conjectures and Monte Carlo. *Ann. Statist. 1*, 799–821.

Hušková, M. and P.K. Sen (1985). On sequentially adaptive asymptotically efficient rank statistics. *Sequential Anal. 4*, 125–151.

Jurečková, J. and P.K. Sen (1996). *Robust Statistical Procedures: Asymptotics and Interrelations.* New York: Wiley.

Keating, J.P., R.L Mason, and P.K. Sen (1993). *Pitman's Measure of Closeness: A Comparison of Statistical Estimators.* Philadelphia: SIAM.

Koenker, R and G. Bassett (1978). Regression quantiles. *Econometrica 46*, 33–50.

LeCam, L. (1986). *Asymptotic Methods in Statistical Decision Theory.* New York: Springer-Verlag.

McCullagh, P. and J.A. Nelder (1989). *Generalized Linear Models, Second edition.* London: Chapman and Hall.

Murphy, S.A and P.K. Sen (1991). Time-dependent coefficients in a Cox-type regression model. *Stoch. Process Appl. 39*, 153–180.

Pedroso de Lima, A.C. and P.K. Sen (1997). A matrix-valued counting process with first-order interactive intensities. *Ann. Appl. Probab. 7*, 494–507.

Puri, M.L. and P.K. Sen (1971). *Nonparametric Methods in Multivariate Analysis.* New York: Wiley.

Puri, M.L. and P.K. Sen (1985). *Nonparametric Methods in General Linear Models.* New York: Wiley.

Roy, S.N. (1953). A heuristic method of test construction and its use in multivariate analysis. *Ann. Math. Statist. 24*, 220–238.

Saleh, A.K.Md.E. and P.K. Sen (1991). Pitman-closeness of some preliminary test and shrinkage estimators. *Comm. Statist. A–Theory Methods 20*, 3643–3657.

Sen, P.K. (1981). *Sequential Nonparametrics: Invariance Principles and Statistical Inference.* New York: Wiley.

Sen, P.K. (1986). On the asymptotic distributional risk shrinkage and preliminary test versions of maximum likelihood estimators. *Sankhyā A 48*, 354–371.

Sen, P.K. (1991). Pitman closeness of statistical estimators: Latent years and the renaissance. In M. Ghosh and P. Pathak (Eds.), *Issues in Statistics: D. Basu Volume*, Inst. Math. Statist. Lect. Notes Monograph Series #17.

Sen, P.K. (1994). Isomorphism of quadratic norm and pc ordering of estimators admitting first order an representation. *Sankhyā A 56*, 465–475.

Sen, P.K. (1997). Estimating functions: Nonparametrics and robustness. In I. Basawa et al. (Eds.), *Selected Proceedings of the Symposium on Estimating Functions*, Volume 32 of *Inst. Math. Statist. Lect. Notes Monograph Series.*

Sen, P.K. (1998). The Hájek convolution theorem and Pitman closeness of regular Bayes estimators. *Tetra Mount. Math. Publ. 17*, 27–35.

Sen, P.K., T. Kubokawa, and A.K.M.E. Saleh (1989). The Stein-paradox in the sense of the Pitman measure of closeness. *Ann. Statist. 17*, 1375–1386.

Sen, P.K. and D. Sengupta (1991). On characterizations of Pitman closeness of some shrinkage estimators. *Commun. Statist. Theory Methods 20*, 3551–3580.

Stein, C. (1956). Inadmissibility of the usual estimator for the mean of a multivariate normal distribution. In *Proc. 3rd Berkeley Symp. Math. Statist. and Prob., Vol. I*, pp. 197–206. Univ. California Press, Berkeley, CA.

Zellner, A. (1986). On assessing prior distributions and Bayesian regression analysis with *g*-prior distributions. In P. Goel and A. Zellner (Eds.), *Bayesian Inference and Decision Techniques*, pp. 233–243. Amsterdam: North Holland.

5

Empirical Bayes Procedures for a Change Point Problem with Application to HIV/AIDS Data

S.-L.T. Normand
K. Doksum

ABSTRACT Let $X(t)$ be a stochastic process which represents a health indicator of an individual. The mean level $\mu(t) = EX(t)$ is constant until the time $t = S$ of infection and after infection it decreases according to some parametric function which may depend on covariate values measured on the individual. We model $X(t) - \mu(t)$ as a stationary Gaussian process and use data from cohort studies to obtain empirical Bayes estimators of the distribution of the change point S.

1 Introduction

We consider experiments in which, for each case in a sample of cases, we have available repeated observations over time of a stochastic process $X(t)$ as well as data on covariates $Z_1, \ldots, Z_d$. The mean level $\mu(t) = EX(t)$ is modelled to be a constant μ until t reaches a change point S. At time S the case becomes subject to stress and for $t \geq S$, the mean level is modelled to follow a parametric time-decreasing function of $Z_1, \ldots, Z_d$ with parameters that need to be estimated. Next, $X(t) - \mu(t)$ is modelled as a stationary Gaussian process. The object is to use the available observations on $X(t)$ and data on covariates $Z_1, \ldots, Z_d$ to produce an estimate of the unknown distribution of the change point S.

Our data example is the San Francisco Mens Health Study, SFMHS Winkelstein et al. (1987). Here S is the unknown time of HIV infection and $X(t)$ is the CD4 T-cell count at time t for infected individuals. The time from infection to detection of HIV is sometimes called the *latency time*. Because the time of detection is known, S is equivalent to the latency time. A number of authors have undertaken modelling of serial CD4 measurements, among these are Munoz et al. (1988); Longini et al. (1989); Berman (1990); Berman (1994); Taylor et al. (1991); Doksum (1991); DeGruttola et al. (1991); Lange et al. (1992); Jewell and Kalbfleisch (1992); Taylor et al. (1994); Vittinghoff et al. (1994); DeGruttola and Tu (1994); Doksum and Normand (1995); Tsiatis et al. (1995); Kuichi et al. (1995); Shi et al. (1996); Satten and Longini (1996); and Faucett and Thomas (1996). These models include a transformation of the raw CD4 counts such as the log, fourth root, or square root transformation.

Our approach is closely related to that of Berman (1990) and Berman (1994) who used observations on $X(t)$ to estimate the distribution of the latency time. We extend his approach to use observable covariates as well as observations on $X(t)$ to estimate the distribution of S in order to make it possible to investigate the effect of covariates on S. The approach is to postulate a prior distribution for S, then use Bayes Theorem to compute the posterior distribution given the covariate values and observations on $X(t)$. Finally, we show how aspects of the prior distribution can be estimated using the data thereby making the approach empirical Bayes.

2 Gaussian Models

Let $X(t)$ denote the level of a health indicator whose mean level $\mu(t)$ starts to decline when the subject is infected at time $t = S$. We have data from two cohorts, one infected and one uninfected, and let $X_P(t)$ and $X_N(t)$ denote the processes for subjects from the infected (positive) and uninfected (negative) groups respectively. Moreover, $\mathbf{Z}_P$ and $\mathbf{Z}_N$ denote vectors whose entries are values of covariates such as age and CD8 counts for the two groups. Let $W(t)$ denote a stationary Gaussian process with mean zero and covariance structure

$$\operatorname{Cov}(W(s), W(t)) = \sigma^2 \rho(t-s), \quad s \leq t,$$

where $\rho(v)$, $v \geq 0$, is a continuous decreasing unknown correlation function with $\rho(0) = 1$. We model $X_N(t)$ and $X_P(t)$ as

$$\begin{aligned} X_N(t) &= \mu_0(\mathbf{Z}_N) + W_1(t), \\ X_P(t) &= \mu(t, \mathbf{Z}_P) + W_2(t), \end{aligned} \tag{1}$$

where W_1 and W_2 are independent and normally distributed as W, $\mu_0 = \mu_0(\mathbf{Z}_N)$ is a parametric function of $\mathbf{Z}_N$ such as $\boldsymbol{\alpha}^T \mathbf{Z}_N$,

$$\mu(t, \mathbf{Z}_P) = \begin{cases} \mu_0(\mathbf{Z}_P), & \text{for } t \leq S \\ \mu_0(\mathbf{Z}_P) - \Delta(\mathbf{Z}_P)(t - S), & \text{for } t > S, \end{cases} \tag{2}$$

S is independent of (W_1, W_2), and $\Delta = \Delta(\mathbf{Z}_P)$ is a parametric function of $\mathbf{Z}_P$ such as $\boldsymbol{\beta}^T \mathbf{Z}_P$. The unknowns μ_0, μ, Δ and σ can be estimated using methods introduced in the next section. For now, we treat μ_0, μ, Δ and σ as known, derive the distribution of the change point in terms of μ_0, μ, Δ and $\sigma > 0$. In Section 3 we present estimates that can be substituted for the unknowns.

Next, we introduce transformations to simplify our notation and calculations. The data consists of $\mathbf{Z}_N$ and $\mathbf{Z}_P$ as well as the values of the processes $X_N(t)$ and $X_P(t)$ at known discrete time points $t_1, \ldots, t_k$. These time points, which we call visit times, are typically between 4 and 8 months apart, and are different for different individuals. Instead of working with the infection time S, it will be more convenient to use the equivalent variable

$$T_0 = t_1 - S = \text{Time from infection to first visit},$$

and to shift the time axis to the left by the amount S,

$$t \to t - S.$$

Because of stationarity and the independence between S and (W_1, W_2), an equivalent model to (1) with (2), is (1) with

$$\mu(t, \mathbf{Z}_P) = \begin{cases} \mu_0, & \text{for } t \leq 0, \\ \mu_0 - \Delta t, & \text{for } t > 0. \end{cases} \tag{3}$$

With this formulation we want the distribution of T_0 given the observable value $X_P(T_0)$ of the health indicator at the first visit time. Further simplifications are obtained by standardizing both health indicator and time using

$$X = \frac{X_P(T_0) - \mu_0}{\sigma} \quad \text{and} \quad T = \frac{\Delta T_0}{\sigma}.$$

The conditional density of X given $T = t$ is given by

$$f(x \mid t) = \varphi(x + t), \quad x \in R, \ t \geq 0, \tag{4}$$

where φ denotes the $N(0,1)$ density. Our goal is to find the conditional density of T given $X = x$, which by Bayes Theorem is given by

$$g(t \mid x) = f(x \mid t) g(t) / f(x),$$

where $g(t)$ is the marginal or "prior" density of T. In this formula, $f(x \mid t)$ is known and $f(x)$ can be estimated from the data of CD4 counts, $X_P(t_1)$, from the sample of HIV positive individuals. Moreover, f determines g. To see this, let M denote "moment generating function". Then

$$\begin{aligned} M_X(u) = E[\exp(uX)] &= \int \exp(ux) \int \varphi(x+t) g(t)\, dt\, dx, \\ &= \int g(t) \left[\int \exp(ux) \varphi(x+t)\, dx \right] dt, \\ &= \int q(t) \exp\left(-ut + \frac{1}{2} u^2 \right) dt, \\ &= \exp\left(\frac{1}{2} u^2 \right) M_{-T}(u). \end{aligned} \tag{5}$$

It follows that the distribution of X determines the distribution of T. Moreover, (5) shows that X is the convolution of $-T$ with a $N(0,1)$ random variable, that is, we can write

$$X = -T + Z,$$

where Z is $N(0,1)$ and independent of T.

It follows that $E(T) = -E(X)$, $E(T^2) = E(X^2 - 1)$, $E(T^3) = E(X^3 - 3X)$, and in general,

$$E(T^m) = (-1)^m EH_m(X), \tag{6}$$

where H_m is the Hermite polynomial. Thus all the moments of g can be estimated using the sample moments of X.

We have seen that f determines g. Conversely, g determines f because

$$f(x) = \int \varphi(x+t)g(t)\,dt.$$

Thus there are no theoretical restrictions on choosing g. One empirical approach would be to select a parametric g where the parameters are chosen by fitting the moments using (6). Note that the estimate of $\theta_m = E(T^m)$ is

$$\tilde{\theta}_m = (-1)^m \frac{1}{n} \sum_{i=1}^{n} H_m(X_i).$$

Here $X_1, \ldots, X_n$ are obtained from the CD4 counts at first visit times of a sample of HIV positive individuals.

A simple choice of g which Berman (1990) has shown to give a good empirical fit in the case of no covariates is the exponential density with mean θ estimated with $\tilde{\theta} = -n^{-1}\Sigma X_i$. With this choice, in Berman's case, (6) approximately holds for $m \geq 1$. In this case the posterior is given by the truncated normal density

$$g(t \mid x) = \frac{\varphi(t + x + \theta^{-1})}{1 - \Phi(x + \theta^{-1})} I_{[0,\infty)}(t),$$

having expectation

$$E(T \mid x) = \frac{\varphi(x + \theta^{-1})}{1 - \Phi(x + \theta^{-1})} - (x + \theta^{-1}),$$

where Φ denotes the $N(0,1)$ distribution function.

A different approach, which allows for an arbitrary prior g, would be to predict T using the class of linear predictors of the form $a + bX$. The linear predictor that minimizes the mean squared prediction error $E[T-(a+bX)]^2$ is

$$E(T) + \frac{E(XT) - E(X)E(T)}{\mathrm{Var}(X)}[X - E(X)].$$

Here $E(T) = -E(X)$, $E(X)$, and $\mathrm{Var}(X)$ can be estimated using the sample mean and variance of $X_1, \ldots, X_n$, where the X_i are defined in Section 3. Moreover, because

$$E(TX) = E[E(TX \mid T)] = E[TE(X \mid T)],$$
$$= E(-T^2) = -E(X^2 - 1),$$

we can estimate $E(TX)$ using $-n^{-1}\Sigma(X_i^2 - 1)$.

This linear approach extends readily to the multivariate predictor $\mathbf{X} = [H_1(X), \ldots, H_m(X)]^T$, in which case the best linear predictor of T is

$$E(T) + \Sigma_{\mathbf{X}T}\Sigma_{\mathbf{XX}}^{-1}[\mathbf{X} - E(\mathbf{X})].$$

The equations

$$E[TH_k(X)] = (-1)^{2k+1}E[H_k(X)], \quad k = 1, 2, \ldots,$$

would yield all the necessary estimates of unknown covariances. Finally, this linear approach also can be used if, for a certain individual, in addition to CD4 at the first visit time, we have available CD4 counts from more than one visit. We can incorporate counts at the other visit times in the vector $\mathbf{X}$.

3 Estimation of Parameters

The processes $X_N(t)$ and $X_P(t)$ consist of the square roots of the CD4 counts at time t calibrated to correct for changes in the technology used to measure CD4. The calibration procedure consists of using the percentage changes in mean negative CD4 counts over time to adjust the CD4 positive counts.

We checked the assumption of a linear decline in $E[X_p(t)]$ over t by plotting the Lowess (Cleveland and Devlin, 1988) estimate of $E[X_P(t)]$. Panels (a) and (b) of Figure 1 display the estimates before and after calibration for 490 seronegative and 344 seropositive men, respectively, from the SFMHS cohort. The SFMHS cohort includes men who did not show up for every visit; thus the curves in (a) and (b) at different time points are to a small extent based on different people. The estimates for a sub-cohort of 187 seronegative and 99 seropositive men who had consistent visits are displayed in panels (c) and (d) of Figure 1. The calibrated curves indicate a linear decline.

The assumption of within-subject Gaussian distributions with variance homogeneous in t for the fourth root of CD4 counts has been shown to be reasonable by other authors (Taylor et al., 1991) and (Taylor et al., 1994). We also conducted a graphical analysis (not shown) and reached the same conclusion.

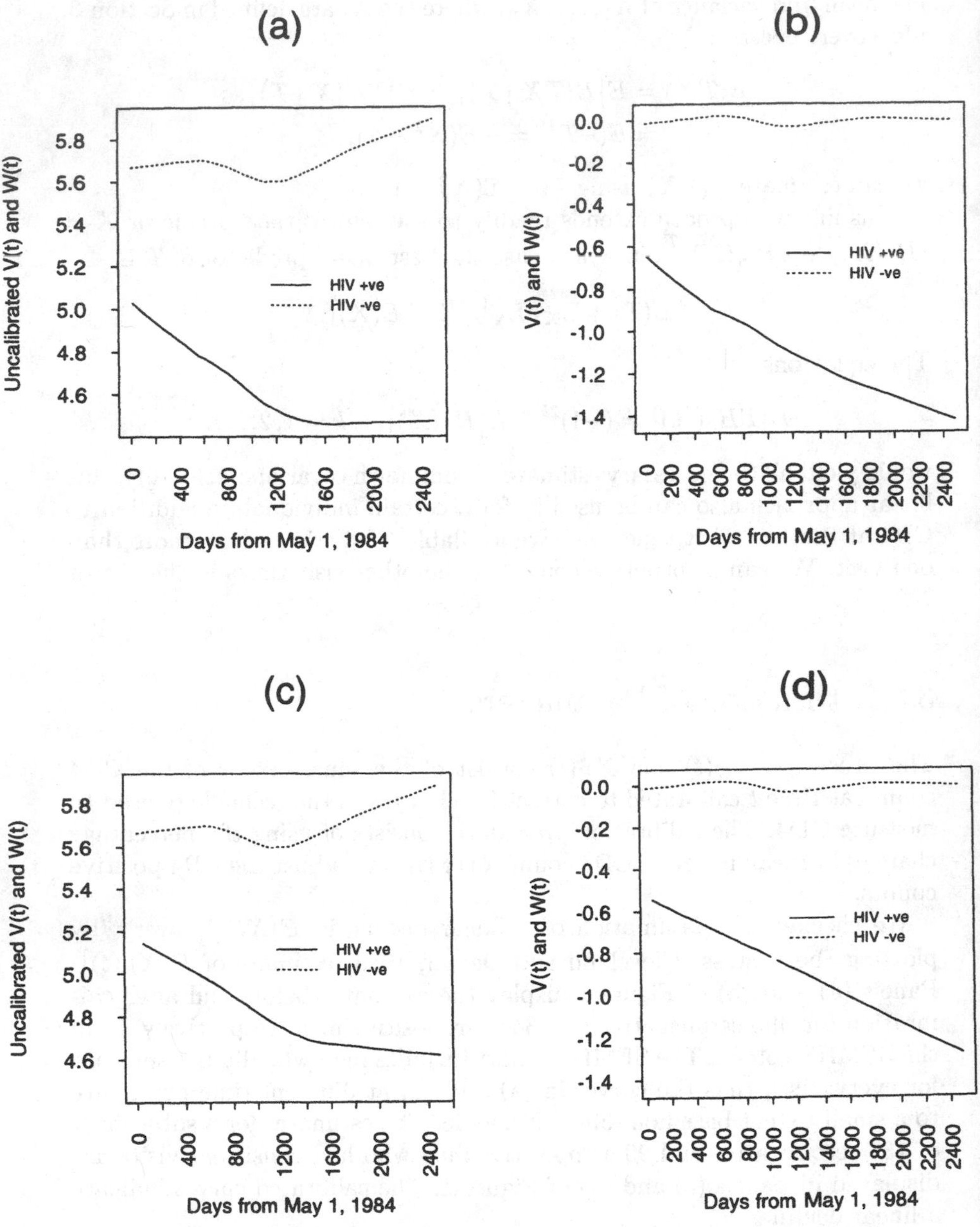

FIGURE 1. Smoothed calibrated CD4 biomarker data. Smoothed $V(t)$ (seropositives) & $W(t)$ (seronegatives) data for the SFMHS full-cohort (panel (b)) and sub-cohort (panel (d)). Panels (a) and (c) display the uncalibrated processes.

For the ith subject in a sample of HIV negative subjects we have available transformed CD4 counts

$$X_N(t_{ij}), \quad j = 1, \ldots, k, \quad \text{where } i = 1, \ldots, n_1 = 490.$$

If we set $\mu_0 = \mu_0(\mathbf{Z}_N) = \boldsymbol{\alpha}^T \mathbf{Z}_N$, then the maximum likelihood estimate of $\boldsymbol{\alpha}$ based $\mathbf{X}_{Nj} = \{X_N(t_{ij}), i = 1, \ldots, n_{1j}\}^T$ is $\widehat{\alpha}_j = (\mathbf{Z}_N^T \mathbf{Z}_N)^{-1} \mathbf{Z}_N^T \mathbf{X}_{Nj}$ where n_{1j} is the number of HIV negative subjects that contributed observations at visit j. Our estimates of $\boldsymbol{\alpha}$, μ_0 and σ^2 are

$$\widehat{\boldsymbol{\alpha}} = \frac{\Sigma_j n_{1j} \boldsymbol{\alpha}_j}{\Sigma_j n_{1j}}, \quad \widehat{\mu}_0 = \widehat{\boldsymbol{\alpha}}^T \mathbf{Z}_N, \quad \hat{\sigma}_N^2 = \frac{\Sigma_j n_{1j} \hat{\sigma}_j^2}{\Sigma_j n_{1j}},$$

where $\hat{\sigma}_j^2$ is the residual sum of squares divided by degrees of freedom for the fit of $\boldsymbol{\alpha}_j^T \mathbf{Z}_N$ to $\mathbf{X}_{Nj}$.

Similarly, for the HIV positive men,

$$X_P(t_{ij}), \quad j = 1, \ldots, k \quad \text{where } i = 1, \ldots, n_2 = 344,$$

are the transformed CD4 counts for individual i. The X_i's referred to in Section 2 are

$$X_i = \frac{X_P(t_{i1}) - \widehat{\mu}_0}{\hat{\sigma}_N}, \quad i = 1, \ldots, n_{P1},$$

where n_{P1} is the number of HIV positive men who contributed observations at the first visit.

We introduce

$$Y_{ij} = -\frac{X_P(t_{ij}) - X_P(t_{ij-1})}{t_{ij} - t_{ij-1}},$$

then Y_{ij} is distributed

$$N\{\Delta, 2\sigma^2 [1 - \rho(t_{ij} - t_{ij-1})] / (t_{ij} - t_{ij-1})^2\}.$$

If we assume $\Delta = \boldsymbol{\beta}^T \mathbf{Z}_P$, then we get an estimator of $\boldsymbol{\beta}$ by using the maximum likelihood estimate appropriate for the correlation function $\rho(t) = \rho_0(t) = (t_{\max} - t)/t_{\max}$, where $t_{\max}$ is the last observed visit time among the HIV positive subjects. This estimator is consistent for general $\rho(t)$ and efficient for $\rho(t) = \rho_0(t)$.

4 Application to the SFMHS Cohort

We considered the three covariates:

1. Z_{i1} = age of subject i at entry.

2. Z_{i2} = fourth root of subject CD8 T-cell count at entry.

TABLE 1. Regression estimates of Gaussian process parameters for the latency time model. Standard errors for parameter estimates calculated assuming the correlation function, $\rho(t)$, is $(t_{\max} - t)/t_{\max}$.

Model 1: Age and Baseline CD8 Full Cohort: 490 Seronegatives and 344 Seropositives						
Variable	$\hat{\alpha}$	$se(\hat{\alpha})$	$\frac{\hat{\alpha}}{se(\hat{\alpha})}$	$\hat{\beta}$	$se(\hat{\beta})$	$\frac{\hat{\beta}}{se(\hat{\beta})}$
Intercept	-1.49	4.20×10^{-1}	-3.35	-1.14×10^{-3}	4.43×10^{-4}	-2.57
CD8(cells/mm^3)$^{1/4}$	2.76×10^{-1}	7.81×10^{-2}	3.54	3.30×10^{-4}	7.96×10^{-5}	4.15
Age - 35 (yrs)	8.16×10^{-3}	5.21×10^{-3}	1.57	1.36×10^{-5}	6.96×10^{-6}	1.95
$\hat{\sigma}^2$	$(0.83)^2$					
Model 2: Age and Average Slope of Calibrated CD8 Full-Cohort: 490 Seronegatives and 344 Seropositives						
Variable	$\hat{\alpha}$	$se(\hat{\alpha})$	$\frac{\hat{\alpha}}{se(\hat{\alpha})}$	$\hat{\beta}$	$se(\hat{\beta})$	$\frac{\hat{\beta}}{se(\hat{\beta})}$
Intercept	-1.18×10^{-2}	3.70×10^{-2}	-0.32	7.00×10^{-4}	4.21×10^{-5}	16.65
Mean CD8 slope	$4.00 \times 10^{+1}$	$5.20 \times 10^{+1}$	0.77	-8.55×10^{-1}	6.23×10^{-2}	-13.71
Age - 35 (yrs)	7.13×10^{-3}	5.12×10^{-3}	1.39	1.31×10^{-5}	6.73×10^{-6}	1.95
$\hat{\sigma}^2$	$(0.81)^2$					

3. The average slope of calibrated CD8 T-cell counts over time; that is, for the ith subject

$$Z_{i3} = \frac{1}{k_i}\Sigma\frac{Z_i(t_{ij}) - Z(t_{ij-1})}{t_{ij} - t_{ij-1}},$$

where $Z(t)$ denotes the CD8 T-cell count at time t, and 2 models:

Model 1: contains the covariates age (Z_1) and the fourth root of the CD8 count (Z_2) at entry.

Model 2: contains age (Z_1) at entry and the average slope of fourth root CD8 counts over time (Z_3).

Model 1 can be used to answer the question: If a subject walks into a clinic today and learns for the first time that he is infected with HIV then, on the basis of todays age, CD4 and CD8 counts, what is the distribution of the time T_0 since infection? Model 2 cannot be used to answer this question because the CD8 slopes are not available for this subject. The second model can be used to analyze the relationship between CD4, CD8 and time since infection.

The results of the regression analysis for the two models are summarized in Table 1. It shows a strong relationship between CD4 T-cell counts and initial CD8 T-cell counts and a much stronger relationship between CD4 and the average slope of CD8. Age is on the boundary of being significant in both models.

TABLE 2. Mean of estimated conditional latency time density. Mean (number of days) estimated using model 1 (baseline CD8) and reported on original time scale.

	CD4	Age 30 Mean Time (Days)	Age 50 Mean Time (Days)
	200	2074	1548
Baseline	300	1448	1111
CD8	400	1015	814
748	500	658	583
	600	308	166
	CD4	Age 30 Mean Time (Days)	Age 50 Mean Time (Days)
	200	1685	1386
Baseline	300	1213	1021
CD8	400	892	772
1204	500	644	586
	600	424	430

TABLE 3. CD4 and covariate values for two subjects with estimated unconditional (prior) mean latency time.

Individual	CD4	Age	Z_2	Z_3	Mean Latency Time Model 1	Model 2
1	500	30	5.04	2.14×10^{-4}	669	913
2	556	50	5.23	3.52×10^{-4}	1,369	1,585

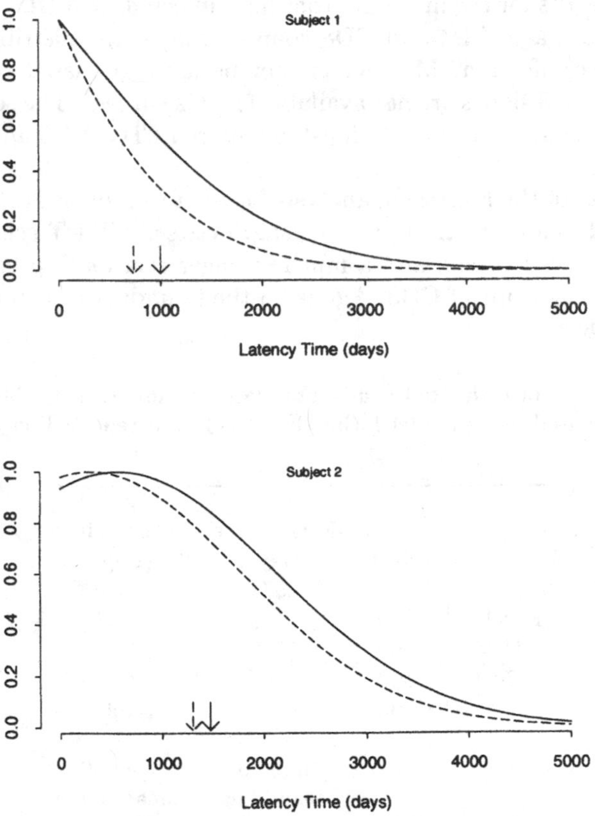

FIGURE 2. Conditional (posterior) distribution of latency time for subjects 1 and 2 estimated using model 1 (dashed line) and model 2 (solid line) given the value of transformed calibrated CD4 value at visit 1. Arrows indicate mean of the estimated posterior distribution.

Next we fitted an exponential distribution to the marginal distribution of T_0 as explained in Section 2 and computed the estimated conditional mean latency time given baseline CD4, baseline CD8 and age. The results in Table 2 indicate that baseline CD8 strongly influences the mean latency time.

Figure 2 displays the posterior densities of the latency time T_0 for two subjects in the study. The relevant values of CD4 counts and covariate values are given in Table 3.

5 Discussion

In this paper we presented methods for predicting the time of an event that happened in the past. Our approach is to use a stochastic model that links the distribution of an event time to the change point of a degradation process and then use data on the degradation process and on covariates to estimate the model parameters.

A key assumption in our model is that the mean of the degradation process declines linearly in time after the change point. However, our methods can be used if this decline is quadratic or some other function of time t as long as it is monotone decreasing. Figure 1 indicates that for the data we consider a linear decline in the mean degradation is a reasonable assumption.

Our model also assumes normality of the degradation process. As discussed in Section 3, this assumption is approximately satisfied for our data example. However, in cases where it is not satisfied, the result giving the posterior distribution of latency time T only gives qualitative insights because this result is sensitive to the normality assumption. In such cases, a nonparametric or semiparametric approach based on

$$g(t \mid x) = \frac{f(x \mid t)g(t)}{f(x)},$$

may give better results. On the other hand, our results involving prediction of the latency time based on moments are more robust to the normality assumption. These questions need further study.

Acknowledgments: Professor Normand's work was partially supported by Grant CA-61141, awarded by the National Cancer Institute, Department of Health and Human Services. Professor Doksum's work was partially supported by Grant CA-56713, awarded by the National Cancer Institute, Dept. of Health and Human Services and Grant DMS-96-25777 awarded by the National Science Foundation. The authors thank Eric Vittinghoff,

Jeremy Taylor, and Mark Segal for useful comments. Thanks are also extended to Marjorie Ng, James Wiley, and Warren Winkelstein for generously providing the SFMHS data. The SFMHS was supported by contract N01-AI-82515 from the National Institute of Allergy and Infectious Diseases.

6 References

Berman, S.M. (1990). A stochastic model for the distribution of HIV latency time based on T4 counts. *Biometrika 77*, 733–741.

Berman, S.M. (1994). Perturbation of normal random vectors by nonnormal translations, and an application to HIV latency time distributions. *Ann. Appl. Prob. 4*, 968–980.

Cleveland, W.S. and S.J. Devlin (1988). Locally weighted regression: An approach to regression analysis by local fitting. *J. Amer. Statist. Assoc. 83*, 597–610.

DeGruttola, V., N. Lange, and U. Dafni (1991). Modeling the progression of HIV infection. *J. Amer. Statist. Assoc. 86*, 569–577.

DeGruttola, V. and X.M. Tu (1994). Modelling the relationship between progression of CD4-lymphocyte count and survival time. In N. Jewell, K. Dietz, and V. Farewell (Eds.), *AIDS Epidemiology: Methodological Issues*, Boston, pp. 275–296.

Doksum, K.A. (1991). Degradation rate models for failure time and survival data. *Centrum voor Wiskunde en Informatica Quarterly 5*, 195–203.

Doksum, K.A. and S.L. Normand (1995). Gaussian models for degradation processes—part I: Methods for the analysis of biomarker data. *Lifetime Data Analysis 1*, 131–144.

Faucett, C.L. and D.C. Thomas (1996). Simultaneously modeling censored survival data and repeatedly measured covariates: A Gibbs sampling approach. *Statist. Med. 15*, 1663–1685.

Jewell, N.P. and J.D. Kalbfleisch (1992). Marker models in survival analysis and applications to issues associated with AIDS. In N. Jewell, K. Dietz, and V. Farewell (Eds.), *IDS Epidemiology: Methodological Issues*, Boston, pp. 211–230.

Kuichi, A.S., J.A. Hartigan, R.R. Holford, P. Rubinstein, and C.E. Stevens (1995). Change points in the series of T4 counts prior to AIDS. *Biometrics 51*, 236–248.

Lange, N., B. Carlin, and A. Gelfand (1992). Hierarchical Bayes models for the progression of HIV infection using longitudinal CD4 T-cell numbers. *J. Amer. Statist. Assoc. 87*, 615–626.

Longini, I.M., W.S. Clark, R.H. Byers, J.W. Ward, W.W. Darrow, G.F. Lem, and H.W. Hethcote (1989). Statistical analysis of the stages of HIV infection using a Markov model. *Statist. Med. 8*, 831–843.

Munoz, A., V. Cary, A.J. Saah, J.P. Phair, L.A. Kingsley, J.L. Fahey, H.M. Ginzburg, and B.F. Polk (1988). Predictors of decline in CD4 lymphocytes in a cohort of homosexual men infected with human immunodeficiency virus. *J. Acquired Immune Deficiency Syndrome 1*, 396–404.

Satten, G.A. and I.M. Longini (1996). Markov chains with measurement error: estimating the "true" course of the progression of human immunodeficiency virus disease (with discussion). *Appl. Statist. C45*, 275–309.

Shi, M., R.E. Weiss, and J.M.G. Taylor (1996). An analysis of pediatric CD4 counts for acquired immune deficiency syndrome using flexible random curves. *Appl. Statist. C 45*, 151–163.

Taylor, J.M.G., W.G. Cumberland, and J.P. Sy (1994). A stochastic model for analysis of longitudinal AIDS data. *J. Amer. Statist. Assoc. 89*, 727–736.

Taylor, J.M.G., S.J. Tan, R. Detels, and J.V. Giorgio (1991). Applications of computer simulation model of the natural history of CD4 T-cell number in HIV-infected individuals. *AIDS 5*, 159–167.

Tsiatis, A.A., V. DeGruttola, and M.S. Wulfson (1995). Modeling the relationship of survival to longitudinal data measured with error: applications to survival and CD4 counts in patients with AIDS. *J. Amer. Statist. Assoc. 90*, 27–37.

Vittinghoff, E., H.M. Malani, and N.P. Jewell (1994). Estimating patterns of CD4 lymphocyte decline using data from a prevalent cohort of HIV infected individuals. *Statist. Med. 13*, 1101–1118.

Winkelstein, W., D.M. Lyman, N. Padian, R. Grant, M. Samuel, R.E. Anderson, W. Lang, J. Riggs, and J.A. Levy (1987). Sexual practices and risk of infection by the human immunodeficiency virus. *J. Amer. Med. Assoc. 257*, 321–325.

6

Bayes and Empirical Bayes Estimates of Survival and Hazard Functions of a Class of Distributions

M. Ahsanullah and S.E. Ahmed

ABSTRACT Bayes and empirical Bayes estimates of the survival and hazard functions of a class of distributions are obtained. These estimates are compared with the corresponding maximum likelihood estimates. A Monte Carlo study is designed to perform this comparison, due to intractability of the estimators distributions from an analytical point of view.

1 Introduction

Suppose X is a random variable with absolutely continuous cumulative distribution function (cdf) $F(x)$. We define the survival function

$$S(t) = P(X > t) = 1 - F(t),$$

and the hazard function

$$h(t) = \frac{f(t)}{1 - F(t)}, \quad F(t) < 1,\ f(t) = \frac{d}{dt}F(t).$$

In this paper we will consider the class of distributions with $F(t)$,

$$F(t) = [g(t)]^{\theta}, \quad -\infty < \gamma < t < \eta,\ \theta > 0, \tag{1}$$

where $g(t)$ is a monotone increasing function of t with $g(\gamma) = 0$ and $g(\eta) = 1$. We will say X is a member of the class C_1 if its distribution function is of the form as given (1). Some members of class C_1 are

$$g(t) = (1 - e^{-t})^k, \quad k > 0,\ \gamma = 0,\ \eta = \infty,$$
$$g(t) = t^k, \quad k > 0, \gamma = 0,\ \eta = 1,$$
$$g(t) = e^{-e^{-t}}, \quad \gamma = -\infty,\ \eta = \infty$$
$$g(t) = \frac{1}{1 + e^{-t}}, \quad \gamma = -\infty,\ \eta = \infty.$$

For members of class C_1, we have

$$h(t) = \frac{\theta g'(t)}{g(t)} \frac{(g(t))^\theta}{(1-g(t))^\theta}.$$

The likelihood function L based on n independent observations, $x_1, \ldots, x_n$ is

$$L(\mathbf{x},\theta) = \theta^n \prod_{i=1}^{n} g'(x_i)\left[g(x_i)\right]^{\theta-1}, \quad \gamma < x_i < \eta. \tag{2}$$

Thus $T_1 = -\sum_{i=1}^{n} \ln g(x_i)$ is a sufficient statistic for θ and T_1 is complete.
Let $u(x) = -\ln g(x)$, then the characteristic function $\phi_1(w)$ of $u(x)$ is

$$\begin{aligned}\phi_1(w) &= \int_\gamma^\eta e^{wu(x)}\theta g'(x)\left[g(x)\right]^{\theta-1} dx,\\ &= \int_0^\infty e^{ws}\theta e^{-\theta s}\, ds,\\ &= \frac{\theta}{\theta - w}, \quad w < \theta.\end{aligned}$$

Thus T_1 is distributed as a gamma with probability density function pdf $f_1(x)$

$$f_1(x) = \frac{x^{n-1}}{\Gamma(n)}\theta^n e^{-\theta x}, \quad 0 < x < \infty. \tag{3}$$

In the next section, we propose some plausible estimators of $S(t)$ and $h(t)$ and study their properties in subsequent section. We propose the estimation problem in a Bayesian framework. The methods of maximum likelihood and minimum variance unbiased estimation are also explored.

2 Estimation Strategies

In this section we consider four estimators of the survival functions and hazard function respectively.

Bayes estimates

In order to obtain the Bayes estimator of the parameter of interest we assume a conjugate prior, $g(\theta, \alpha, \beta)$ for θ as

$$g(\theta, \alpha, \beta) = \frac{\beta^\alpha}{\Gamma(\alpha)}\theta^{\alpha-1}e^{-\beta\theta}, \quad \theta > 0, \ \alpha > 0, \ \beta > 0.$$

Thus, the joint pdf $f(x)$ of $(x_1, \dots, x_n)$ is given by

$$\begin{aligned} f(\mathbf{x}) &= \int f(\mathbf{x} \mid \theta) g(\theta, \alpha, \beta)\, d\theta, \\ &= \int_0^\infty \frac{\beta^\alpha}{\Gamma(\alpha)} \theta^{n+\alpha-1} e^{-\beta\theta} \prod_{i=1}^n g'(x_i) \left[g(x_i)\right]^{\theta-1} d\theta, \\ &= \frac{\Gamma(n+\alpha)}{\Gamma(\alpha)} \frac{\beta^\alpha}{(\beta+T_1)^{n+\alpha}} \prod_{i=1}^n \frac{g'(x_i)}{g(x_i)}, \end{aligned} \tag{4}$$

where $T_1 = -\sum_{i=1}^n \ln g(x_i)$. The conditional pdf of θ given $\mathbf{x}$ is given by

$$f(\theta \mid \mathbf{x}) = \frac{(\beta+T_1)^{n+\alpha}}{\Gamma(n+\alpha)} \theta^{(n+\alpha-1)} e^{-\theta(\beta+T_1)}. \tag{5}$$

The corresponding Bayesian estimates of $S(t)$ and $h(t)$ are respectively

$$\begin{aligned} \widehat{S}_B(t) &= 1 - \int_0^\infty \left[g(t)\right]^\theta \frac{(\beta+T_1)^{n+\alpha}}{\Gamma(n+\alpha)} \theta^{n+\alpha-1} e^{-\theta(\beta+T_1)}\, d\theta, \\ &= 1 - \left[\frac{\beta+T_1}{\beta+T_1-\ln g(t)}\right]^{n+\alpha}, \\ \hat{h}_B(t) &= \frac{g'(t)}{g(t)} \int_0^\infty \theta \frac{(g(t))^\theta}{1-(g(t))^\theta} \frac{(\beta+T_1)^{n+\alpha}}{\Gamma(n+\alpha)} \theta^{n+\alpha-1} e^{-\theta(\beta+T_1)}\, d\theta, \\ &= \frac{\int_0^\infty w(\theta) e^{-\theta(\beta+T_1)+(n+\alpha-1)ln\theta}\, d\theta}{\int_0^\infty e^{-\theta(\beta+T_1)+(n+\alpha-1)\theta}\, d\theta}, \end{aligned} \tag{6}$$

where

$$w(\theta) = \frac{\theta g'(t)}{g(t)} \frac{(g(t))^\theta}{1-(g(t))^\theta}.$$

Since $\hat{h}_B(t)$ is difficult to obtain in a simplified form, we use Lindley's (1980) approximation of the ratio of two integrals,

$$\frac{\int w(\theta) e^{q(\theta)}\, d\theta}{\int e^{q(\theta)}\, d\theta} \approx w(\theta_0) - \frac{1}{2} \frac{1}{\partial^2 q/\partial\theta^2} \left[\frac{\partial^2 w}{\partial\theta^2} - \frac{(\partial w/\partial q)(\partial q/\partial\theta^3)}{\partial^2 q/\partial\theta^2}\right]_{\theta=\theta_0},$$

where θ_0 is the solution of $\partial q(\theta)/\partial\theta = 0$. Here $\theta_0 = (n+\alpha-1)/(\beta+T_1)$.

It can be shown that

$$\frac{\partial w}{\partial\theta} = w(\theta)c(\theta), \quad c(\theta) = \frac{1}{\theta} + c_1(\theta), \quad c_1(\theta) = \frac{\ln g(t)}{1-[g(t)]^\theta}$$

$$\frac{\partial^2 w}{\partial\theta^2} = w(\theta)c^2(\theta) + w(\theta)\left\{-\frac{1}{\theta^2} + [g(t)]^\theta \left[\frac{\ln g(t)}{1-[g(t)]^\theta}\right]^2\right\}.$$

On simplification we obtained

$$\hat{h}_\beta(t) \approx h(t)\big|_{\theta=\theta_0} \times \left[1 + \frac{\theta}{n+\alpha-1}\left(\frac{1}{\theta} + 3c_1(\theta) + \theta c_1^2(\theta)\{1 + [g(t)]^\theta\}\right)\right]_{\theta=\theta_0}. \quad (7)$$

Empirical Bayes estimates

We will estimate α and β from m past samples. Suppose we have m past sample each of size n,

$$\begin{array}{cccc} X_{11} & X_{12} & \dots & X_{1n} \\ X_{21} & X_{22} & \dots & X_{2n} \\ \dots & \dots & \dots & \dots \\ X_{m1} & X_{m2} & \dots & X_{mn} \end{array}$$

and a current sample $X_{(m+1)1}, X_{(m+2)2}, \dots, X_{(m+n)n}$. For $j = 1, 2, \dots, m+1$, let $T_j = -\sum_{i=1}^n \ln g(x_{ji})$, and $W_j = n/T_j$. It can be shown that W_j has inverted gamma with pdf as

$$f(w_j \mid \theta) = \frac{(n\theta)^n}{\Gamma(n)} \frac{1}{w_j^{n+1}} e^{-n\theta/w_j}, \quad w_j > 0.$$

The marginal pdf of W_j is

$$\begin{aligned} f(w_j) &= \int_0^\infty f(w_j \mid \theta) g(\theta, \alpha, \beta)\, d\theta, \\ &= \frac{\beta n^n}{B(\alpha, n)} \frac{(\beta w_j)^{\alpha-1}}{(n + \beta w_j)^{n+\alpha}}, \quad w_j > 0. \end{aligned}$$

It can be easily shown that

$$\begin{aligned} E(W_j) &= \frac{n}{n-1} \frac{\alpha}{\beta}, \\ E(W_j^2) &= \frac{n^2}{(n-1)(n-2)} \frac{\alpha(\alpha+1)}{\beta^2}. \end{aligned}$$

Using the methods of moments, we have

$$\widehat{\alpha} = \frac{A^2}{B - A^2} \quad \text{and} \quad \hat{\beta} = \frac{A}{B - A^2},$$

where

$$A = \frac{n-1}{mn} \sum_{j=1}^m w_j \quad \text{and} \quad B = \frac{(n-1)(n-2)}{mn^2} \sum_{j=1}^m w_j^2, \quad n > 2.$$

By substituting $\widehat{\alpha}$, $\hat{\beta}$ and $\theta_0^* = (n + \widehat{\alpha} - 1)/(\hat{\beta} + T_{m+1})$ for α, β, and θ_0 in (6) and (7), we obtain the corresponding empirical Bayes estimates $\widehat{S}_{\text{EB}}(t)$ and $\hat{h}_{\text{EB}}(t)$ of $S(t)$ and $h(t)$ respectively.

Maximum likelihood (ML) *estimates*

The maximum likelihood estimates of θ can be obtained from (2) by solution of $\partial \ln L/\partial\theta = 0$. The solution of $\partial \ln L/\partial\theta = 0$ is $\hat{\theta}_{\mathrm{ML}} = n/T_1$, $T_1 = -\sum_{i=1}^{n} \ln g(x_i)$.

Thus, the ML estimates of $S(t)$ and $h(t)$ are

$$\widehat{S}_{\mathrm{ML}}(t) = \left[1 - g(t)\right]^{\hat{\theta}_{\mathrm{ML}}}, \tag{8}$$

and

$$\hat{h}_{\mathrm{ML}}(t) = \hat{\theta}_{\mathrm{ML}} \frac{g'(t)}{1-g(t)} \frac{(g(t))^{\hat{\theta}_{\mathrm{ML}}}}{1-(g(t))^{\hat{\theta}_{\mathrm{ML}}}}. \tag{9}$$

Minimum variance unbiased (MVU) *estimates*

Using the complete sufficient statistic, T_1 it can easily be shown by using Rao–Blackwell theorem that the MVU estimates of $S(t)$ and $h(t)$ are respectively

$$\widehat{S}_{\mathrm{MVU}}(t) = 1 - \left[1 + \frac{\ln g(t)}{T_1}\right]^{n-1}, \quad -\ln g(t) < T_1 < \infty. \tag{10}$$

The results follow by noting that T_1 is a complete sufficient statistic and

$$\begin{aligned} E\widehat{S}_{\mathrm{MVU}}(t) &= 1 - \int_{-\ln g(t)}^{\infty} \left[1 + \frac{\ln g(t)}{T_1}\right]^{n-1} \frac{1}{\Gamma(n)} \theta^n e^{\theta T_1} T_1^{n-1}\, dT_1, \\ &= 1 - \int_0^{\infty} \frac{1}{\Gamma(n)} u^{n-1} \theta^n e^{-\theta\{u - \ln[g(t)]\}}\, du, \\ &= 1 - \left[g(t)\right]^{\theta}, \\ &= S(t). \end{aligned}$$

3 Numerical Results

We took $g(x) = 1 - e^{-x^4}$, $x \geq 0$ and $\theta = 4$ to compare the various estimates of $S(t)$ and $h(t)$. Using the $g(x)$ and θ, we obtained n ($n = 20$, 30 and 50) observations and calculated the ML and MUV estimates as given by (8), (9), and (10). We repeated the estimates M ($M = 10^5$) times and calculated the estimated risk (ER) given by

$$\mathrm{ER}(\delta) = \frac{1}{M} \sum_{i=1}^{M} (\hat{\delta}_i - \delta)^2,$$

where $\hat{\delta}_i$ is an estimate of δ. For given values of α ($\alpha = 1.5, 2.5$) and β ($\beta = 2.5, 3.5$) we obtained the Bayes estimates using (6) and (7). For empirical

TABLE 1. Estimated Risk (ER) of estimated values of $S(t)$.

t	α	β	n	ER$[\widehat{S}_{\mathrm{ML}}(t)]$	ER$[\widehat{S}_{\mathrm{MVU}}(t)]$	ER$[\widehat{S}_{\mathrm{B}}(t)]$	ER$[\widehat{S}_{\mathrm{EB}}(t)]$	
							$m = 15$	$m = 30$
1.2	1.5	2.5	20	.019	.015	.017	.021	.018
			30	.015	.013	.015	.016	.015
			50	.011	.010	.011	.012	.011
1.2	2.5	3.5	20	.022	.015	.020	.023	.020
			30	.018	.013	.016	.019	.017
			50	.014	.017	.013	.014	.013
1.6	1.5	2.5	20	.0004	.002	.0003	.0004	.0003
			30	.0003	.0002	.0003	.0003	.0003
			50	.0003	.0001	.0002	.0003	.0002
1.6	2.5	3.5	20	.0005	.0002	.0004	.0005	.0004
			30	.0004	.0002	.0003	.0004	.0003
			50	.0003	.0001	.0003	.0003	.0003

TABLE 2. Estimated risk (ER) of estimated values of $h(t)$.

t	α	β	n	ER$[\hat{h}_{\mathrm{ML}}(t)]$	ER$[\hat{h}_B(t)]$	$ER[\hat{h}_{\mathrm{EB}}(t)]$	
						$m = 15$	$m = 30$
1.2	1.5	2.5	20	.0347	.0308	.0377	.0326
			30	.0272	.0253	.0302	.0268
			50	.0212	.0196	.0229	.0204
1.2	2.5	3.5	20	.0409	.0355	.0423	.0357
			30	.0318	.0295	.0340	.0302
			50	.0245	.0227	.0251	.0233
1.6	1.5	2.5	20	.0028	.0021	.0031	.0022
			30	.0028	.0019	.0029	.0021
			50	.0020	.0016	.0020	.0019
1.6	2.5	3.5	20	.0012	.0011	.0013	.0011
			30	.0010	.0009	.0010	.0009
			50	.0005	.0007	.0008	.0007

Bayes estimates for the given values of α and β, the past m estimates $\theta_j = w_j, j = 1, \ldots, m$ of θ were obtained from inverted gamma distribution. One observation was obtained for θ were obtained from $g(\theta, \alpha, \beta)$ and using this θ. One observation of $T = T_{m+1}$ was obtained from (9). Following the method outline in Subsection 2.3, we obtained $\widehat{\alpha}$ and $\hat{\beta}$ using w_j's and substituting $\widehat{\alpha}$ and $\hat{\beta}$ and $\theta_0^* = (n + \widehat{\alpha} - 1)/(\hat{\beta} + T_{m+1})$ for α, β and θ_0 in (6) and (7), we obtained the corresponding empirical estimates of $S(t)$ and $h(t)$. Finally based on M repetitions we obtained the corresponding estimated risks of these estimates. The results are given in Tables 1 and 2. The estimated risk of Bayes and empirical Bayes estimated were calculated based on M $(= 10^5)$ repetitions. IMSL subroutines were used to generated random numbers. The tables give the ER for $S(t)$ and $h(t)$.

It is seen from the these tables that for fixed values of the prior parameters α and β, the estimated risks of the estimates (ML, MVU, B, EB) decreases as n increases. Generally Bayes estimate of $S(t)$ and $h(t)$ are more efficient than their corresponds MLE's and EBE's. Also, MLE's of $S(t)$ and $h(t)$ are better than EBE for $m = 15$. As m increases EBE's tend to be more efficient than the MLE's in the sense of smaller estimated risks.

Acknowledgments: The work of Professor S. E. Ahmed was supported by grants from the Natural Sciences and Engineering Research Council of Canada.

4 References

Lindley, D.V. (1980). Approximate Bayesian methods. In J. Bernardo, M. DeGroot, D. Lindley, and A. Smith (Eds.), *Bayesian Statistics*, Valencia, 1979, pp. 223–245. Univ. Press, Valencia.

7

Bayes and Empirical Bayes Procedures for Selecting Good Populations From a Translated Exponential Family

R.S. Singh

ABSTRACT

This paper presents Bayes and Empirical Bayes procedures based on two-tail test criteria with product distance loss function for selecting good populations coming from a translated exponential family. This family has widely been used in reliability theory, quality control and engineering sciences. In the event the structural form of the prior distribution is unknwon, non-parametric empirical Bayes selection procedures are developed. These procedures are shown to be asymptotically optimal. Speed of convergence for asymptotic optimality are investigated.

1 Introduction

The empirical Bayes approach, as originally formulated by Robbins (1956), is applicable to situations where one confronts repeatedly and independently the same decision problems involving the same unknown prior distribution. Robbins showed that in such situations the accumulated historical data acquired as the sequence of problems progresses may be used to obtain approximations to the minimum Bayes risk procedure, which would be available for use only if the prior distribution were known. Such procedures, named "empirical Bayes procedures" by Robbins, may be asymptotically optimal in the sense that their Bayes risks may converge to the minimum Bayes risk which would have been achieved if the prior distribution were known and the minimum Bayes risk procedure based on this knowledge were used.

This approach has been applied rather extensively to various statistical problems by many authors, including, Robbins (1963), Robbins (1964), Johns and Van Ryzin (1971), Johns and Van Ryzin (1972), Nogami (1988), Singh (1976), Singh (1979), Singh (1995), Gupta and Liang (1993), Wei (1989), Wei (1991) and Datta (1994).

In this paper we consider situations where one has a sequence of independent populations $\pi_1, \pi_2, \ldots$, and at each stage i, $i = 1, 2, \ldots$ one is to select the population at hand if it is a "good" population and reject it otherwise, based on a data set X from this population (and possibly that acquired from the previous populations) and on a certain selection criterion with a penalty function for a wrong decision. Whereas a vast literature on EB selection procedures using a one-tail test criterion exists, very little has been done on EB selection procedures using the two-tail test criterion. Recently, Wei (1989, 1991) provided EB two-tail procedures for one parameter exponential family for the continuous case and for the discrete case, whereas Singh and Wei (1999) looked at this problem for a one parameter scale exponential family.

In this paper we apply the EB approach to the two-tail test selection problem when the populations are from an one-parameter translated exponential family given by the densities

$$f(x \mid \theta) = \begin{cases} ke^{-k(x-\theta)} & \text{for } x > \theta, \\ 0 & \text{elsewhere,} \end{cases} \tag{1}$$

where $\theta \in \Theta \subseteq (-\infty, \infty)$. This family has been found to be very useful in many areas of application, including survival analysis, life testing and reliability theory. In these areas the parameter θ is more commonly referred to as the "threshold parameter"; and in life-testing it is interpreted as the "minimum guaranteed life-time". For further applications of this density, see Johnson et al. (1994).

The general set up is as follows. Based on an observation X from a population whose density is given by (1), one is to select the population as a "good population" if

$$H_o\colon \theta_1 \leq \theta \leq \theta_2,$$

is true, and reject it as a "bad population" if

$$H_1\colon \theta < \theta_1 \text{ or } \theta > \theta_2,$$

is true, where $\theta_1 < \theta_2$ are the boundary points of the standard chosen appropriately by the practitioner. The penalty function is

$$L_o(\theta) = c(\theta - \theta_1)(\theta - \theta_2)I(\theta < \theta_1 \text{ or } \theta > \theta_2),$$

for selecting the population incorrectly, and it is

$$L_1(\theta) = c(\theta - \theta_1)(\theta - \theta_2)I(\theta \leq \theta_1 \leq \theta_2),$$

for rejecting it incorrectly, where $c > 0$ is a constant of proportionality and $I(A)$ stands for the indicator function of the set A. That is, if the population $\pi(\theta)$ is accepted (or rejected) correctly then there is no penalty, otherwise it is proportional to the product of the distances the true θ is away from θ_1 and θ_2 in each case. Such a loss function for two-tail test problems has been used in the literature by many authors, including Samuel (1963) and Wei (1989, 1991).

Our plan is as follows. In Section 2 we discuss Bayes and empirical Bayes approaches to this selection problem. In Section 3 we develop empirical Bayes selection procedures for situations where the prior distribution is unknown. In Section 4 we establish asymptotic optimality of the EB selection procedures and give rates of convergence. We conclude the paper with a remark in Section 5 suggesting how the work here can be extended to the case where one has to deal with the selection of many populations simultaneously at each stage.

2 Bayes and Empirical Bayes Approach

Bayes approach

In the Bayes approach we assume that the parameter θ is a random variable distributed according to a prior distribution G with support in the parameter space Θ. Based on an observation $X \sim f(\cdot \mid \theta)$ let

$$\delta(x) = P\left(\text{accepting } H_0 \mid X = x\right), \tag{2}$$

be a randomized decision rule when $X = x$. Then, the Bayes risk of δ w.r.t. G is

$$\begin{aligned} R(\delta, G) &= E\{L_0(\theta)\delta(X) + L_1(\theta)[1 - \delta(X)]\} \\ &= \int \{L_0(\theta)\delta(X) + L_1(\theta)[1 - \delta(X)]\} f(x \mid \theta)\, dx\, dG(\theta), \end{aligned}$$

where E stands for the expectation operator w.r.t. all the r.v's involved.

Now by little algebraic manipulation we can show that L_o and L_1 defined in Section 1 can be re-written as

$$L_o(\theta) = c\left[(\theta - a)^2 - b^2\right] I\left(|\theta - a| > b\right),$$

and

$$L_1(\theta) = c\left[b^2 - (\theta - a)^2\right] I\left(|\theta - a| \leq b\right),$$

where

$$a = \frac{(\theta_1 + \theta_2)}{2} \text{ and } b = \frac{(\theta_2 - \theta_1)}{2}.$$

Noting these observations, it can be shown, after some simplification, that $R(\delta, G)$ can be expressed as

$$R(\delta, G) = c \int \alpha(x)\delta(x) + C_G, \tag{3}$$

which is similar to (1) of Johns and Van Ryzin (1971), where

$$C_G = \iint L_1(\theta)\, dG(\theta),$$
$$\alpha(x) = \int [(\theta - a)^2 - b^2] f(x \mid \theta)\, dG(\theta). \tag{4}$$

Throughout the remainder of this paper we assume that $\int \theta^2 dG(\theta) < \infty$. This restriction is necessary to ensure that $R(\delta, G) < \infty$. From (3) we see that $R(\delta, G)$ depends on δ only through the first term on the r.h.s. of (3). Therefore, the Bayes optimal procedure which minimizes the Bayes risk $R(\delta, G)$ over all possible randomized decision rules $\delta(\cdot)$ on $(-\infty, \infty)$ onto $[0, 1]$ is, from (3), given by

$$\delta_G(X) = I\big[\alpha(X) \leq 0\big]. \tag{5}$$

That is, when $X = x$ is observed from the population under study then the minimum Bayes risk procedure is to select the population as a good population if $\alpha(x) \leq 0$; otherwise reject it as a bad population.

We denote the (minimum) Bayes risk attained by δ_G by

$$R(G) = \min_{\delta} R(\delta, G) = R(\delta_G, G). \tag{6}$$

Note that the Bayes optimal procedure δ_G is available for use in practice if and only if G is known or, alternatively, we put a known prior, as usual, on Θ. This is undoubtedly a very strong requirement. When G is unknown δ_G is not available for use. To handle this problem we take Robbins' nonparametric empirical Bayes approach.

Empirical Bayes approach

Consider the situation where populations are incoming sequentially and independently, and, at each stage one has to select a population as a good population or reject it as a bad population. Assuming that the present one is the $(n+1)$st population in the sequence, we represent the past n such selection problems and the present one by the $n+1$ independent pairs $(X_1, \theta_1), \ldots, (X_n, \theta_n)$ and $(X_{n+1} = X, \theta_{n+1} = \theta)$, where X_i given θ_i, has the conditional density $f(\cdot \mid \theta_i)$ given by (1); $\theta_1, \theta_2, \ldots$ continue to be unobservable, having the same unknown prior distribution G; and $X_1, \ldots, X_n$ and X are i.i.d. according to the marginal density $f(\cdot) = \int f(\cdot \mid \theta)\, dG(\theta)$.

In this note we obtain an EB selection procedure for selecting (or rejecting) the present population based on $X_1, \ldots, X_n$ and X, first by forming a statistic $\alpha_n(x) = \alpha_n(X_1, \ldots, X_n; x)$ which estimates appropriately the unknown function $\alpha(x)$ and then adopting the (EB) selection procedure, following (5), $\delta_n(X) = I[\alpha_n(X) \leq 0]$ as against the (*unknown*) Bayes optimal procedure $\delta_G(X)$. These selection procedures δ_n are asymptotically optimal in the sense of Robbins (1956) and Robbins (1964), i.e. for which $\lim_{n\to\infty} R(\delta_n, G) = R(\delta_G, G) = R(G)$. Thus for large n, δ_n could be considered as good as δ_G, the best procedure, but unavailable for use. We also investigate the speed of convergence and show that, for any $\varepsilon > 0$, procedures δ_n can be constructed for which $R(\delta_n, G) - R(G) = O(n^{-1/2+\varepsilon})$ under certain moment conditions on G. It is pointed out how the results can be extended to the case where one has to deal with m varieties of populations simultaneously at each stage.

3 Development of Empirical Bayes Procedures

We first examine the function $\alpha(\cdot)$ which from (5) is the key function involved in the expression for the optimal Bayes selection procedure $\delta(\cdot)$. We note from (4) that after some simplification, we can rewrite $\alpha(x)$ as

$$\alpha(x) = (a^2 - b^2)f(x) - 2a\,V(x) + J(x), \tag{7}$$

where

$$f(x) = \int f(x \mid \theta)\, dG(\theta), \tag{8}$$

is the marginal pdf of X,

$$V(x) = \int \theta f(x \mid \theta)\, dG(\theta), \tag{9}$$

and

$$J(x) = \int \theta^2 f(x \mid \theta)\, dG(\theta). \tag{10}$$

From (7) we see that good approximations to $\alpha(\cdot)$ can be achieved by getting good estimates of $f(\cdot)$, $V(\cdot)$ and $J(\cdot)$.

Throughout the remainder of this paper, let $\chi = \{x\colon f(x) > 0\}$. Then, as noted in Singh (1995) for $x \in \chi$ the conditional (given θ) cdf of X can be written as $F(x \mid \theta) = \int_{-\infty}^{x} f(t \mid \theta)\, dt = I(x > \theta) - k^{-1} f(x \mid \theta)$. Hence, since $F(x) = \int_{-\infty}^{x} F(x \mid \theta)\, dG(\theta) = \int_{-\infty}^{x} [I(x > \theta) - k^{-1} f(x \mid \theta)]\, dG(\theta)$, we have

$$G(\cdot) = F(\cdot) + k^{-1} f(\cdot). \tag{11}$$

This identity has been helpful in obtaining nonparametric estimates of the prior distribution G utilizing the data $X_1, X_2, \ldots, X_n$, e.g, see Singh (1995).

Remark 1. From (4) and (1) we note that the function $\alpha(\cdot)$ can also be written as

$$\alpha(x) = ke^{-kx}\big[(a^2 - b^2)w_o(x) - 2aw_1(x) + w_2(x)\big], \tag{7'}$$

where, for $j = 0$, 1, and 2,

$$w_j(x) = \int \theta^j e^{k\theta} I(\theta < x)\, dG(\theta).$$

If the prior distribution G was explicitly known, then we would have evaluated exact values of $w_j(\cdot)$ and used the minimum Bayes risk test statistic $\delta_G(\cdot)$ in (5), via $\alpha(\cdot)$ in (7) for the selection purpose. On the other hand, if G was not known but the structural form of G up to possibly some (unknown) parameters was known, then in view of (1) $\delta_G(\cdot)$ would be known up to the (unknown) parameters of G. In such situations it might have been possible to use the data $X_1, X_2, \ldots, X_n$ and X to estimate these parameters and hence to estimate the Bayes procedure δ_G. This would have been what is popularly known as the standard Morris–Efron-type EB approach to the problem. In the absence of any knowledge of the structural form of G, this approach, however, is not applicable here. Even if we had realized the values of θ in the past experiences of the problem, then we could have possibly utilized these realized values of θ from the past experiences to formulate an empirical distribution of θ and could have used the Bayes test versus this empirical distribution. But again this is not the case here. In view of these difficulties, we have no choice but to apply Robbins' nonparametric empirical Bayes approach to our problem here. Note that even though the parametric forms of the conditional c.d.f. $F(\cdot \mid \theta)$ and conditional pdf $f(\cdot \mid \theta)$ are known in view of (1) the structural forms of the unconditional c.d.f. $F(\cdot)$ and the unconditional pdf $f(\cdot)$ remain unknown since the structural form of G is completely unknown. As such, we have no choice but to estimate $F(\cdot)$ and $f(\cdot)$, (and hence $G(\cdot)$ from (11)) nonparametrically.

To obtain appropriate approximations to the unknown functions $f(x)$, $V(x)$ and $J(x)$ involved in the expression (7) for $\alpha(\cdot)$ using the data X_1, $X_2, \ldots, X_n$, we proceed as follows. As in Singh (1974), Singh (1977) for kernel estimates of density, let $s > 0$ be an integer and let K be any Borel-measurable real valued function of bounded variation on the real line such that $\int y^t K(y)\, dy = 1$ for $t = 0$, and 0 for $t = 1, 2, \ldots, s-1$. Further, let $\int |y^s K(y)|\, dy < \infty$ and $\lim_{|y|\to\infty} |yK(y)| = 0$. Let $h = h_n$ be a sequence of numbers such that $h \to 0$ and $nh \to \infty$ as $n \to \infty$. Then

$$f_n(x) = (nh)^{-1} \sum_{j=1}^{n} K\big((X_j - x)/h\big), \tag{12}$$

is a higher order kernel estimate of a density given in Singh (1974), Singh (1977), which has subsequently been used in the empirical Bayes context for obtaining higher rates of convergence by many authors.

For our density $f(x)$, it follows from Singh (1974) [also see Singh (1977), Singh (1995)] that if $h \to 0$ and $nh \to \infty$ then

$$\sup_x \big| f_n(x) - f(x) \big| = o(1) \text{ w.p.1}, \tag{13}$$

and if h is of the order $O[n^{-1/(2s+2)}]$, then

$$\sup_x \left| f_n(x) - f(x) \right| = O(n^{-s/(s+1)} \log\log n)^{1/2} \text{ w.p.1.} \tag{14}$$

Also, it is well known that, if $F_n(x) = n^{-1} \sum_{j=1}^{n} I(X_j \leq x)$ is the empirical distribution function of $X_1, \ldots, X_n$, then

$$\sup_x \left| F_n(x) - F(x) \right| = O(n^{-1} \log\log n)^{1/2} \text{ w.p.1.} \tag{15}$$

Further if

$$G_n(\cdot) = F_n(\cdot) + k^{-1} f_n(\cdot), \tag{11$'$}$$

then from Theorem 3.1 of Singh (1995), for $h \to 0$ and $nh \to \infty$ as $n \to \infty$,

$$\sup_x \left| G_n(x) - G(x) \right| = o(1) \text{ w.p.1,} \tag{16}$$

and for $h = O[n^{-s/(s+1)}]$,

$$\sup_x \left| G_n(x) - G(x) \right| = O[n^{-s/(s+1)} \log\log n]^{1/2} \text{ w.p.1.} \tag{17}$$

Remark 2. We have tacitly provided in (11)$'$ a sequence of estimators G_n of the prior distribution G. The result in (16) shows that these estimators are uniformly strongly consistent for G, and that in (17) provides the speed of convergence for such consitency.

In view of the above estimators of f and G, we estimate $V(x)$ and $J(x)$ for $x \in X$ by

$$V_n(x) = \int_{-\infty}^{x} t f(x \mid t)\, dG_n(t), \tag{18}$$

and

$$J_n(x) = \int_{-\infty}^{x} t^2 f(x \mid t)\, dG_n(t), \tag{19}$$

respectively; and finally, in view of (7) we estimate $\alpha(x)$ by

$$\alpha_n(x) = (a^2 - b^2) f_n(x) - 2a V_n(x) + J_n(x). \tag{20}$$

In view of these results and (5) our proposed EB selection procedure for the $(n+1)$st population is

$$\delta_n(x) = I\left[\alpha_n(x) \leq 0\right], \tag{21}$$

i.e, once $X = x$ is observed from the $(n+1)$st population w.p.1 we select this population as a good population if $\alpha_n(x) \leq 0$, otherwise we reject it w.p.1.

Note that the sequence δ_n is completely nonparametric in the sense that it does not depend on the knowledge or the structure of the prior distribution G.

4 Asymptotic Optimality of the EB Procedures and Rates of Convergence

In this section, we prove two theorems which give the main results of this paper concerning the properties of the EB selection procedures δ_n. The first theorem proves the asymptotic optimality of δ_n whereas the second one provides the speed of convergence. The symbols c_o, $c_1, \dots$ appearing below in the proofs are absolute constants. Again, all the convergences are w.r.t. $n \to \infty$, unless stated otherwise.

Theorem 1. *Let the window-width function h in the definition of the kernel estimates f_n of f be such that $h \to 0$ and $nh \to \infty$ as $n \to \infty$. Let the sequence $\delta_n(x)$ of* EB *selection procedures be given by* (21) *via* (20). *Then*

$$R(\delta_n, G) - R(G) = o(1). \tag{22}$$

for all G on Θ for which $\int \theta^2 \, dG(\theta) < \infty$.

Proof. Note that from (3) and (4) we can rewrite $R(\delta, G)$ as

$$R(\delta, G) = E\{[(\theta - a)^2 - b^2]\delta(X)\} + C_G,$$

where the expectation operator E is w.r.t. all the r.v.'s involved, including θ. Similarly,

$$R(\delta_n, G) = E\{[(\theta - a)^2 - b^2]\delta_n(X)\} + C_G.$$

Therefore, the excess risk δ_n over that of δ_n is

$$\begin{aligned} o \le R(\delta_n, G) - R(G) &= E\{[(\theta - a)^2 - b^2][\delta_n(x) - \delta_G(x)]\} \\ &= E([(\theta - a)^2 - b^2]E\{[\delta_n(x) - \delta_G(x)] \mid X = x\}). \end{aligned}$$

Throughout the remainder of the proof, the argument x displayed in various functions is in the set χ. It can be shown, e.g., see Singh (1995), that

$$\begin{aligned} |\delta_n(x) - \delta_G(x)| &= |I[\alpha_n(x) \le 0] - I[\alpha(x) \le 0]| \\ &\le |\alpha(x)|^{-1} |\alpha_n(x) - \alpha(x)|. \end{aligned}$$

Now, from (7) and (20),

$$|\alpha_n(x) - \alpha(x)| \le c_o[l_1(x) + l_2(x) + l_3(x)],$$

where

$$\begin{aligned} l_1(x) &= |f_n(x) - f(x)|, \\ l_2(x) &= \left| \int_{-\infty}^{x} t f(x \mid t) \, d[G_n(t) - dG(t)] \right|, \\ l_3(x) &= \left| \int_{-\infty}^{x} t^2 f(x \mid t) \, d[G_n(t) - dG(t)] \right|. \end{aligned}$$

We note that $l_1(x) = o(1)$ w.p.1 from (13). Also, by integration by parts

$$l_2(x) = [G_n(x) - G(x)]kx - \int_{-\infty}^{x} (1+tk)f(x \mid t)[G_n(t) - G(t)]\,dt.$$

Therefore,

$$|l_2(x)| \le \sup_x |G_n(x) - G(x)| \left[k|x| + \int_{-\infty}^{x} (1+|t|k)f(x \mid t)\,dt\right].$$

Since, $\int_{-\infty}^{x}(1+|t|k)f(x \mid t)\,dt < \infty$, from (16) $l_2(x) = o(1)$ w.p.1. Similarly, using the integration by parts technique it can be shown that $l_3(x) = o(1)$ w.p.1. Thus we conclude that

$$|\delta_n(x) - \delta_G(x)| = o(1). \tag{23}$$

Now since $|\delta_n(\cdot) - \delta_G(\cdot)| \le 1$, (23) followed by the Lebesgue dominated convergence theorem, shows that $E[\delta_n(x) - \delta_G(x) \mid X = x] = o(1)$ for every $x \in \chi$. Now, since $E(\theta^2) < \infty$, another use of the dominated convergence theorem shows that the r.h.s. of (22) is $o(1)$. This completes the proof of the theorem. □

Remark 3. We note that in the process of the proof of Theorem 1, we also proved that *the empirical Bayes selection procedures δ_n are strongly consistent estimators of the Bayes optimal selection procedure δ_G*. Our next theorem gives a minimum rate of convergence for asymptotic optimality of $\{\delta_n\}$ under some moment conditions on the prior distribution G.

Theorem 2. *Let $s > 0$ be an arbitrary but fixed integer and for this s, let the kernel function K in the definition of f_n in (12) satisfy the conditions on the kernel defined in Section 3. Further, let the window-width function h involved in f_n be of the order $O[n^{-s/(2s+2)}]$. Then for any prior distribution G on Θ for which*

$$\int |\theta|^{3/(1-\gamma)}\,dG(\theta) < \infty, \tag{24}$$

for a $0 < \gamma < 1$,

$$R(\delta_n, G) - R(G) = O[n^{-s/(s+1)} \log\log n]^{\gamma/2}. \tag{25}$$

Proof. First of all we note, by the arguments used to arrive at (3), that the Bayes risks of the selection procedures $\{\delta_n\}$ can be expressed as

$$R(\delta_n, G) = c\int \alpha(x)\delta_n(x)\,dx + C_G. \tag{26}$$

Therefore, the excess risk δ_n over that of the minimum Bayes risk procedure δ_G is from (3),

$$R(\delta_n, G) - R(G) = c\int B_n(x)\,dx, \tag{27}$$

where

$$B_n(x) = \alpha(x)E\big[\delta_n(x) - \delta(x)\big]. \tag{28}$$

But, from Lemma 1 of Johns and Van Ryzin (1972) or, alternatively, from the proof of Theorem 1 above, it follows that for $\gamma > 0$

$$\big|B_n(x)\big| \le \big|\alpha(x)\big|^{1-\gamma} E\big|\alpha_n(x) - \alpha(x)\big|^{\gamma}, \tag{29}$$

$$\le c_1\big[\big|\alpha(x)\big|^{1-\gamma} l_1^{\gamma}(x) + l_2^{\gamma}(x) + l_3^{\gamma}(x)\big], \tag{30}$$

where the second inequality follows from (7) and (20) with $l_1(x)$, $l_2(x)$ and $l_3(x)$ as defined in the proof of Theorem 1.

Also it is shown in the proof of Theorem 1 that

$$\big|l_2(x)\big| \le \sup_x \big|G_n(x) - G(x)\big| \left[k|x| + \int_{-\infty}^{x} (1 + |t|k) f(x \mid t)\, dt\right], \tag{31}$$

$$\le c_2\big(|x| + 1\big) \sup_x \big|G_n(x) - G(x)\big|. \tag{32}$$

Similary, using the integration by parts technique, it can be shown that

$$l_3(x) \le c_3\big(x^2 + |x| + 1\big) \sup_x \big|G_n(x) - G(x)\big|. \tag{33}$$

Therefore, from (27) and (28) followed by (29)–(31), (14), (17) and (4) we have for $0 < \gamma < 1$,

$$R(\delta_n, G) - R(G) = O[n^{-s/(s+1)} \log\log n]^{\gamma/2} \sum_{j=0}^{2}\sum_{i=0}^{2} A_{ij}, \tag{34}$$

where, for i, $j = 0$, 1, 2,

$$A_{ij} = \int |x|^{i\gamma} \big[|\theta|^j f(x \mid \theta)\, dG(\theta)\big]^{1-\gamma} dx.$$

Now we show that A_{ij}'s for $i, j = 0$, 1 and 2 are finite for every $0 < \gamma < 1$ for which (24) holds. Note that, by Holder's inequality, for $0 < \gamma < 1$,

$$\int_{|x|\le 1} |x|^{i\gamma} \big[|\theta|^j f(x \mid \theta)\, dG(\theta)\big]^{1-\gamma} dx$$

$$\le \left(\int_{|x|\le 1} |x|^i\, dx\right)^{\gamma} \left[\int_{|x|\le 1} \int |\theta|^j f(x \mid \theta)\, dG(\theta)\, dx\right]^{1-\gamma}$$

$$< \infty \quad \text{whenever } E|\theta|^j < \infty.$$

Also, by Holder's inequality and for $\eta > i + 1$

$$\int_{|x|\ge 1} |x|^{i\gamma} \left[\int |\theta|^j f(x \mid \theta)\right]^{1-\gamma} dx$$

$$\le \left[\int_{|x|\ge 1} |x|^{\eta\gamma/(1-\gamma)} \int |\theta|^j f(x \mid \theta)\, dG(\theta)\, dx\right]^{1-\gamma} \left[\int_{|x|\ge 1} |x|^{(i-\eta)}\, dx\right]^{\gamma}$$

$< \infty$ whenever the $[j + \eta\gamma(1+\gamma)^{-1}]$th moment of G is finite.

Thus, since $\eta > i+1$ is arbitrary, taking $\eta = i+\gamma^{-1}$, we see that for $i, j = 0$, 1 and 2, $A_{ij} < \infty$ whenever the $[j + (i\gamma + 1)(1-\gamma)^{-1}]$th moment of G is finite. This latter condition on G is satisfied by (24) for every combination of i, $j = 0, 1, 2$. The proof of theorem is now complete from (34). □

5 Concluding Remarks and Extension of the Results

In this paper, we have provided asymptotically optimal nonparametric empirical Bayes selection procedures for accepting a population as a "good" population or rejecting it as a "bad" population under the two-tail test criterion with product-distance loss function when the population density belongs to the translated exponential family given by (1). We have shown that for any $\varepsilon > 0$ under certain moment conditions on the prior distribution, the rates of convergence could be arbitrarily close to $O(n^{-1/2+e})$.

The work here can also be extended to the situation where one has to deal with m different (or same) varieties of populations simultaneously and one has to accept the good ones and rejects the bad ones at each stage. Let $\pi_1, \ldots, \pi_m$ denote the m populations to be tested for their acceptance and let π_j have the density $f(y_j \mid \theta_j) = \exp[-(y_j - \theta_j)]I(y_j > \theta_j)$. The population π_j is acceptable as a good population if $\theta_j \in [\theta_{1j}, \theta_{2j}]$, where θ_{1j} and θ_{2j} are the boundary points of the standard set by the practitioner for the jth variety of the population. The random vector $\theta = (\theta_1, \ldots, \theta_m)$ has unknown and unspecified prior distribution on $\Theta^m, \Theta = (-\infty, \infty)$. The total risk accross the m selection problems is $\sum_{j=1}^{m} E\{L_0(\theta_j)\delta_j(X_j) + L_1(\theta_j)[1 - \delta_j(X_j)]\}$, where for $j = 1, \ldots, m$ and $i = 0, 1$, $L_i(\theta_j) = (1 - i)(\theta_j - \theta_{1j})(\theta_j - \theta_{2j})I(\theta_j < \theta_{1j})$ or $(\theta_j > \theta_{2j}) + i(\theta_j - \theta_{1j})(\theta_{2j} - \theta_j)I(\theta_{1j} \leq \theta_j \leq \theta_{2j}$ and $L_0(\theta_j)$ [and $L_1(\theta_j)$] is the penalty due to incorrectly accepting (rejecting) the population π_j, and X_j is the randomly selected data from π_j. The problem can be handled along the lines discussed in this paper by treating each population π_j separately and by developing the EB selection procedure δ_{jn} in the way discussed in this paper.

Acknowledgments: The research is in part supported by Natural Science and Engineering Research Council of Canada, grant No. A4631.

6 References

Datta, G. (1994). Empirical Bayes estimation in a threshold model. *Sankhya, Ser. A 54*, 106–117.

Gupta, S.S. and T. Liang (1993). Bayes and empirical Bayes rules for selecting fair multinomial populations. In S. Mitra, K. Parthasarathy, and P. Rao (Eds.), *Statistics and Probability*. Wiley Eastern Ltd. Pub.

Johns, Jr., M.V. and J. Van Ryzin (1971). Convergence rates in empirical Bayes two action problems I: Discrete case. *Ann. Math. Statist. 42*, 1521–1539.

Johns, Jr., M.V. and J. Van Ryzin (1972). Convergence rates in empirical Bayes two-action problem II: Continuous case. *Ann. Math. Statist. 43*, 934–947.

Johnson, N., S. Kotz, and N. Balakrishnan (1994). *Continuous Univariate Distributions*, Volume Vol. 1, second ed. New York: John Wiley & Sons.

Nogami, Y. (1988). Convergence rates for empirical Bayes estimation in the uniform $u(0,\theta)$ distribution. *Ann. Statist. 16*, 1335–1341.

Robbins, H. (1956). An empirical Bayes approach to statistics. In *Proc. 2nd Berkeley Symp. Math. Statist.*, Volume 1, pp. 157–164.

Robbins, H. (1963). The empirical Bayes approach to testing statistical hypothesis. *Rev. Int. Statist. Inst. 31*, 195–208.

Robbins, H. (1964). The empirical Bayes approach to statistical decision problems. *Ann. Math. Statist. 35*, 1–10.

Samuel, E. (1963). An empirical Bayes approach to the testing of certain parametric hypothesis. *Ann. Math. Statist. 34*, 1370–1385.

Singh, R.S. (1974). Estimation of derivatives of average of μ-densities and sequence compound estimation in exponential families RM-319. Technical report, Michigan State University.

Singh, R.S. (1976). Empirical Bayes estimation with convergence rates in noncontinuous Lebesgue exponential families. *Ann. Statist. 4*, 431–439.

Singh, R.S. (1977). Improvement on some known nonparametric uniformly consistent estimates of derivatives of a density. *Ann. Statist. 5*, 394–399.

Singh, R.S. (1979). Empirical Bayes estimation in Lebesgue-exponential families with rates near the best possible rate. *Ann. Statist. 9*, 890–902.

Singh, R.S. (1995). Empirical Bayes linear loss hypothesis testing in a nonregular exponential family. *J. Statist. Plann. Inter. 43*, 107–120.

Singh, R.S. and L. Wei (1999). Nonparametric empirical Bayes two-tail test procedure in scale exponential family. *J. Nonpar. Statist.*. to appear.

Wei, L. (1989). Empirical Bayes two-sided test problem for continuous one-parameter exponential families. *System Sci. Math. 2*, 369–384.

Wei, L. (1991). Empirical Bayes two-sided test problem about one-parameter discrete exponential families. *Chinese J. Appl. Prob.* 7, 299–310.

8

Shrinkage Estimation of Regression Coefficients From Censored Data With Multiple Observations

S.E. Ahmed

ABSTRACT This paper considers the preliminary test and Stein-type estimation of regression parameters in exponential regression failure time distribution. We consider a situation where the lifetime data may be right censored with multiple observations taken at each regression vector. We propose improved estimators of the regression vector when it is suspected that the true regression parameter vectors may be restricted to a linear subspace. The large sample risk properties of the proposed estimators are derived. The relative merits of the proposed estimators are discussed.

1 Preliminaries and Introduction

In some clinical trials and medical investigations scientists are interested in investigating the relationship of lifetime variable (or a transformation of lifetime) to other factors when the lifetime data may be right censored. As an example, this type of censoring may occur in investigations planned for a fixed duration when the subjects participated in the study may not have responded by the termination time of the study. Further, censoring also occurs when a competing risk precludes observation of the desired end point of interest in the study. As an example of the model considered in this paper, let us consider a group of elderly men with advanced lung cancer who have undergone surgery. The measurement of interest is the duration of time, following surgery, until the patient attains a specified level of functioning. Oncologists are interested in estimating the effect of tumor size on duration of time to recovery.

In many industrial life test experiments, multiple observations may be obtained at each separate regression vector. Suppose observations are taken at m distinct vectors $\mathbf{x}_i = (x_{i1}, x_{i2}, \ldots, x_{ip})$ for the independent variables $(i = 1, 2, \ldots, m)$, with observations on n_i individuals at x_i. Suppose that the sample at $\mathbf{x}_i$ to be right or Type II censored. Thus, the data consist of the r_i smallest lifetimes

$$t_{i1} \leq t_{i2} \leq \cdots \leq t_{ir_i} \quad (r_i \leq n_i) \text{ at each } x_i.$$

Further, let

$$T_i = \sum_{j=1}^{r_i} t_{ij} + (n_i - r_i) t_{ir_i}.$$

Suppose we have lifetime measurements from a population modeled by exponential distribution with mean $1/\lambda$. Then we have the following *probability density function* (pdf) $f_T(t) = \lambda e^{-\lambda t}$, with $t \geq 0$, $\lambda > 0$. This distribution has been widely used in many areas. Further, an exponential regression model is useful in situations when individual observations have constant hazard rates that may depend on concomitant variables. Let $\mathbf{x}_i = (x_{i1}, \ldots, x_{ip})$ be a vector of regressor variables. The probability density function of T, given $\mathbf{x}_i$ is

$$f_T(t \mid \mathbf{x}_i) = \frac{1}{\theta_{\mathbf{x}_i}} e^{-t/\theta_{\mathbf{x}_i}}, \quad t \geq 0,\ \theta > 0. \tag{1}$$

We consider the most useful form of $\theta_{\mathbf{x}_i}$ given by

$$\theta_{\mathbf{x}_i} = e^{\mathbf{x}_i \boldsymbol{\beta}}, \tag{2}$$

where $\boldsymbol{\beta} = (\beta_1, \ldots, \beta_p)'$ is a vector of regression parameters. We refer to Lawless (1982) and others, for a detailed study on the above form of regression model. For the model (1) with $\theta_{\mathbf{x}_i} = e^{\mathbf{x}_i \boldsymbol{\beta}}$, the quantities $T_1, T_2, \ldots, T_m$ are sufficient for $\boldsymbol{\beta}$. Also $(T_i \mid \theta_{\mathbf{x}_i})$ has a one parameter gamma distribution with index parameter r_i. Thus, the joint pdf of $T_1, T_2, \ldots, T_m$ is

$$\prod_{i=1}^{m} \frac{T_i^{r_i - 1}}{\theta_{\mathbf{x}_i}^{r_i} (r_i - 1)!} \exp\left(-\frac{T_i}{\theta_{\mathbf{x}_i}}\right).$$

Note that the lifetime model given in relation (1) is a proportional hazards model, which can be viewed as a location-scale model with a log transformation on random variable T_i. Usually, we deal with log lifetimes, $Y_i = \log T_i$. Thus, the joint pdf of $Y_1, Y_2, \ldots, Y_m$ is

$$f(y_1, y_2, \ldots, y_m) = \left\{\prod_{i=1}^{m} \frac{\exp[r_i(y_i - \mathbf{x}_i\beta)]}{(r_i - 1)!}\right\} \exp\left[-\sum_{i=1}^{m} \exp(y_i - \mathbf{x}_i\beta)\right]. \tag{3}$$

The log likelihood function is

$$\ell(\beta) = \sum_{i=1}^{m} \frac{r_i(y_i - \mathbf{x}_i\beta)}{(r_i - 1)!} - \sum_{i=1}^{m} \exp(y_i - \mathbf{x}_i\beta). \tag{4}$$

The *unrestricted maximum likelihood* estimators are obtained by solving the following system of equations

$$\frac{\partial \ell(\boldsymbol{\beta})}{\partial \beta_s} = -\sum_{i=1}^{m} r_i \mathbf{x}_{is} + \sum_{i=1}^{m} \mathbf{x}_{is} \exp[(y_i - \mathbf{x}_i \boldsymbol{\beta})] = 0, \quad \text{for } s = 1, \ldots, p. \quad (5)$$

Also, the second derivative of $\ell(\boldsymbol{\beta})$ is given by

$$\frac{\partial^2 \ell(\boldsymbol{\beta})}{\partial \beta_s \partial \beta_t} = -\sum_{i=1}^{m} \mathbf{x}_{is} \mathbf{x}_{it} \exp[(y_i - \mathbf{x}_i \boldsymbol{\beta})] = 0, \quad \text{for } s,\ t = 1, \ldots, p. \quad (6)$$

These equations are solved by an iterative procedure such as the Newton–Raphson method. The observed information matrix is given by the following relation

$$\mathfrak{I}^o = \left[\frac{-\partial^2 \ell(\boldsymbol{\beta})}{\partial \beta_s \partial \beta_t}\right]_{\widehat{\boldsymbol{\beta}}}$$

$$= \left[\sum_{i=1}^{m} x_{is} x_{it} e^{(y_i - \mathbf{x}_i \widehat{\boldsymbol{\beta}})}\right] \quad \text{where } s,\ t = 1, \ldots, p. \quad (7)$$

For inference purposes when there is censoring, one may use the asymptotic distribution of $\widehat{\boldsymbol{\beta}}$, i.e., $\widehat{\boldsymbol{\beta}} \sim \mathcal{N}(\boldsymbol{\beta}, {\mathfrak{I}^o}^{-1})$.

Uncertain Prior Information (UPI)

Suppose the true regression parameter vector $\boldsymbol{\beta}$ may be partitioned as $\boldsymbol{\beta}' = (\boldsymbol{\beta}_1', \boldsymbol{\beta}_2')$, where $\boldsymbol{\beta}_1$ is a regression parameter vector of order $a \times 1$ and $\boldsymbol{\beta}_2$ is of order $b \times 1$ with $a + b = p$. In fact, this partition is a special case of more general partition $\mathbf{R}\boldsymbol{\beta} = \mathbf{r}$ where $\mathbf{R}$ is a $b \times p$ matrix of known constants with rank b and $\mathbf{r}$ is vector of order b.

In the present investigation we are interested in estimating $\boldsymbol{\beta}_1$, a subvector of $\boldsymbol{\beta}$ when it is plausible that $\boldsymbol{\beta}_2$ is close to a specified vector $\boldsymbol{\beta}_2^o$. This information may be regarded as *uncertain prior information*:

$$\text{UPI}: \boldsymbol{\beta}_2 = \boldsymbol{\beta}_2^o. \quad (8)$$

The statistical objective is to estimate $\boldsymbol{\beta}_1$, when UPI is available. Now, we pose the commonly asked question "Does the reduction of dimension of the parameter space play a role in obtaining superior estimators"? We will deal with this question in a logical manner by proposing various estimation schemes for $\boldsymbol{\beta}_1$.

The *maximum likelihood* estimator of $\boldsymbol{\beta}_1$ can be obtained by solving the system of equations in (5). Note that the *ml* estimator of $\boldsymbol{\beta}_1$ relies on sample data only. Such an estimator is generally called an *unrestricted ml* estimator.

However, it may be advantageous to use the available information to obtain improved estimates. The UPI given in (8) may be explicitly incorporated into the estimation scheme by modifying the parameter space. In this case, the new (restricted) parameter space is a subspace of the original one (reduced in dimension). Generally speaking, the reduction in dimensionality provides efficient parameter estimates. However, in the case of an incorrect restriction opposite conclusions will hold.

Let $\widehat{\boldsymbol{\beta}}_1^{\mathrm{R}}$ be the *restricted maximum likelihood* estimator of $\boldsymbol{\beta}_1$ when the UPI in (8) is correct. The estimator is found by solving the system of equations $\partial\ell(\boldsymbol{\beta})/\partial\beta_r = 0$, $(s = 1, \ldots, a)$ with $\boldsymbol{\beta}_2$ equal to $\boldsymbol{\beta}_2^{\mathrm{o}}$. It can be shown that

$$\widehat{\boldsymbol{\beta}}_1^{\mathrm{R}} = \widehat{\boldsymbol{\beta}}_1^{\mathrm{U}} + \mathfrak{I}_{11}^{-1}\mathfrak{I}_{12}(\widehat{\boldsymbol{\beta}}_2^{\mathrm{U}} - \boldsymbol{\beta}_2^{\mathrm{o}}) + o_p(n^{-1/2}), \tag{9}$$

where $\widehat{\boldsymbol{\beta}}^{U}$ is given by (5) and $n = \sum_{i=1}^{m} n_i$. We further assume that for large n the ratio $n_i/n \to \pi \in (0,1)$ in the remaining discussions. Further, $\mathfrak{I}^{\mathrm{o}}$ is the observed information matrix and may be partitioned as

$$\mathfrak{I}^{\mathrm{o}} = \begin{pmatrix} \mathfrak{I}_{11} & \mathfrak{I}_{12} \\ \mathfrak{I}_{21} & \mathfrak{I}_{22} \end{pmatrix}. \tag{10}$$

Lemma 1. *Under the usual regularity conditions and as n increases we have following:*

(a) $n^{1/2}(\widehat{\boldsymbol{\beta}}_1^{\mathrm{U}} - \boldsymbol{\beta}_1) \xrightarrow{\mathcal{L}} \mathcal{N}(\mathbf{0}, \mathfrak{I}_{11.2}^{-1})$, $\mathfrak{I}_{11.2} = \mathfrak{I}_{11} - \mathfrak{I}_{12}\mathfrak{I}_{22}^{-1}\mathfrak{I}_{21}$,

(b) $n^{1/2}(\widehat{\boldsymbol{\beta}}_1^{\mathrm{R}} - \boldsymbol{\beta}_1) \xrightarrow{\mathcal{L}} \mathcal{N}(\mathbf{0}, \mathfrak{I}_{11}^{-1})$.

Here $\xrightarrow{\mathcal{L}}$ means convergence in distribution.

The proof that $\widehat{\boldsymbol{\beta}}_1^{\mathrm{R}}$ is more efficient than $\widehat{\boldsymbol{\beta}}_1^{\mathrm{U}}$ under the UPI in (8) is relatively simple, since the difference between the covariance matrices of $\widehat{\boldsymbol{\beta}}_1^{\mathrm{U}}$ and $\widehat{\boldsymbol{\beta}}_1^{\mathrm{R}}$ is a positive semi-definite matrix.

We have just established that the restricted estimator $\widehat{\boldsymbol{\beta}}_1^{\mathrm{R}}$ is more efficient (or, at least, no less efficient) than the unrestricted estimator when the model satisfies the restriction. But what happens when it does not satisfy the restriction. It is easy to see that the restricted estimator will, in general, be biased.

Even though $\widehat{\boldsymbol{\beta}}_1^{\mathrm{R}}$ is biased when the constraint in (8) does not hold, it is still of interest to find how well it performs. The analog of the covariance matrix for biased estimator is the *mean squared matrix* and we use this matrix to compute the risk of the estimator. Further, in a large sample situation, it is seen that for fixed alternatives $\widehat{\boldsymbol{\beta}}_1^{\mathrm{R}}$ has an unbounded risk. Hence, in the large sample set-up there is not much to do. To obtain some interesting and meaningful results, we shall therefore restrict ourselves to contiguous alternatives. More specifically, we consider a sequence $\{K_n\}$ of contiguous alternatives defined by

$$K_n \colon \boldsymbol{\beta}_2 = \boldsymbol{\beta}_2^o + \frac{\boldsymbol{\delta}}{n^{1/2}}, \tag{11}$$

where $\boldsymbol{\delta} = (\delta_1, \ldots, \delta_b)' \in \mathbb{R}^b$. Note that $\boldsymbol{\delta} = \mathbf{0}$ implies $\boldsymbol{\beta}_2 = \boldsymbol{\beta}_2^o$, so (8) is a particular case of $\{K_n\}$. We consider the computation of bias, mean squared error matrices and risks of the restricted estimators under the local alternatives.

We are mainly interested in estimating the unknown parameter vector $\boldsymbol{\beta}_1$ by means of an estimator $\widetilde{\boldsymbol{\beta}}_1$. A loss function $\text{Ł}(\widetilde{\boldsymbol{\beta}}_1, \boldsymbol{\beta}_1)$ will show the loss incurred by making wrong decision about $\boldsymbol{\beta}_1$ using the estimator $\widetilde{\boldsymbol{\beta}}_1$. We confine ourselves to loss functions of form

$$\text{Ł}(\widetilde{\boldsymbol{\beta}}_1, \boldsymbol{\beta}_1; \boldsymbol{\Gamma}) = n(\widetilde{\boldsymbol{\beta}}_1 - \boldsymbol{\beta}_1)'\boldsymbol{\Gamma}(\widetilde{\boldsymbol{\beta}}_1 - \boldsymbol{\beta}_1), \tag{12}$$

where $\boldsymbol{\Gamma}$ is positive semi-definite weighting matrix. Then, the expected loss functions

$$\mathcal{E}[\text{Ł}(\widetilde{\boldsymbol{\beta}}_1, \boldsymbol{\beta}_1; \boldsymbol{\Gamma})] = R(\widetilde{\boldsymbol{\beta}}_1, \boldsymbol{\beta}_1; \boldsymbol{\Gamma}) \equiv R(\widetilde{\boldsymbol{\beta}}_1, \boldsymbol{\beta}_1) \equiv R(\widetilde{\boldsymbol{\beta}}_1), \tag{13}$$

are called *risk functions.* We have

$$R(\widetilde{\boldsymbol{\beta}}_1) = n\, \text{trace}\{\boldsymbol{\Gamma}[E(\widetilde{\boldsymbol{\beta}}_1 - \boldsymbol{\beta}_1)(\widetilde{\boldsymbol{\beta}}_1 - \boldsymbol{\beta}_1)']\} = \text{trace}(\boldsymbol{\Gamma}\boldsymbol{\Upsilon}), \tag{14}$$

where $\boldsymbol{\Upsilon}$ is the covariance matrix of $\widetilde{\boldsymbol{\beta}}_1$. Further, $\widetilde{\boldsymbol{\beta}}_1$ is an *inadmissible estimator* of $\boldsymbol{\beta}_1$ if there exists an alternative estimator $\widehat{\boldsymbol{\beta}}_1$ such that

$$\mathcal{R}(\widehat{\boldsymbol{\beta}}_1, \boldsymbol{\beta}_1) \le \mathcal{R}(\widetilde{\boldsymbol{\beta}}_1, \boldsymbol{\beta}_1) \quad \text{for all} \quad (\boldsymbol{\beta}_1, \boldsymbol{\Gamma}), \tag{15}$$

with strict inequality for some $\boldsymbol{\beta}_1$. We also say that $\widehat{\boldsymbol{\beta}}_1$ dominates $\widetilde{\boldsymbol{\beta}}_1$. If, instead of (15) holding for every n, we have

$$\lim_{n\to\infty} \mathcal{R}(\widehat{\boldsymbol{\beta}}_1, \boldsymbol{\beta}_1) \le \lim_{n\to\infty} \mathcal{R}(\widetilde{\boldsymbol{\beta}}_1, \boldsymbol{\beta}_1) \quad \text{for all} \quad \boldsymbol{\beta}_1, \tag{16}$$

with strict inequality for some $\boldsymbol{\beta}_1$, then $\widetilde{\boldsymbol{\beta}}_1$ is termed an *asymptotically inadmissible estimator* of $\boldsymbol{\beta}_1$. However, the expression in (16) may be difficult to obtain. Hence we consider the *asymptotic distributional risk* (adr) for a sequence $\{K_n\}$ of local alternatives defined in relation (11).

Assume that under local alternatives, $n^{1/2}(\widetilde{\boldsymbol{\beta}}_1 - \boldsymbol{\beta}_1)$ has a limiting distribution given by

$$F(\mathbf{y}) = \lim_{n\to\infty} P[n^{1/2}(\widetilde{\boldsymbol{\beta}}_1 - \boldsymbol{\beta}_1) \le \mathbf{y}], \tag{17}$$

which is called the *asymptotic distribution function* of $\widetilde{\boldsymbol{\beta}}_1$. Further, let

$$\boldsymbol{\Upsilon} = \int \cdots \int \mathbf{y}\mathbf{y}'\, dF(\mathbf{y}), \tag{18}$$

be the dispersion matrix which is obtained from F, then

$$R(\widetilde{\beta}_1;\beta_1) = \text{trace}(\mathbf{\Gamma}\mathbf{\Upsilon}). \tag{19}$$

We may be able to compute asymptotic risk by replacing $\mathbf{\Upsilon}$ with the limit of actual dispersion matrix of $n^{1/2}(\widetilde{\beta}_1 - \beta_1)$. This may require extra regularity conditions to suit the problem in hand. This point has been explained in various other contexts by Saleh and Sen (1985); Sen (1986), and others.

Lemma 2. *Under $\{K_n\}$ and the usual regularity conditions as n increases,* $n^{1/2}(\widehat{\beta}_1^{\mathrm{R}} - \beta_1) \xrightarrow{\mathcal{L}} \mathcal{N}(\mathfrak{I}_{11}^{-1}\mathfrak{I}_{12}\delta, \mathfrak{I}_{11}^{-1})$.

We define the *asymptotic distribution bias* (adb) of an estimator $\widetilde{\beta}_1$ of β_1 as

$$\text{adb}(\widetilde{\beta}_1) = \lim_{n\to\infty}\left[n^{1/2}(\widetilde{\beta}_1 - \beta_1)\right]. \tag{20}$$

The adb of the $\widehat{\beta}_1^{\mathrm{R}}$ is given by

$$\text{adb}(\widehat{\beta}_1^{\mathrm{R}}) = \mathfrak{I}_{11}^{-1}\mathfrak{I}_{12}\delta. \tag{21}$$

Unless $\mathfrak{I}_{12}$ is a zero matrix or δ is a null vector, $\widehat{\beta}_1^{\mathrm{R}}$ will be a biased estimator. The magnitude of the bias will depend on the matrices $\mathfrak{I}_{11}$, $\mathfrak{I}_{12}$ and the vector δ. Similar results are available for all sorts of restrictions, not just for linear restrictions, and for various models in addition to linear regression models. Most importantly, the imposition of false restrictions on some of the parameters of a model generally cause all of the parameter estimates to be biased. Moreover, the bias does not vanish as the sample size gets larger.

In order to obtain simple and meaningful analysis of the adb of the estimator, let us transform this vector-valued bias into a scalar (quadratic) form. To achieve this, define the *quadratic* adb (qadb) of an estimator $\widetilde{\beta}_1$ of parameter vector β_1 by

$$\text{qadb}(\widetilde{\beta}_1) = \left[\text{adb}(\widetilde{\beta}_1)\right]^t \mathfrak{I}_{11.2}\left[\text{adb}(\widetilde{\beta}_1)\right],$$

where we use $\mathbf{v}^t$ to denote the transpose of a vector $\mathbf{v}$. Hence,

$$\text{qadb}(\widehat{\beta}_1^{\mathrm{R}}) = \Delta^*, \quad \Delta^* = \delta'\mathfrak{I}^*\delta, \quad \mathfrak{I}^* = \mathfrak{I}_{21}\mathfrak{I}_{11}^{-1}\mathfrak{I}_{11.2}\mathfrak{I}_{11}^{-1}\mathfrak{I}_{12}.$$

Thus, $\widehat{\beta}_1^{\mathrm{R}}$ has no control on its quadratic bias since $\Delta^* \in [0, \infty)$.

Under local alternatives and usual regularity conditions, we obtain the adr functions of $\widehat{\beta}_1^{\mathrm{U}}$ and $\widehat{\beta}_1^{\mathrm{R}}$ given in the following theorem.

Theorem 1.

$$\text{adr}(\widehat{\beta}_1^{\mathrm{U}}) = \text{trace}(\mathbf{\Gamma}\mathfrak{I}_{11.2}^{-1}), \tag{22}$$

$$\text{adr}(\widehat{\beta}_1^{\mathrm{R}}) = \text{trace}(\mathbf{\Gamma}\mathfrak{I}_{11}^{-1}) + \delta'\mathfrak{I}^{\bullet}\delta, \quad \mathfrak{I}^{\bullet} = \mathfrak{I}_{21}\mathfrak{I}_{11}^{-1}\mathbf{\Gamma}\mathfrak{I}_{11}^{-1}\mathfrak{I}_{12}. \tag{23}$$

Proof. By plugging, $g(\Lambda) = 1$ and 0 respectively, in Lemma 4 of the Appendix we get the expressions (22) and (23). □

Again, we exclude the case $\mathfrak{J}_{12} = \mathbf{0}$, since in this situation, $\mathfrak{J}^{\bullet} = \mathbf{0}$ and $\mathfrak{J}_{11.2} = \mathfrak{J}_{11}$. Then the adr of $\widehat{\boldsymbol{\beta}}_1^{R}$ is equal to the adr of $\widehat{\boldsymbol{\beta}}_1^{U}$. Therefore, in the remaining discussions we establish the asymptotic dominance properties of $\widehat{\boldsymbol{\beta}}_1^{R}$ relative to $\widehat{\boldsymbol{\beta}}_1^{U}$ when $\mathfrak{J}_{12}$ is not a null matrix.

The adr of $\widehat{\boldsymbol{\beta}}_1^{U}$ does not depend on the UPI, i.e., on $\boldsymbol{\delta}$ while $\widehat{\boldsymbol{\beta}}_1^{R}$ does. Let us consider a special choice of $\boldsymbol{\Gamma} = \mathfrak{J}_{11.2}$, then Theorem 1 holds for the Mahalanobis distance as a loss function. For this special choice of $\boldsymbol{\Gamma}$ we have $\text{trace}(\boldsymbol{\Gamma}\mathfrak{J}_{11.2}^{-1}) = a$, and $\text{trace}(\boldsymbol{\Gamma}\mathfrak{J}_{11}^{-1}) = a - \text{trace}(\mathfrak{J}^{\circ})$, where $\mathfrak{J}^{\circ} = \mathfrak{J}_{12}\mathfrak{J}_{22}^{-1}\mathfrak{J}_{21}\mathfrak{J}_{11}^{-1}$. Thus, we note that the $\text{adr}(\widehat{\boldsymbol{\beta}}_1^{U}) = a$, which is constant and independent of $\boldsymbol{\delta} \in \mathbb{R}^b$.

Now, we compare $\widehat{\boldsymbol{\beta}}_1^{R}$ with $\widehat{\boldsymbol{\beta}}_1^{U}$ when the UPI is correct, $\text{adr}(\widehat{\boldsymbol{\beta}}_1^{U}) - \text{adr}(\widehat{\boldsymbol{\beta}}_1^{R}) = \text{trace}(\mathfrak{J}^{\circ}) > 0$. Hence, when the restriction is correctly specified $\widehat{\boldsymbol{\beta}}_1^{R}$ strictly dominates $\widehat{\boldsymbol{\beta}}_1^{U}$. However, when $\boldsymbol{\delta}$ moves away from the null vector, the adr of $\widehat{\boldsymbol{\beta}}_1^{R}$ monotonically increases and goes to ∞ as $\boldsymbol{\delta}'\mathfrak{J}^{\bullet}\boldsymbol{\delta} \to \infty$. Thus, $\widehat{\boldsymbol{\beta}}_1^{R}$ may not behave well when the assumed pivot is different from the specified value of $\boldsymbol{\beta}_2$. It is seen that $\text{adr}(\widehat{\boldsymbol{\beta}}_1^{R}) \leq \text{adr}(\widehat{\boldsymbol{\beta}}_1^{U})$ if $\boldsymbol{\delta}'\mathfrak{J}^{\bullet}\boldsymbol{\delta} \leq \text{trace}(\mathfrak{J}^{\circ})$ and strict inequality holds if $\boldsymbol{\delta}$ is a null vector. Thus under the null hypothesis, $\widehat{\boldsymbol{\beta}}_1^{R}$ is superior to $\widehat{\boldsymbol{\beta}}_1^{U}$. This clearly indicates that the performance of $\widehat{\boldsymbol{\beta}}_1^{R}$ will strongly depend on the reliability of the UPI. The performance of $\widehat{\boldsymbol{\beta}}_1^{U}$ is always steady throughout $\Delta \in [0, \infty)$.

Further, by Courant's theorem

$$\text{ch}_{\min}(\mathfrak{J}^{\circ}) \leq \frac{\boldsymbol{\delta}'\mathfrak{J}^{\bullet}\boldsymbol{\delta}}{\boldsymbol{\delta}'\mathfrak{J}_{22.1}\boldsymbol{\delta}} \leq \text{ch}_{\max}(\mathfrak{J}^{\circ}),$$

where $\text{ch}_{\min}(*)$ and $\text{ch}_{\max}(*)$ mean the smallest and largest eigenvalues of $(\bullet)$ respectively. Thus, $\text{adr}(\widehat{\boldsymbol{\beta}}_1^{R})$ intersects with the $\text{adr}(\widehat{\boldsymbol{\beta}}_1^{U})$ at

$$\Delta_1 = \frac{\text{trace}(\mathfrak{J}^{\circ})}{\text{ch}_{\max}(\mathfrak{J}^{\circ})} \quad \text{and} \quad \Delta_2 = \frac{\text{trace}(\mathfrak{J}^{\circ})}{\text{ch}_{\min}(\mathfrak{J}^{\circ})},$$

respectively. Note that $\Delta_1 \in (0, 1]$ and $\Delta_2 \in (0, 1]$, however, $\Delta_1 < \Delta_2$. The equality will hold only if $\mathfrak{J}^{\circ}$ is a diagonal matrix with same diagonal elements. Thus, for

$$\Delta \in \left[0, \frac{\text{trace}(\mathfrak{J}^{\circ})}{\text{ch}_{\max}(\mathfrak{J}^{\circ})}\right],$$

$\widehat{\boldsymbol{\beta}}_1^{R}$ has smaller adr than that of $\widehat{\boldsymbol{\beta}}_1^{U}$. Alternatively, for

$$\Delta \in \left[\frac{\text{trace}(\mathfrak{J}^{\circ})}{\text{ch}_{\min}(\mathfrak{J}^{\circ})}, \infty\right],$$

$\widehat{\beta}_1^{\mathrm{U}}$ has smaller adr. Clearly, when Δ moves away from the null vector beyond the value trace$(\mathfrak{J}^{\circ})/\,\mathrm{ch}_{\min}(\mathfrak{J}^{\circ})$, the adr of $\widehat{\beta}_1^{\mathrm{R}}$ increases and becomes unbounded. This clearly indicates that the performance of $\widehat{\beta}_1^{\mathrm{R}}$ will strongly depend on the reliability of the UPI.

The consequences of incorporating UPI depend on the ***quality or reliability*** of information introduced into the estimation process. Practitioners frequently find themselves in a situation like the one we have considered. Applied workers wish to estimate β_1 and do not know whether or not $\beta_2 = \beta_2^{\mathrm{o}}$. There is a dire need to consider some alternative estimators to $\widehat{\beta}_1^{\mathrm{U}}$ and $\widehat{\beta}_1^{\mathrm{R}}$. In the following sections we provide some alternative estimation strategies to restricted and unrestricted estimators.

2 Preliminary Test Estimation

Let $\Lambda_{\beta_2=\beta_2^{\circ}}$ be a suitable test statistic for the preliminary (null) hypothesis that $H_{\mathrm{P}}: \beta_2 = \beta_2^{\mathrm{o}}$, with c_α as the critical value for the test of size α.

It seems natural to define an estimator of the following form.

$$\widehat{\beta}_1^{\mathrm{P}} = \begin{cases} \widehat{\beta}_1^{\mathrm{U}} & \text{if } \Lambda_{\beta_2=\beta_2^{\circ}} \geq c_\alpha; \\ \widehat{\beta}_1^{\mathrm{R}} & \text{if } \Lambda_{\beta_2\neq\beta_2^{\circ}} < c_\alpha. \end{cases}$$

Thus, $\widehat{\beta}_1^{\mathrm{P}}$ will be the unrestricted estimator $\widehat{\beta}_1^{\mathrm{U}}$ when the Λ test does reject the null the hypothesis and will be the restricted estimator when the Λ test does not reject hypothesis that the restriction is satisfied. This is called the ***preliminary test estimator*** (PTE) of parameter vector β_1. A preliminary test is first performed on the validity of the UPI in the form of parametric restrictions. Then on the basis of the outcome of the test, the restricted or unrestricted estimator is selected. Thus, the selection procedure depends on the result of the preliminary test H_{P}. Bancroft (1944) first proposed the idea. This may be partly motivated by the remarks made by Berkson (1942). Furthermore, the PTE may also be defined in the following computationally attractive form:

$$\begin{aligned} \widehat{\beta}_1^{\mathrm{P}} &= \widehat{\beta}_1^{\mathrm{R}} I(\Lambda \leq c_\alpha) + \widehat{\beta}_1^{\mathrm{U}} I(\Lambda > c_\alpha) \\ &= \widehat{\beta}_1^{\mathrm{U}} - (\widehat{\beta}_1^{\mathrm{U}} - \widehat{\beta}_1^{\mathrm{R}}) I(\Lambda \leq c_\alpha), \end{aligned} \tag{24}$$

where $I(A)$ is the indicator function of the set A.

It is important to note that the estimators obtained by the preliminary test rule perform better than the estimators based on sample data only in a relatively narrow part of the parameter space induced by the UPI. Suppose the restriction holds, then the estimator we would like to select is the restricted estimator, $\widehat{\beta}_1^{R}$. Keeping in mind that $\alpha\%$ of the time, the Λ test will incorrectly reject the null hypothesis and $\widehat{\beta}_1^{P}$ will equal to the restricted estimator $\widehat{\beta}_1^{U}$ instead. Thus, $\widehat{\beta}_1^{P}$ must be less efficient than $\widehat{\beta}_1^{R}$ when the UPI does in fact hold. On the other hand, when the UPI does not hold, one may or may not want to use the unrestricted estimator $\widehat{\beta}_1^{U}$. Depending on how much power the test statistic has, $\widehat{\beta}_1^{P}$ will make a choice between $\widehat{\beta}_1^{U}$ and $\widehat{\beta}_1^{R}$. Since $\widehat{\beta}_1^{R}$ is not unbiased, $\widehat{\beta}_1^{P}$ will not be an unbiased estimator of β_1. Further, it may be more or less efficient than unrestricted estimator.

In the following, we will investigate the large sample theory of the $\widehat{\beta}_1^{P}$ for the problem in hand. First, we develop a large sample test statistic to test the H_P. One can use the likelihood ratio statistic

$$\Lambda = -2\log\left[\frac{L(\widehat{\beta}_1^{R}, \beta_2^{o})}{L(\widehat{\beta}_1^{U}, \widehat{\beta}_2^{U})}\right], \tag{25}$$

for this purpose. Obviously, large values of Λ give evidence against H_P and significance levels can be calculated via the asymptotic approximation $\Lambda \sim \chi_b^2$ (a chi-square distribution with b degrees of freedom). Further,

$$\Lambda = n(\widehat{\beta}_2^{U} - \beta_2^{o})'\mathfrak{I}_{22.1}(\widehat{\beta}_2^{U} - \beta_2^{o}) + o_p(1), \tag{26}$$

with

$$\mathfrak{I}_{22.1} = \mathfrak{I}_{22} - \mathfrak{I}_{21}\mathfrak{I}_{11}^{-1}\mathfrak{I}_{12}. \tag{27}$$

Theorem 2. *Under $\{K_n\}$ and the usual regularity conditions and as n increases Λ follows a noncentral chi-squared distribution with b degrees of freedom and noncentrality parameter $\Delta = \delta'\mathfrak{I}_{22.1}\delta$.*

To obtain the asymptotic distribution of $\widehat{\beta}_1^{P}$, let us consider the matrix $\mathbf{C}$

$$\mathbf{C} = \mathfrak{I}^{o^{-1}} = \begin{pmatrix}\mathbf{C}_1\\ \mathbf{C}_2\end{pmatrix} = \begin{pmatrix}\mathbf{C}_{11} & \mathbf{C}_{12}\\ \mathbf{C}_{21} & \mathbf{C}_{22}\end{pmatrix}. \tag{28}$$

Theorem 3. *Under $\{K_n\}$ and the assumed regularity conditions, as $n \to \infty$,*

$$\lim_{n\to\infty} P_{K_n}\left[n^{1/2}(\widehat{\beta}_1^{P} - \beta_1) \le \mathbf{x}\right]$$
$$= G_b(\mathbf{x} + \mathfrak{I}_{12}\delta; \mathbf{0}, \mathfrak{I}_{11}^{-1})\Phi_b(\chi_{b,\alpha}^2; \Delta)$$
$$+ \int_{E(\delta)} G_b(\mathbf{x} - \mathbf{C}_{12}\mathbf{C}_{22}^{-1}\mathbf{z}; \mathbf{0}, \mathbf{C}_{11.2})\, dG_b(\mathbf{z}; \mathbf{0}, \mathbf{C}_{22}),$$

where $G_\nu(\mathbf{x}, \boldsymbol{\mu}, \boldsymbol{\Sigma})$ is the ν-variate normal distribution function with mean vector $\boldsymbol{\mu}$ and covariance matrix $\boldsymbol{\Sigma}$. Here $\Phi_\nu(y; \Delta)$ stands for the noncentral *chi-square distribution function with ν degrees of freedom and noncentrality parameter Δ. And*

$$E(\boldsymbol{\delta}) = \left\{\mathbf{z} \in \mathbb{R}^b : (\mathbf{z}+\boldsymbol{\delta})'\mathfrak{J}_{22.1}(\mathbf{z}+\boldsymbol{\delta}) \geq \chi^2_{b,\alpha}\right\}.$$

The probability density function is given by

$$f(\mathbf{x},\boldsymbol{\delta}) = g_b(\mathbf{x} + \mathfrak{J}_{11}^{-1}\mathfrak{J}_{12}\boldsymbol{\delta}; \mathbf{0}, \mathfrak{J}_{11}^{-1})\Phi_b(\chi^2_{b,\alpha};\Delta) + \int_{E(\boldsymbol{\delta})} g_b(\mathbf{x} - \mathbf{C}_{12}\mathbf{C}_{22}^{-1}\mathbf{z}; \mathbf{0}, \mathbf{C}_{11.2})\, dG_b(\mathbf{z}; \mathbf{0}, \mathbf{C}_{22}),$$

where g_ν stands for the multinormal probability density function.

The above theorem is a direct multivariate generalization of Saleh and Sen (1978, Theorem 3.2), and in view of the similarity the details of the proof are omitted.

By plugging $g(\Lambda) = I(\Lambda > \chi^2_{b,\alpha})$ in Lemma 3 in the appendix, the adb of $\widehat{\boldsymbol{\beta}}_1^{\mathrm{P}}$ is given by

$$\mathrm{adb}(\widehat{\boldsymbol{\beta}}_1^{\mathrm{P}}) = \Phi_{b+2}(\chi^2_{b,\alpha};\Delta)\mathfrak{J}_{11}^{-1}\mathfrak{J}_{12}\boldsymbol{\delta}, \quad \Delta = \boldsymbol{\delta}'\mathfrak{J}_{22.1}\boldsymbol{\delta}. \tag{29}$$

Hence

$$\mathrm{qadb}(\widehat{\boldsymbol{\beta}}_1^{\mathrm{P}}) = \left[\Phi_{b+2}(\chi^2_{b,\alpha};\Delta)\right]^2\Delta^*.$$

Interestingly but not surprisingly, $\widehat{\boldsymbol{\beta}}_1^{\mathrm{P}}$ offers a good control over the qadb. In fact $\mathrm{qadb}(\widehat{\boldsymbol{\beta}}_1^{\mathrm{P}})$ vanishes as $\Delta \to \infty$, since $\Phi_{b+2}(\chi^2_{b,\alpha};\Delta) \to 0$ as $\Delta \to \infty$.

Theorem 4. *Under $\{K_n\}$ in* (11) *and the assumed regularity conditions, as $n \to \infty$, the asymptotic covariance of $\widehat{\boldsymbol{\beta}}_1^{\mathrm{P}}$ is*

$$\mathfrak{J}_{11.2} - \Phi_{b+2}(\chi^2_{b,\alpha};\Delta)(\mathfrak{J}_{11.2} - \mathfrak{J}_{11}) + \mathfrak{J}_{11}^{-1}\mathfrak{J}_{12}\boldsymbol{\delta}\boldsymbol{\delta}'\mathfrak{J}_{21}\mathfrak{J}_{11}^{-1}\left[2\Phi_{b+2}(\chi^2_{b,\alpha};\Delta) - \Phi_{b+4}(\chi^2_{b,\alpha};\Delta)\right]. \tag{30}$$

Consequently,

$$\mathrm{adr}(\widehat{\boldsymbol{\beta}}_1^{\mathrm{P}}) = \mathrm{trace}(\boldsymbol{\Gamma}\mathfrak{J}_{11.2}^{-1}) + \boldsymbol{\delta}'\mathfrak{J}^{\bullet}\boldsymbol{\delta}\left[2\Phi_{b+2}(\chi^2_{b,\alpha};\Delta) - \Phi_{b+4}(\chi^2_{b,\alpha};\Delta)\right] - \Phi_{b+2}(\chi^2_{b,\alpha};\Delta)\left[\mathrm{trace}(\boldsymbol{\Gamma}\mathfrak{J}_{11.2}^{-1}) - \mathrm{trace}(\boldsymbol{\Gamma}\mathfrak{J}_{11}^{-1})\right]. \tag{31}$$

The expression (30) is readily obtained by setting $g(\Lambda) = I(\Lambda > \chi^2_{b,\alpha})$ in Lemma 4 of the appendix.

Again, here we exclude the case $\mathfrak{J}_{12} = \mathbf{0}$, since in this situation, $\mathfrak{J}^{\bullet} = \mathbf{0}$ and $\mathfrak{J}_{11.2} = \mathfrak{J}_{11}$. Then adr of $\widehat{\boldsymbol{\beta}}_1^{\mathrm{P}}$ is reduced to the adr of $\widehat{\boldsymbol{\beta}}_1^{\mathrm{U}}$. Therefore, in the remaining discussions we establish the asymptotic properties of $\widehat{\boldsymbol{\beta}}_1^{\mathrm{P}}$ when $\mathfrak{J}_{12}$ is not null.

Noting that $\Phi_{b+4}(\chi^2_{b,\alpha};\Delta) \le \Phi_{b+2}(\chi^2_{b,\alpha};\Delta) \le \Phi_{b+2}(\chi^2_{b,\alpha};0) = 1-\alpha$, for $\alpha \in (0,1)$ and $\Delta > 0$. The left hand-side of the above relation converges to 0 as $\Delta \to \infty$. Also, as $\|\boldsymbol{\delta}\| \to \infty \implies \Delta \to \infty$, then $\Phi_{b+4}(\chi^2_{b,\alpha};\Delta)$, $\boldsymbol{\delta}'\mathfrak{J}^{\bullet}\boldsymbol{\delta}\Phi_{b+2}(\chi^2_{b,\alpha};\Delta)$ and $\boldsymbol{\delta}'\mathfrak{J}^{\bullet}\boldsymbol{\delta}\Phi_{b+4}(\chi^2_{b,\alpha};\Delta)$ approach 0, and the adr of $\widehat{\boldsymbol{\beta}}_1^{\mathrm{P}}$ approaches the adr of $\widehat{\boldsymbol{\beta}}_1^{\mathrm{U}}$. The adr of $\widehat{\boldsymbol{\beta}}_1^{\mathrm{P}}$ is smaller than the adr of $\widehat{\boldsymbol{\beta}}_1^{\mathrm{U}}$ near the null hypothesis which keeps on increasing, crosses the adr of $\widehat{\boldsymbol{\beta}}_1^{\mathrm{U}}$, reaches to maximum then decreases monotonically to the adr of $\widehat{\boldsymbol{\beta}}_1^{\mathrm{U}}$. Hence a preliminary test approach controls the magnitude of the risk. The dominating condition is given by

$$\mathrm{adr}(\widehat{\boldsymbol{\beta}}_1^{\mathrm{P}}) \le \mathrm{adr}(\widehat{\boldsymbol{\beta}}_1^{\mathrm{U}}) \quad \text{if} \quad \boldsymbol{\delta}'\mathfrak{J}^{\bullet}\boldsymbol{\delta} \le \frac{\mathrm{trace}(\mathfrak{J}^{\circ})h(\Delta)}{2h(\Delta)-g(\Delta)},$$

where

$$h(\Delta) = \Phi_{b+2}(\chi^2_{b,\alpha};\Delta), \quad g(\Delta) = \Phi_{b+4}(\chi^2_{b,\alpha};\Delta).$$

There are points in the parameter space for which $\widehat{\boldsymbol{\beta}}_1^{\mathrm{P}}$ is inferior to $\widehat{\boldsymbol{\beta}}_1^{\mathrm{U}}$ and a sufficient condition is $\boldsymbol{\delta}'\mathfrak{J}^{\bullet}\boldsymbol{\delta} \in \{\mathrm{trace}(\mathfrak{J}^{\circ})h(\Delta)/[2h(\Delta)-g(\Delta)], \infty\}$. By the application of Courant's theorem it can be shown that $\mathrm{adr}(\widehat{\boldsymbol{\beta}}_1^{\mathrm{P}})$ intersects with the $\mathrm{adr}(\widehat{\boldsymbol{\beta}}_1^{\mathrm{U}})$ at

$$\Delta_1^{\circ} = \frac{h(\Delta)\,\mathrm{trace}(\mathfrak{J}^{\circ})}{\mathrm{ch}_{\max}(\mathfrak{J}^{\circ})[2h(\Delta)-g(\Delta)]} \quad \text{and} \quad \Delta_2^{\circ} = \frac{h(\Delta)\,\mathrm{trace}(\mathfrak{J}^{\circ})}{\mathrm{ch}_{\min}(\mathfrak{J}^{\circ})[2h(\Delta)-g(\Delta)]},$$

respectively.

At $\boldsymbol{\delta} = 0$, the $\mathrm{adr}(\widehat{\boldsymbol{\beta}}_1^{\mathrm{P}})$ assumes the value

$$\mathrm{trace}(\boldsymbol{\Gamma}\mathfrak{J}_{11.2}^{-1})\left[1-\Phi_{b+2}(\chi^2_{b,\alpha};0)\right] + \mathrm{trace}(\boldsymbol{\Gamma}\mathfrak{J}_{11}^{-1})\Phi_{b+2}(\chi^2_{b,\alpha};0).$$

Thus, the reduction in the risk is

$$\Phi_{b+2}(\chi^2_{b,\alpha};0)\left[\mathrm{trace}(\boldsymbol{\Gamma}\mathfrak{J}_{11}^{-1}) - \mathrm{trace}(\boldsymbol{\Gamma}\mathfrak{J}_{11.2}^{-1})\right].$$

We find that performance of the preliminary test estimators, which combine sample information with UPI, heavily depend on the correctness of the UPI. The gain in the risk is substantial over the usual procedure when the UPI is nearly correct. However, $\widehat{\boldsymbol{\beta}}_1^{\mathrm{P}}$ combines the information in a superior way than that of $\widehat{\boldsymbol{\beta}}_1^{\mathrm{R}}$ in the sense that its risk is bounded.

We note at $\boldsymbol{\delta} = \mathbf{0}$ the risk difference between $\widehat{\boldsymbol{\beta}}_1^{\mathrm{P}}$ and $\widehat{\boldsymbol{\beta}}_1^{\mathrm{R}}$ is

$$\mathrm{adr}(\widehat{\boldsymbol{\beta}}_1^{\mathrm{P}}) - \mathrm{adr}(\widehat{\boldsymbol{\beta}}_1^{\mathrm{R}}) = \mathrm{trace}(\mathfrak{J}^{\circ})\left[1-\Phi_{b+2}(\chi^2_{b,\alpha};0)\right] > 0.$$

Thus, under the null hypothesis $\widehat{\boldsymbol{\beta}}_1^{\mathrm{R}}$ performs better than $\widehat{\boldsymbol{\beta}}_1^{\mathrm{P}}$. Alternatively when Δ drifts away from the origin the opposite conclusion holds.

Finally, keeping in mind the discussions above we draw the following conclusion: None of the three estimators is inadmissible with respect to each other. However, at $\Delta = 0$, the estimators may be ordered according to the magnitude of their adr as follows: $\widehat{\beta}_1^{\mathrm{R}} \succ \widehat{\beta}_1^{\mathrm{P}} \succ \widehat{\beta}_1^{\mathrm{U}}$.

Obviously, such behavior of the preliminary test estimator is due to the dichotomous function $I(\Lambda \leq c_\alpha)$. This suggests to replace $I(\Lambda \leq c_\alpha)$ by a smooth function. This leads to constructive estimation procedure based on the most celebrated Stein-type rule. This estimator is generally called a shrinkage estimator.

3 Shrinkage Estimation

The *shrinkage estimator* (SE) (Stein, 1956; James and Stein, 1961) is defined by

$$\widehat{\beta}_1^{\mathrm{S}} = \widehat{\beta}_1^{\mathrm{R}} + \left[1 - \frac{(b-2)}{\Lambda}\right](\widehat{\beta}_1^{\mathrm{U}} - \widehat{\beta}_1^{\mathrm{R}}), \quad b \geq 3. \tag{32}$$

For some insight to shrinkage estimation we refer to Brandwein and Strawderman (1990); Stigler (1990); Ahmed (1992); Ahmed and Saleh (1993) among others.

The limiting distribution of $\widehat{\beta}_1^{\mathrm{S}}$ is given in the following theorem.

Theorem 5. *Under $\{K_n\}$ and the assumed regularity conditions, as $n \to \infty$,*

$$n^{1/2}(\widehat{\beta}_1^{\mathrm{S}} - \beta_1) \xrightarrow{\mathcal{L}} \frac{(b-2)\mathfrak{J}_{11}^{-1}\mathfrak{J}_{12}(\mathbf{C}_2\mathbf{V}+\delta)}{(\mathbf{C}_2\mathbf{V}+\delta)'\mathfrak{J}_{22.1}(\mathbf{C}_2\mathbf{V}+\delta)} + \mathbf{C}_1\mathbf{V},$$

where the random vector $\mathbf{V}$ has a p-variate normal distribution with null mean vector and dispersion matrix $\mathfrak{J}^{\circ}$.

By setting $g(\Lambda) = [1 - (b-2)\Lambda^{-1}]$ in Lemma 3 in the Appendix, we get

$$\operatorname{adb}(\widehat{\beta}_1^{\mathrm{S}}) = (b-2)\mathfrak{J}_{11}^{-1}\mathfrak{J}_{12}\delta\mathcal{E}\left[\chi_{b+2}^{-2}(\Delta)\right]. \tag{33}$$

Here,

$$\mathcal{E}\left[\chi_\nu^{-2j}(\Delta)\right] = \int_0^\infty x^{-2j}\, d\Phi_\nu(x;\Delta).$$

Alternatively,

$$\operatorname{qadb}(\widehat{\beta}_1^{\mathrm{S}}) = (b-2)^2\Delta^*\left\{\mathcal{E}\left[\chi_{b+2}^{-2}(\Delta)\right]\right\}^2. \tag{34}$$

The quadratic bias of $\widehat{\beta}_1^{\mathrm{S}}$ starts from 0 at $\Delta^* = 0$, increases to a point then decreases towards 0, since $\mathcal{E}\left[\chi_{b+2}^{-2}(\Delta)\right]$ is decreasing log-convex function of Δ.

Theorem 6. *Under $\{K_n\}$ in (11) and the assumed regularity conditions, as $n \to \infty$*

$$\operatorname{adm}(\widehat{\beta}_1^{S}) = \mathfrak{J}_{11.2} + (b^2-4)\mathcal{E}\left[\chi_{b+4}^{-4}(\Delta)\right]\mathfrak{J}_{11}^{-1}\mathfrak{J}_{12}\delta\delta'\mathfrak{J}_{21}\mathfrak{J}_{11}^{-1} - (b-2)\mathfrak{J}^{*}\left\{\mathcal{E}\left[\chi_{b+2}^{-2}(\Delta)\right] + \Delta\mathcal{E}\left[\chi_{b+4}^{-4}(\Delta)\right]\right\}, \quad (35)$$

where $\mathfrak{J}^{*} = \mathfrak{J}_{11}^{-1}\mathfrak{J}_{12}\mathfrak{J}_{22.1}\mathfrak{J}_{21}\mathfrak{J}_{11}^{-1}$. *Hence,*

$$\begin{aligned}\operatorname{adr}(\widehat{\beta}_1^{S}) &= \operatorname{trace}(\mathbf{\Gamma}\mathfrak{J}_{11.2}^{-1}) + \delta'\mathfrak{J}^{\bullet}\delta(b^2-4)\mathcal{E}\left[\chi_{b+4}^{-4}(\Delta)\right] \\ &\quad - (q-2)\operatorname{trace}(\mathfrak{J}^{\bullet}\mathfrak{J}_{22.1}^{-1})\left\{2\mathcal{E}\left[\chi_{b+2}^{-2}(\Delta)\right] - (b-2)\mathcal{E}\left[\chi_{b+2}^{-4}(\Delta)\right]\right\}. \quad (36)\end{aligned}$$

We get the expression for adm *of* $\widehat{\beta}_1^{S}$ *by setting* $g(\Lambda) = \left[1-(b-2)\Lambda^{-1}\right]$ *in Lemma 4 in the Appendix.*

We now wish to compare $\widehat{\beta}_1^{S}$ with $\widehat{\beta}_1^{R}$ and $\widehat{\beta}_1^{P}$. Near $\delta = \mathbf{0}$ both estimators ($\widehat{\beta}_1^{R}$ and $\widehat{\beta}_1^{P}$) dominate $\widehat{\beta}_1^{S}$ at the expense of poor performance in the rest of the parameter space. However, it is seen that for large values of α the preliminary estimator may not be able to retain this property. Nevertheless, $\widehat{\beta}_1^{S}$ and $\widehat{\beta}_1^{P}$ both enjoy a common asymptotic property, i.e., as $\Delta \to \infty$, their risks converge to a common limit, i.e., the risk of $\widehat{\beta}_1^{U}$. In fact, the risk of $\widehat{\beta}_1^{S}$ is always under this asymptotic value.

Now, let us consider the performance of $\widehat{\beta}_1^{S}$ in case of the Mahalanobis distance. We compare $\widehat{\beta}_1^{S}$ with $\widehat{\beta}_1^{U}$. The adr expressions $\widehat{\beta}_1^{U}$ and $\widehat{\beta}_1^{S}$ reveal that $\widehat{\beta}_1^{S}$ will dominate $\widehat{\beta}_1^{U}$ if

$$\delta'\mathfrak{J}^{\bullet}\delta \le \left[\frac{\operatorname{trace}(\mathfrak{J}^{\circ})}{(b+2)}\right]\left\{\frac{2\mathcal{E}[\chi_{b+2}^{-2}(\Delta)] - (b-2)\mathcal{E}[\chi_{b+2}^{-4}(\Delta)]}{\mathcal{E}[\chi_{b+4}^{-4}(\Delta)]}\right\}. \quad (37)$$

Then by the use of Courant's theorem, the above condition can be expressed as:

$$\Delta \le \left[\frac{\operatorname{trace}(\mathfrak{J}^{\circ})}{\operatorname{ch}_{\max}(\mathfrak{J}^{\circ})}\right]\left\{\frac{2\mathcal{E}[\chi_{b+2}^{-2}(\Delta)] - (b-2)\mathcal{E}[\chi_{b+2}^{-4}(\Delta)]}{(b^2-4)\mathcal{E}[\chi_{b+4}^{-4}(\Delta)]}\right\}, \quad \forall\Delta > 0, \quad (38)$$

which in turn requires that $b > 2$ and $\operatorname{ch}_{\max}(\mathfrak{J}^{\circ})/\operatorname{trace}(\mathfrak{J}^{\circ}) \le 2/(b+2)$. Thus, $\operatorname{adr}(\widehat{\beta}_1^{S}) \le \operatorname{adr}(\widehat{\beta}_1^{U})$ if the following three conditions are met.

1. $p_{\min} = \min(a, b)$,
2. $\operatorname{ch}_{\max}(\mathfrak{J}^{\circ}) < \frac{1}{2}\operatorname{trace}(\mathfrak{J}^{\circ})$,
3. $0 \le (b-2) \le \min\left[2\operatorname{trace}(\mathfrak{J}^{\circ})/\operatorname{ch}_{\max}(\mathfrak{J}^{\circ}) - 4, 2(b-2)\right]$.

4 Positive-Part Shrinkage Estimation

Noting that the $\widehat{\beta}_1^{\mathrm{S}}$ defined in relation (32) is not a convex combination of $\widehat{\beta}_1^{\mathrm{R}}$ and $\widehat{\beta}_1^{\mathrm{U}}$. Hence, the proposed estimator $\widehat{\beta}_1^{\mathrm{S}}$ may change the sign of $\widehat{\beta}_1^{\mathrm{U}}$. To avoid this strange behavior of $\widehat{\beta}_1^{\mathrm{S}}$, we truncate $\widehat{\beta}_1^{\mathrm{S}}$ which leads to a convex combination of $\widehat{\beta}_1^{\mathrm{U}}$ and $\widehat{\beta}_1^{\mathrm{R}}$ and is called ***positive-part shrinkage*** (PPS) estimator. The positive-part estimator can be defined as

$$\widehat{\beta}_1^{\mathrm{S}+} = \widehat{\beta}_1^{\mathrm{R}} + \left[1 - \frac{(b-2)}{\Lambda}\right]^+ (\widehat{\beta}_1^{\mathrm{U}} - \widehat{\beta}_1^{\mathrm{R}}), \tag{39}$$

where $z^+ = \max(0, z)$. Further,

$$\begin{aligned}
\widehat{\beta}_1^{\mathrm{S}+} &= \widehat{\beta}_1^{\mathrm{R}} + \left[1 - \frac{(b-2)}{\Lambda}\right]^+ (\widehat{\beta}_1^{\mathrm{U}} - \widehat{\beta}_1^{\mathrm{R}}) \\
&= \widehat{\beta}_1^{\mathrm{R}} + \left[1 - \frac{(b-2)}{\Lambda}\right] I(\Lambda > b-2)(\widehat{\beta}_1^{\mathrm{U}} - \widehat{\beta}_1^{\mathrm{R}}) \\
&= \widehat{\beta}_1^{\mathrm{R}} + \left[1 - \frac{(b-2)}{\Lambda}\right] (\widehat{\beta}_1^{\mathrm{U}} - \widehat{\beta}_1^{\mathrm{R}}) \\
&\qquad - \left[1 - \frac{(b-2)}{\Lambda}\right] I(\Lambda \leq b-2)(\widehat{\beta}_1^{\mathrm{U}} - \widehat{\beta}_1^{\mathrm{R}}).
\end{aligned}$$

Thus, $\widehat{\beta}_1^{\mathrm{S}+}$ may be written as

$$\widehat{\beta}_1^{\mathrm{S}+} = \widehat{\beta}_1^{\mathrm{S}} - \left[1 - \frac{(b-2)}{\Lambda}\right] I(\Lambda \leq b-2)(\widehat{\beta}_1^{\mathrm{U}} - \widehat{\beta}_1^{\mathrm{R}}).$$

We emphasize here that $\widehat{\beta}_1^{\mathrm{S}+}$ is particularly important to control the over-shrinking inherent in $\widehat{\beta}_1^{\mathrm{S}}$. Since the Stein-type estimator has the problem of not being convex combination of unrestricted and restricted estimator and it also changes the sign of the unrestricted estimator, we recommend that this estimator should be used as a tool for developing the PPS estimator and should not be used as an estimator in its own right.

The adb of $\widehat{\beta}_1^{\mathrm{S}+}$ is given by

$$\begin{aligned}
\mathrm{adb}(\widehat{\beta}_1^{\mathrm{S}+}) = \mathfrak{J}_{11}^{-1}\mathfrak{J}_{12}\delta\big(&\Phi_{b+2}(b-2;\Delta) \\
&+ \mathcal{E}\{\chi_{b+2}^{-2}(\Delta) I[\chi_{b+2}^2(\Delta) > (b-2)]\}\big).
\end{aligned} \tag{40}$$

The above relation is established by replacing $g(\Lambda) = [1 - (b-2)\Lambda^{-1}]^+$ in Lemma 3 in the appendix. Finally,

$$\mathrm{qadb}(\widehat{\beta}_1^{\mathrm{S}+}) = \Delta^*\big(\Phi_{b+2}(b-2;\Delta) + \mathcal{E}\{\chi_{b+2}^{-2}(\Delta) I[\chi_{b+2}^2(\Delta) > (b-2)]\}\big)^2.$$

The graph of qadb of $\widehat{\beta}_1^{S+}$ more or less follows the same pattern as $\widehat{\beta}_1^{S}$, however the quadratic bias curve of $\widehat{\beta}_1^{S+}$ remains below the curve of $\widehat{\beta}_1^{S}$ for all values of Δ.

Theorem 7. *Under $\{K_n\}$ in* (11) *and the assumed regularity conditions, as $n \to \infty$ the* adm *of $\widehat{\beta}_1^{S+}$ is given by*

$$\begin{aligned}
\operatorname{adm}(\widehat{\beta}_1^{S+}) &= \operatorname{adm}(\widehat{\beta}_1^{S}) \\
&\quad + (b-2)\mathfrak{I}^*\big(2\mathcal{E}\{\mathcal{E}[\chi_{b+2}^{-2}(\Delta)]I[\chi_{b+2}^2(\Delta) \le (b-2)]\} \\
&\qquad - (b-2)\mathcal{E}\{\mathcal{E}[\chi_{b+2}^{-4}(\Delta)]I[\chi_{b+2}^2(\Delta) \le (b-2)]\}\big) \\
&\quad - \mathfrak{I}^*\Phi_{b+2}(b-2;\Delta) \\
&\quad + \mathfrak{I}_{11}^{-1}\mathfrak{I}_{12}\delta\delta'\mathfrak{I}_{21}\mathfrak{I}_{11}^{-1}[2\Phi_{b+2}(b-2;\Delta) - \Phi_{b+4}(b-2;\Delta)] \\
&\quad - (b-2)\mathfrak{I}_{11}^{-1}\mathfrak{I}_{12}\delta\delta'\mathfrak{I}_{21}\mathfrak{I}_{11}^{-1} \\
&\qquad \times \big(2\mathcal{E}\{\chi_{b+2}^{-2}(\Delta)I[\chi_{b+2}^2(\Delta) \le (b-2)]\} \\
&\qquad - 2\mathcal{E}\{\chi_{b+4}^{-2}(\Delta)I[\chi_{b+4}^2(\Delta) \le (b-2)]\} \\
&\qquad + (b-2)\mathcal{E}\{\chi_{b+4}^{-4}(\Delta)I[\chi_{b+4}^2(\Delta) \le (b-2)]\}\big). \qquad (41)
\end{aligned}$$

Consequently,

$$\begin{aligned}
\operatorname{adr}(\widehat{\beta}_1^{S+}) &= \operatorname{adr}(\widehat{\beta}_1^{S}) \\
&\quad + (b-2)\operatorname{trace}(\mathfrak{I}^\bullet\mathfrak{I}_{22.1}^{-1})\big(2\mathcal{E}\{\chi_{b+2}^{-2}(\Delta)I[\chi_{b+2}^2(\Delta) \le (b-2)]\} \\
&\qquad - (b-2)\mathcal{E}\{\chi_{b+2}^{-4}(\Delta)I[\chi_{b+2}^2(\Delta) \le (b-2)]\}\big) \\
&\quad - \operatorname{trace}(\mathfrak{I}^\bullet\mathfrak{I}_{22.1}^{-1})\Phi_{b+2}(b-2;\Delta) \\
&\quad + \delta'\mathfrak{I}^\bullet\delta[2\Phi_{b+2}(b-2;\Delta) - \Phi_{b+4}(b-2;\Delta)] \\
&\quad - (b-2)\delta'\mathfrak{I}^\bullet\delta\big(2\mathcal{E}\{\chi_{b+2}^{-2}(\Delta)I[\chi_{b+2}^2(\Delta) \le (b-2)]\} \\
&\qquad - 2\mathcal{E}\{\chi_{b+4}^{-2}(\Delta)I[\chi_{b+4}^2(\Delta) \le (b-2)]\} \\
&\qquad + (b-2)\mathcal{E}\{\chi_{b+4}^{-4}(\Delta)I[\chi_{b+4}^2(\Delta) \le (b-2)]\}\big). \qquad (42)
\end{aligned}$$

The proof of expression (41) follows by setting $g(\Lambda) = [1-(b-2)\Lambda^{-1}]^+$ in Lemma 4 of the appendix.

Finally, we compare the adr performance of $\widehat{\beta}_1^{S+}$ and $\widehat{\beta}_1^{S}$. We may conclude from relations (36) and (42) that $\operatorname{adr}(\widehat{\beta}_1^{S+}) \le \operatorname{adr}(\widehat{\beta}_1^{S})$, for all δ, with strict inequality for some δ.

Therefore, $\widehat{\beta}_1^{S+}$ asymptotically dominates $\widehat{\beta}_1^{S}$ under local alternatives. Hence, $\widehat{\beta}_1^{S+}$ is also superior to $\widehat{\beta}_1^{U}$.

Finally, it is important to remark here that generally Stein-type estimators are superior to ml estimators for $b \ge 3$ while ml estimator is admissible for $b = 1$ and $b = 2$. In these cases the use of shrinkage estimation may be limited due to this dimensional restriction. As such, in bivariate situation we will be unable to use UPI in the estimation procedure. In this situation, we recommend the use of estimators based on preliminary test rule.

5 Recommendations and Concluding Remarks

Several estimation strategies for using the UPI with sample information are presented for estimating the regression parameter vector in a censored model. It is concluded that positive-part shrinkage estimator dominates the usual shrinkage estimator and they both perform well relative to usual classical estimator of the parameter vector in the entire parameter space. In contrast, the performance of the restricted and preliminary test estimators heavily depends on the quality of the UPI. The risk of the restricted estimator is unbounded when the parameter moves away from the assumed subspace of the restriction. From a practical point of view it is suggested to use Mahalanobis distance as a loss function which simplifies the risk expressions.

For $b \geq 3$, we recommend $\widehat{\beta}_1^{S+}$ over $\widehat{\beta}_1^{S}$ and $\widehat{\beta}_1^{U}$. We re-emphasize here that $\widehat{\beta}_1^{S}$ should not be used since it is not a convex combination of $\widehat{\beta}_1^{R}$ and $\widehat{\beta}_1^{U}$ and also changes the sign of $\widehat{\beta}_1^{U}$. One may agree on adjusting the magnitudes of the $\widehat{\beta}_1^{U}$, but change of sign is somewhat a serious matter and it would make a practitioner rather uncomfortable. For this reason we have introduced $\widehat{\beta}_1^{S}$ as a tool for developing $\widehat{\beta}_1^{S+}$.

It is noted that the application of shrinkage estimators are subject to the dimensional condition. If $b \leq 2$, then we recommend to use estimators based on preliminary test procedure. However, $\widehat{\beta}_1^{P}$ does not have the asymptotic minimax character. One may improve on $\widehat{\beta}_1^{P}$ by replacing $\widehat{\beta}_1^{U}$ by $\widehat{\beta}_1^{S}$ or $\widehat{\beta}_1^{S+}$ in relation (24). It can be shown that the resulting estimators will dominate the $\widehat{\beta}_1^{P}$ but now we have the restriction $b \geq 3$. Further, it can be shown that this estimator does not dominate $\widehat{\beta}_1^{U}$.

In the end we would like to mention that Shao and Strawderman (1994) were able to obtain an estimator which dominates the positive-part shrinkage estimator. However, their estimator is also inadmissible, and there is further scope for research in this direction.

Acknowledgments: I would like to thank the referee for very constructive comments and suggestions on earlier version of this paper. This research was supported by grant from the Natural Sciences and Engineering Research Council of Canada.

6 Appendix

Lemma 3. *Define* $\widetilde{\beta}_1 = \widehat{\beta}_1^{\mathrm{R}} + g(\Lambda)(\widehat{\beta}_1^{\mathrm{U}} - \widehat{\beta}_1^{\mathrm{R}})$, *an estimator of* β_1. *Under* $\{K_n\}$ *in* (11) *and the assumed regularity conditions, as* $n \to \infty$ *the* adb *of* $\widetilde{\beta}_1$ *is given by*

$$\mathrm{adb}(\widetilde{\beta}_1) = \delta(1 - \mathcal{E}\{g[\chi^2_{b+2}(\Delta)]\}).$$

Proof.

$$\mathcal{E}[n^{1/2}(\widetilde{\beta}_1 - \beta_1)] = \mathcal{E}[n^{1/2}(\widehat{\beta}_1^{\mathrm{R}} - \beta_1)] + \mathcal{E}\{n^{1/2}[g(\Lambda)(\widehat{\beta}_1^{\mathrm{U}} - \widehat{\beta}_1^{\mathrm{R}})]\}$$
$$\to \delta - \delta\mathcal{E}\{g[\chi^2_{b+2}(\Delta)]\} = \delta(1 - \mathcal{E}\{g[\chi^2_{b+2}(\Delta)]\}). \quad \square \qquad (43)$$

Lemma 4. . *Under* $\{K_n\}$ *in* (11) *and the assumed regularity conditions, as* $n \to \infty$ *the* adm *of* $\widetilde{\beta}_1$ *is*

$$\mathrm{adm}(\widetilde{\beta}_1) = \mathfrak{J}_{11.2}^{-1} + \mathfrak{J}_{11.2}^{-1}\mathcal{E}g^2[\chi^2_{b+2}(\Delta)]$$
$$+ \mathfrak{J}_{11}^{-1}\mathfrak{J}_{12}\delta\delta'\mathfrak{J}_{21}\mathfrak{J}_{11}^{-1}\{[1 - 2\mathcal{E}g[\chi^2_{b+2}(\Delta)] + \mathcal{E}g^2[\chi^2_{b+4}(\Delta)]\}.$$

Proof. As in the proof of Lemma 3, let $\widetilde{\beta}_1 = \widehat{\beta}_1^{\mathrm{R}} + g(\Lambda)(\widehat{\beta}_1^{\mathrm{U}} - \widehat{\beta}_1^{\mathrm{R}})$. Then by definition the adm of $\widetilde{\beta}_1$ is given by the expectation

$$\mathcal{E}[n(\widetilde{\beta}_1 - \beta_1)(\widetilde{\beta}_1 - \beta_1)']$$
$$= \mathcal{E}[n(\widehat{\beta}_1^{\mathrm{R}} - \beta_1)(\widehat{\beta}_1^{\mathrm{R}} - \beta_1)']$$
$$+ \mathcal{E}\{g^2(\Lambda)[n(\widehat{\beta}_1^{\mathrm{U}} - \widehat{\beta}_1^{\mathrm{R}})(\widehat{\beta}_1^{\mathrm{U}} - \widehat{\beta}_1^{\mathrm{R}})']\}$$
$$+ 2\mathcal{E}\{g(\Lambda)[n(\widehat{\beta}_1^{\mathrm{R}} - \beta_1)(\widehat{\beta}_1^{\mathrm{U}} - \widehat{\beta}_1^{\mathrm{R}})']\}$$
$$\to \mathfrak{J}_{11.2}^{-1} + \mathfrak{J}_{11}^{-1}\mathfrak{J}_{12}\delta\delta'\mathfrak{J}_{21}\mathfrak{J}_{11}^{-1} + \mathfrak{J}_{11.2}^{-1}\mathcal{E}g^2[\chi^2_{b+2}(\Delta)]$$
$$+ \mathfrak{J}_{11}^{-1}\mathfrak{J}_{12}\delta\delta'\mathfrak{J}_{21}\mathfrak{J}_{11}^{-1}\mathcal{E}g^2[\chi^2_{b+4}(\Delta)]$$
$$- 2\mathfrak{J}_{11}^{-1}\mathfrak{J}_{12}\delta\delta'\mathfrak{J}_{21}\mathfrak{J}_{11}^{-1}\mathcal{E}g[\chi^2_{b+2}(\Delta)]$$
$$= \mathfrak{J}_{11.2}^{-1} + \mathfrak{J}_{11.2}^{-1}\mathcal{E}g^2[\chi^2_{b+2}(\Delta)]$$
$$+ \mathfrak{J}_{11}^{-1}\mathfrak{J}_{12}\delta\delta'\mathfrak{J}_{21}\mathfrak{J}_{11}^{-1}\{[1 - 2\mathcal{E}g[\chi^2_{b+2}(\Delta)] + \mathcal{E}g^2[\chi^2_{b+4}(\Delta)]\}. \quad \square$$

Consequently, the expression for the adr is obtained by computing the quantity

$$\mathrm{adr}(\widetilde{\beta}_1) = \mathrm{trace}(\mathbf{\Gamma}\,\mathrm{adm}(\widetilde{\beta}_1)).$$

7 References

Ahmed, S.E. (1992). Large-sample pooling procedure for correlation. *J. Roy. Statist. Soc. Ser. D 41*, 425–438.

Ahmed, S.E. and A.K.Md.E. Saleh (1993). Improved estimation for the component mean-vector. *J. Japan Statist. Soc. 23*, 145–159.

Bancroft, T.A. (1944). On biases in estimation due to the use of preliminary tests of significance. *Ann. Math. Statist. 15*, 190–204.

Berkson, J. (1942). Test of significance considered as evidence. *J. Amer. Statist. Assoc. 37*, 325–335.

Brandwein, A.C. and W.E. Strawderman (1990). Stein estimation: The spherically symmetric case. *Statist. Sci. 5*, 356–369.

James, W. and C. Stein (1961). Estimation with quadratic loss. In *Proc. 4th Berkeley Symp. Math. Statist. and Prob., Vol. I*, pp. 361–379. Univ. California Press, Berkeley, CA.

Lawless, J.F. (1982). *Statistical Models and Methods for Lifetime Data.* New York: Wiley.

Saleh, A.K.Md.E. and P.K. Sen (1978). Non-parametric estimation of location parameter after a preliminary test regression. *Ann. Statist. 6*, 154–168.

Saleh, A.K.Md.E. and P.K. Sen (1985). On shrinkage M-estimator of location parameters. *Comm. Statist. A–Theory Methods 14*, 2313–2329.

Sen, P.K. (1986). On the asymptotic distributional risk shrinkage and preliminary test versions of maximum likelihood estimators. *Sankhyā A 48*, 354–371.

Shao, P.Y.-S. and W.E. Strawderman (1994). Improving on the James-Stein positive-part estimator. *Ann. Statist. 22*, 1517–1538.

Stein, C. (1956). Inadmissibility of the usual estimator for the mean of a multivariate normal distribution. In *Proc. 3rd Berkeley Symp. Math. Statist. and Prob., Vol. I*, pp. 197–206. Univ. California Press, Berkeley, CA.

Stigler, S.M. (1990). The 1988 Neyman Memorial Lecture: a Galtonian perspective on shrinkage estimators. *Statist. Sci. 5*, 147–155.

9

Bayesian and Likelihood Inference for the Generalized Fieller–Creasy Problem

M. Yin and M. Ghosh

ABSTRACT
The famous Fieller–Creasy problem involves inference about the ratio of two normal means. It is quite challenging from either a frequentist or a likelihood perspective. Bayesian analysis with noninformative priors usually provides ideal solutions for this problem. In this paper, we find a second order matching prior and a one at a time reference prior which work well for Fieller–Creasy problem in the more general setting of two location-scale models with smooth symmetric density functions. The properties of the posterior distributions are investigated for some particular cases including the normal, t, and the double exponential. The Bayesian procedure is implemented via Markov Chain Monte Carlo (MC^2). Our simulation study indicates that the second order matching priors, in general, perform better than the reference priors in terms of matching the target coverage probabilities in a frequentist sense.

1 Introduction

The original Fieller–Creasy problem involves inference about the ratio of two normal means. To be specific, consider two independent samples $(X_1, \ldots, X_n)$ and $(Y_1, \ldots, Y_n)$, where the X_i are iid $N(\mu, \sigma^2)$ and the Y_j are iid $N(\theta\mu, \sigma^2)$. Initially, we assume $\sigma^2 (> 0)$ to be known, although Fieller (1954) considered the problem also for a general unknown variance-covariance matrix. The objective is inference about θ.

Fieller (1954) proposed a confidence interval for θ based on the pivot $(\overline{Y} - \theta\overline{X})/(1+\theta^2)^{1/2}$ which is $N(0, \sigma^2/n)$. Barnard and Sprott (1983) have provided further justification of the same for the general location-scale model where inference concerns the ratio of two location parameters. Noting that $P[|\overline{Y} - \theta\overline{X}|/(1+\theta^2)^1/2 \le z_{\alpha/2}\sigma/n^{1/2}] = 1-\alpha$, where $z_{\alpha/2}$ is the upper $100\alpha/2\%$ point of the $N(0,1)$ distribution, one gets the interval

$$(\overline{X}^2 - z^2_{\alpha/2}\sigma^2/n)^{-1}[\overline{X}\,\overline{Y} \pm z_{\alpha/2}(\sigma^2/n)^{1/2}(\overline{X}^2 + \overline{Y}^2 - z^2_{\alpha/2}\sigma^2/n)^{12}], \quad (1)$$

for θ provided $\overline{X}^2 + \overline{Y}^2 > z^2_{\alpha/2}\sigma^2/n$.

The above procedure suffers from a serious drawback. Since $P_{\mu,\theta}(\overline{X}^2 + \overline{Y}^2 < z^2_{\alpha/2}\sigma^2/n) > 0$, a confidence interval for θ as in (1) may not exist for a given set of data. An alternative asymptotically valid procedure based on the ratio $\overline{Y}/\overline{X}$ provides the confidence interval

$$\overline{Y}/\overline{X} \pm z_{\alpha/2}\frac{\sigma}{\sqrt{n}}\frac{(\overline{X}^2 + \overline{Y}^2)^{1/2}}{\overline{X}^2}, \tag{2}$$

for θ which has asymptotic coverage probability $1-\alpha$. This is based on the fact that $\sqrt{n}(\overline{Y}/\overline{X} - \theta) \xrightarrow{d} N[0, \sigma^2(1+\theta^2)/\mu^2]$ as $n \to \infty$ and $\overline{X}$ and $\overline{Y}$ are consistent estimators of μ and $\theta\mu$. It may be noted that the confidence interval (1) also provides an asymptotically valid procedure, but (2) has the advantage that unlike (1) it always produces an answer.

Creasy (1954) viewed the frequency distributions of $\overline{X}$ and $\overline{Y}$ as fiducial distributions of μ and $\theta\mu$ respectively, and used the usual Jacobian calculus to find the distribution of the ratio θ which she referred to as the fiducial distribution of θ. This interpretation was severely criticized by the discussants of her paper who pointed out that there cannot be any fiducial interpetation of the same. Judging from a Bayesian perspective, Creasy's procedure amounts to putting two independent flat distributions for μ and $\theta\mu$. But, as we shall see later, such a prior is not particularly helpful for the construction of frequentist confidence intervals.

Also, by now, there is substantive literature covering both the Bayesian and likelihood based inference for this problem. Section 2 provides a brief review of the likelihood based analysis for this problem based on the work of Barndorff-Nielsen (1983), Cox and Reid (1987), McCullagh and Tibshirani (1990), Liseo (1993), and Berger et al. (1999) for the normal case.

The main thrust of this paper, however, is to provide a Bayesian analysis for the more general problem of inference about the ratio of two location parameters in the two-sample location-scale problem. Other than the normal, two important members of this general class of distributions are the t and double exponential distributions.

Bayesian analysis for the original Fieller–Creasy problem based on noninformative priors began with Kappenman et al. (1970), and was addressed subsequently in Bernardo (1977), Sendra (1982), Mendoza (1996), Stephens and Smith (1992), Liseo (1993), Phillipe and Robert (1994), Reid (1995) and Berger et al. (1999). All these papers considered either Jeffreys' prior or reference priors. A Bayesian analysis based on proper priors is given in Carlin and Louis (2000).

A somewhat different criterion for developing noninformative priors is based on matching the posterior coverage probability of a Bayesian credible set with the corresponding frequentist coverage probability upto a certain order. For a recent review of various approaches to the development of probability matching priors, we refer to Ghosh and Mukerjee (1998).

In Section 3 of this paper, we develop first and second order probability matching priors as well as two-group and one-at-a-time reference priors. The proposed priors remain the same for every member of the location-scale family. In Section 4, we study the properties of posteriors for two important subclasses of distributions including the normal, t and double exponential distributions. For the normal case, we have corrected a result of Liseo (1993). Implementation of Bayes procedures via Markov Chain Monte Carlo is discussed in Section 5.

2 Likelihood Based Analysis

Let $x_1, \ldots, x_n$ and $y_1, \ldots, y_n$ be two independent samples with pdf's (with respect to Lebesgue measure) $\sigma^{-1} f(x - \mu/\sigma)$ and $\sigma^{-1} f(y - \theta\mu/\sigma)$, where $f(z) = f(-z)$. Here $\theta \in (-\infty, \infty)$, $\mu \in (-\infty, \infty) - \{0\}$ and $\sigma (> 0)$ are all unknown. The parameter of interest is θ, the ratio of the location parameters. The assumption that the parameter space of μ does not contain 0 is not very restrictive since the ratio of the two location parameters is undefined when both are zeros.

The per unit Fisher information matrix is given by

$$I(\theta, \mu, \sigma) = \sigma^{-2} \begin{bmatrix} c_1\mu^2 & c_1\theta\mu & 0 \\ c_1\theta\mu & c_1(1+\theta^2) & 0 \\ 0 & 0 & c_2 \end{bmatrix}, \tag{3}$$

where $c_1 = \int_{-\infty}^{\infty} [f'(x)/f(x)]^2 f(x)\, dx$ (the Fisher information number) and $c_2 = 2\left[\int_{-\infty}^{\infty} x^2 (f'(x)/f(x))^2 f(x)\, dx - 1\right]$. It is assumed that both c_1 and c_2 are finite. In particular, for the normal case $c_1 = 1$ and $c_2 = 4$.

Next, following Cox and Reid (1987), we find the orthogonal parametric transformation $\theta_1 = \theta$, $\theta_2 = \mu(1+\theta^2)^{1/2}, \theta_3 = \sigma$. The resultant per unit Fisher information matrix is given by

$$I(\theta_1, \theta_2, \theta_3) = \theta_3^{-2}\, \mathrm{Diag}\left[c_1\theta_2^2(1+\theta_1^2)^{-2}, c_1, c_2\right]. \tag{4}$$

Now the log profile likelihood is given by

$$\begin{aligned} l_p(\theta_1) = -2n \log \widehat{\theta}_3(\theta_1) &+ \sum_{i=1}^{n} \log f\left\{\left[x_i - \frac{\widehat{\theta}_2(\theta_1)}{\sqrt{1+\theta_1^2}}\right] \Big/ \widehat{\theta}_3(\theta_1)\right\} \\ &+ \sum_{i=1}^{n} \log f\left\{\left[y_i - \frac{\theta_1\widehat{\theta}_2(\theta_1)}{\sqrt{1+\theta_1^2}}\right] \Big/ \widehat{\theta}_3(\theta_1)\right\}, \end{aligned} \tag{5}$$

while the log conditional profile likelihood cf. (Cox and Reid, 1987) is given by

$$l_{cp}(\theta_1) = l_p(\theta_1) - \frac{1}{2}\log\left[n^2 c_1 c_2 \widehat{\theta}_3^{-4}(\theta_1)\right]. \tag{6}$$

In general, closed form expressions for $l_{\mathrm{p}}(\theta_1)$ and $l_{\mathrm{cp}}(\theta_1)$ are not available. However, for the special case of a normal distribution, $l_{\mathrm{p}}(\theta_1)$ and $l_{\mathrm{cp}}(\theta_1)$ are given respectively by

$$\begin{aligned} l_{\mathrm{p}}(\theta_1) &= -n \log S(\theta_1) - K, \\ l_{\mathrm{cp}}(\theta_1) &= -(n-1) \log S(\theta_1) - K, \end{aligned} \tag{7}$$

where in the above and what follows, $K (> 0)$ is a generic constant, and

$$S(\theta_1) = \sum_{i=1}^{n} (x_i - \bar{x})^2 + \sum_{i=1}^{n} (y_i - \bar{y})^2 + \frac{n(\bar{y} - \theta_1 \bar{x})^2}{1 + \theta_1^2}.$$

But $S(\theta_1)$ is maximized at $\theta_1 = -\bar{x}/\bar{y}$. Hence, both $l_{\mathrm{p}}(\theta_1)$ and $l_{\mathrm{cp}}(\theta_1)$ are bounded away from 0 when $|\theta_1| \to \infty$. Consequently, a confidence interval for θ_1 by inverting either the profile likelihood test statistic or conditional profile likelihood test statistic could potentially be the entire real line. This fact is mentioned in Liseo (1993) in the case when σ (> 0) is known. In that case

$$l_{\mathrm{p}}(\theta_1) = l_{\mathrm{cp}}(\theta_1) = -S(\theta_1)/(2\sigma^2) - K. \tag{8}$$

Another adjustment of the profile likelihood is proposed in McCullagh and Tibshirani (1990) based on the idea of unbiased estimating functions. Suppose θ is the parameter of interest and ψ is the nuisance parameter. Denote the score function derived from the profile log-likelihood by $U(\theta) = (\partial)/(\partial\theta) l_{\mathrm{p}}(\theta)$, where $l_{\mathrm{p}}(\theta) = l\left[\theta, \widehat{\psi}(\theta)\right]$, $\widehat{\psi}(\theta)$ being the MLE of ψ for fixed θ. A property of the regular maximum likelihood score function is that it has zero mean, and its variance is the negative of the expected derivative matrix, expectation being computed at the true parameter value. McCullagh and Tibshirani (1990) propose adjusting $U(\theta)$ so that these properties hold when expectations and derivatives are computed at $\left[\theta, \widehat{\psi}(\theta)\right]$ rather than at the true parameter point.

To this end, let $\widetilde{U}(\theta) = [U(\theta) - m(\theta)] w(\theta)$, where $m(\theta)$ and $w(\theta)$ are so chosen that $E_{\theta, \widehat{\psi}(\theta)} \widetilde{U}(\theta) = 0$ and $V_{\theta, \widehat{\psi}(\theta)} [\widetilde{U}(\theta)] = -E_{\theta, \widehat{\psi}(\theta)} [(\partial/\partial\theta) \widetilde{U}(\theta)]$. This leads to the solutions

$$\begin{aligned} m(\theta) &= E_{\theta, \widehat{\psi}(\theta)} U(\theta), \\ w(\theta) &= \left\{ -E_{\theta, \widehat{\psi}(\theta)} \frac{\partial^2}{\partial\theta^2} l_{\mathrm{p}}(\theta) + \frac{\partial}{\partial\theta} m(\theta) \right\} \Big/ V_{\theta, \widehat{\psi}(\theta)} [U(\theta)] \end{aligned} \tag{9}$$

Then the adjusted profile log-likelihood is given by

$$l_{\mathrm{Ap}}(\theta) = \int^{\theta} \widetilde{U}(t)\, dt, \tag{10}$$

and $\exp\left(l_{\mathrm{Ap}}(\theta)\right)$ is called the adjusted profile likelihood. However, for the normal case it turns out that $m(\theta_1) = 0$ and $w(\theta_1) = K + O_p(n^{-1})$, where

$K(>0)$ is a generic constant, so that asymptotically (as $n \to \infty$) $l_{\mathrm{Ap}}(\theta)$ behaves like $l_{\mathrm{p}}(\theta)$ or $l_{\mathrm{cp}}(\theta)$.

Yet another likelihood based approach for this problem is the integrated likelihood approach proposed in Berger et al. (1999). Following their prescription, one uses the conditional prior $\pi(\theta_2, \theta_3 \mid \theta_1)$ which in this case is given by

$$\pi(\theta_2, \theta_3 \mid \theta_1) \propto \theta_3^{-2},$$

based on the positive square root of the determinant of the appropriate submatrix of the Fisher information matrix. Then the integrated likelihood $L_I(\theta_1)$ is given by

$$\begin{aligned}
L_I(\theta_1) &= \int_0^\infty \int_{-\infty}^\infty L(\theta_1, \theta_2, \theta_3)\theta_3^{-2}\, d\theta_2\, d\theta_3 \\
&\propto \int_0^\infty \int_{-\infty}^\infty \theta_3^{-2n-2} \\
&\quad \times \exp\Bigg\{-\frac{1}{\theta_3^2}\Bigg[\sum_1^n (x_i - \overline{x})^2 + \sum_1^n (y_i - \overline{y})^2 \\
&\quad + n\left(\overline{x} - \frac{\theta_2}{\sqrt{1+\theta_1^2}}\right)^2 + n\left(\overline{y} - \frac{\theta_1\theta_2}{\sqrt{1+\theta_1^2}}\right)^2\Bigg]\Bigg\}\, d\theta_2\, d\theta_3 \\
&\propto \int_0^\infty \theta_3^{-2n-1} \exp\left[-\frac{1}{\theta_3^2} S(\theta_1)\right] d\theta_3 \\
&\propto S^{-n}(\theta_1),
\end{aligned}$$

which is the same as the profile likelihood. Thus, any likelihood-based procedure may not always yield any meaningful confidence interval for θ_1. We now turn to the Bayesian approach for this problem.

3 Noninformative Priors

In this section we derive two classes of noninformative priors, namely probability matching priors and reference priors. Bayesian analysis based on these priors are discussed in next sections.

Probability matching priors

Probability matching priors π are those for which the coverage probability of Bayesian credible intervals for θ matches the corresponding frequentist coverage probability asymptotically upto a certain order.

Specifically, let $\theta^{(1-\alpha)}(\pi, x, y)$ denote the $(1-\alpha)$th posterior quantile of θ under a prior π. First and second order probability matching priors refer respectively to the cases when the corresponding frequentist probability $P_{\theta,\mu,\sigma}\left[\theta \le \theta^{(1-\alpha)}(\pi, x, y)\right] = 1 - \alpha + o(n^{-u})$ with $u = 1/2$ and 1.

For validity of the derivation of the probability matching priors, we assume the regularity conditions of Johnson (1970). Then from Tibshirani (1989, Eq. (2.4)), the class of first order matching priors is characterized by:

$$\pi(\theta_1, \theta_2, \theta_3) \propto |\theta_2| \theta_3^{-1} (1 + \theta_1^2)^{-1} g(\theta_2, \theta_3),$$

where $g(\cdot)$ is an arbitrary positive function of θ_2 and θ_3, differentiable in each argument such that the corresponding posterior is proper.

The function g being arbitrary, there are infinitely many first order probability matching priors. Second order probability matching priors narrow down the selection within this class, and are found from Mukerjee and Ghosh (1997, Eq. (2.10)). To find this equation for this problem, first we write the likelihood function

$$L(\theta_1, \theta_2, \theta_3) \\ = \theta_3^{-2n} \prod_{i=1}^{n} \left\{ f\left[\frac{x_i - \theta_2(1+\theta_1^2)^{-1/2}}{\theta_3}\right] f\left[\frac{y_i - \theta_1\theta_2(1+\theta_1^2)^{-1/2}}{\theta_3}\right] \right\}. \quad (11)$$

Then, after simplification,

$$L_{111} = \frac{1}{n} E\left[\frac{\partial \log L(\theta)}{\partial \theta_1}\right]^3 = 0,$$

$$L_{112} = \frac{1}{n} E\left[\frac{\partial^3 \log L(\theta)}{\partial \theta_1^2 \partial \theta_2}\right],$$

and

$$L_{113} = \frac{1}{n} E\left[\frac{\partial^3 \log L(\theta)}{\partial \theta_1^2 \partial \theta_3}\right] = (2c_1 + c_3) \frac{\theta_2^2}{(1+\theta_1^2)^2 \theta_3^3},$$

where

$$c_3 = -\int_{-\infty}^{\infty} x \left[\frac{d^3 \log f(x)}{dx^3}\right] f(x)\, dx.$$

Now, from Mukerjee and Ghosh (1997, Eq. (2.10)), a second order probability matching prior is found as a solution of

$$\theta_3 \frac{\partial}{\partial \theta_2} g(\theta_2, \theta_3) + c\theta_2 \frac{\partial}{\partial \theta_3} g(\theta_2, \theta_3) = 0, \quad (12)$$

where $c = (2c_1 + c_3)/c_2$. A general class of solutions to (12) is given by $g(\theta_2, \theta_3) = h(c\theta_2^2 - \theta_3^2)$, where h is an arbitrary function differentiable in both θ_2 and θ_3. A recommended choice is $g(\theta_2, \theta_3) =$ constant. This is primarily for simplicity, but we have found that some of the other choices of h, for example a quadratic function, may lead to improper posteriors.

Also, with this choice, the second order matching prior is independent of the choice of c, and thus does not vary from one distribution to another.

The corresponding second order matching prior is now

$$\pi^m(\theta_1,\theta_2,\theta_3) \propto |\theta_2|\theta_3^{-1}(1+\theta_1^2)^{-1}. \tag{13}$$

Remark 1. Due to invariance of probability matching priors (Datta and Ghosh, 1996; Mukerjee and Ghosh, 1997), the second order matching prior in the (θ,μ,σ) parametrization reduces to

$$\pi^m(\theta,\mu,\sigma) \propto |\mu|/\sigma.$$

Reference priors

Due to the orthogonality of the parameters, following Bernardo (1979), Berger and Bernardo (1989, 1992) and Datta and Ghosh (1995), using rectangular compacts for θ_1, θ_2 and θ_3, the two-group reference prior is given by

$$\pi^{R_2}(\theta_1,\theta_2,\theta_3) \propto \theta_3^{-2}(1+\theta_1^2)^{-1},$$

while the one-at-a-time reference prior is given by

$$\pi^R(\theta_1,\theta_2,\theta_3) \propto \theta_3^{-1}(1+\theta_1^2)^{-1}.$$

Each one is a first order probability matching prior by proper choice of $g(\theta_2,\theta_3)$, but none is a second order probability matching prior. Jeffreys' prior, namely the positive square root of the determinant of the Fisher information matrix, is given by

$$\pi^J(\theta_1,\theta_2,\theta_3) \propto \theta_3^{-3}|\theta_2|(1+\theta_1^2)^{-1},$$

is also first order probability matching prior, but is not a second order probability matching prior.

Remark 2. We may also note that due to invariance see (Datta and Ghosh, 1996), the reference priors and Jeffreys priors in the (θ,μ,σ) parametrization are given, respectively, by $\pi^{R_2}(\theta,\mu,\sigma) \propto \sigma^{-2}(1+\theta^2)^{-1/2}$, $\pi^R(\theta,\mu,\sigma) \propto \sigma^{-1}(1+\theta^2)^{-1/2}$, $\pi^J(\theta,\mu,\sigma) \propto |\mu|\sigma^{-3}$.

Remark 3. In the special case when σ (or θ_3) is known, the unique second order probability matching prior is given by: $\pi^m(\theta_1,\theta_2) \propto |\theta_2|(1+\theta_1^2)^{-1}$ which is the same as Jeffreys' prior. This is also noticed by Mukerjee and Dey (1993) in the normal case. The reference prior, on the other hand, is given by $\pi^R(\theta_1,\theta_2) \propto (1+\theta_1^2)^{-1}$ which is a first order probability matching prior, but is not a second order probability matching prior.

Remark 4. The one-at-a-time reference prior was also derived by Mendoza (1996) in the normal case.

4 Propriety of Posteriors

A detailed study of posteriors under the proposed priors requires specification of f. We concentrate on two important families of distributions:

(i) the power family, and

(ii) t-family.

The former includes the normal and the double exponential distributions as special cases. The latter includes the normal distribution as a limit. For each family, we investigate the propriety of posteriors for a general class of priors which includes the reference prior π^R, Jeffrey's prior π^J, and the second order matching prior π^m. In what follows, we shall write $x = (x_1, \ldots, x_n)$, $y = (y_1, \ldots, y_n)$.

We begin with the power family. The following general theorem can be proved.

Theorem 1. *Let $f(z) = k(\delta)\exp[-|z|^\delta]$, $\delta \geq 1$, where $k(\delta)$ is the normalizing constant. Consider the class of priors $\pi_{a,\alpha}(\theta, \mu, \sigma) \propto |\mu|^a \sigma^{-\alpha} \times (1+\theta^2)^{-(1-a)/2}$, $0 \leq a \leq 1$, $\alpha \geq 0$. Then the posterior $\pi_{a,\alpha}(\theta, \mu, \sigma \mid x, y)$ is proper if $2n + \alpha > 3$.*

The proof of the theorem is given in appendix.

Remark 5. It may be noted that the propriety of posteriors does not depend on $\delta(\geq 1)$. For $0 < \alpha \leq 1$, one needs $n \geq 2$ for proper posteriors, while for $\alpha > 1$, one needs $n \geq 1$ for proper posteriors.

Next we consider the t-family. The following theorem is proved.

Theorem 2. *Assume f be a t density function with d.f $\nu > 0$. Then with probability one, the posterior distributions are proper for the class of priors given in Theorem 1, if $(n-2)\nu > 3 + \alpha/2$.*

The proof of the theorem is also given in appendix.

A detailed analysis is given in the normal case. We compare the Bayesian analysis with the likelihood based analysis. This generalizes the findings of Liseo (1993); see also (Reid, 1995).

For the one-at-a-time reference prior π^R, the marginal posterior distribution of θ is :

$$\pi^R(\theta \mid x, y) \propto S^{-(n-1/2)}(\theta)(1+\theta^2)^{-1}.$$

Under the second order matching prior π^m, after some algebra, the marginal posterior distribution of θ is

$$\pi^m(\theta \mid x, y) \propto g(\theta; x, y)(1+\theta^2)^{-1},$$

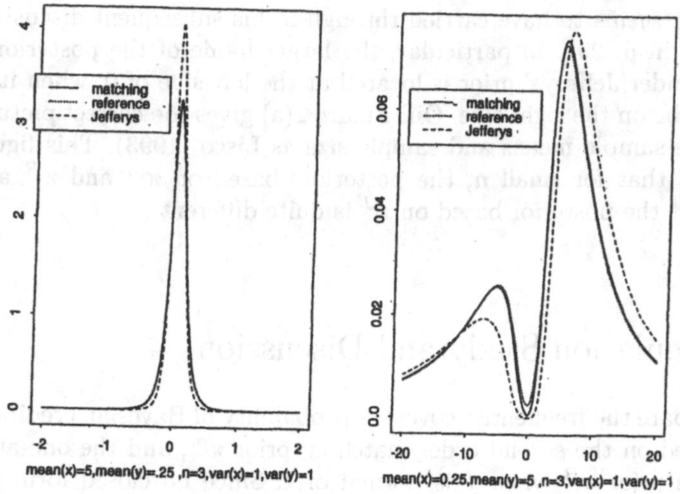

FIGURE 1. Posterior distributions for the reference prior, the second order matching prior and the Jeffreys prior.

where

$$g(\theta; x, y) = \frac{1}{(n-1)(\sum_{i=1}^n x_i^2 + \sum_{i=1}^n y_i^2)^{n-1}} + 2d(\theta)\frac{1}{S^{n-1}(\theta)}\int_0^{d(\theta)} \frac{1}{(1+w^2)^n}\,dw,$$

and

$$d(\theta) = \frac{n^{1/2}|\overline{x} + \theta\overline{y}|}{(1+\theta^2)^{1/2}S^{1/2}(\theta)}.$$

It is easy to see that $g(\theta)$ is an increasing function of $d(\theta)$. Like $l_p(\theta)$, $g(\theta)$ is maximized at $\theta = \overline{y}/\overline{x}$ and is minimized at $\theta = -\overline{x}/\overline{y}$. However, due to the factor $(1+\theta^2)^{-1}$, $\pi^m(\theta \mid x, y)$ tends to zero as $|\theta| \to \infty$, and thus a credible interval for θ under the second order matching prior will never become the entire real line. Substituting $n-1$ by n in $\pi^m(\theta \mid x, y)$, we get the posterior distribution for Jeffreys' prior. Like $l_p(\theta)$ and $l_{cp}(\theta)$, all the posterior distributions are bimodal, but when sample size is moderately large, or θ is small, one of the modes is far out in the tail. This is evident in Figure 1(b).

Liseo (1993) derived profile likelihood, modified likelihood, posterior distributions of θ under the reference prior and Jeffreys' prior when f is normal and $\sigma = 1$. In such a case, Jeffreys' prior is also the unique second order matching prior. Liseo's derivation of the posterior distribution of θ under Jeffrey' prior, however, seems to involve an algebraic error. The error occurs in $\pi_J(\theta \mid x, y)$ in (Liseo, 1993, p. 297) where $+1$ should be -1. Also, this error seems to have carried through in his subsequent discussion and Figure 1 in p. 298. In particular, the larger mode of the posterior distribution under Jeffreys' prior is located at the left side of 0, when it should actually be on the right side. Our Figure 1(a) gives the correct picture with the same sample means and sample size as Liseo (1993). This figure also indicates that for small n, the posteriors based on π^m and π^R are very close, but the posterior based on π^J is quite different.

5 Simulation Study and Discussion

We compare the frequentist coverage probability of Bayesian credible intervals based on the second order matching prior π^m, and the one-at-a-time reference prior π^R when f is normal or t. Since no closed form posteriors are available, the posterior quantiles are obtained via application of the Markov Chain Monte Carlo (MC^2) numerical integration. We provide below some of the implementational details for the t-likelihood.

To this end first represent a t-density with location parameter μ, scale parameter σ and degrees of freedom ν as the gamma scale mixture of a normal density. Specifically, writing $f_{\mu,\sigma,\nu}(t)$ for such a pdf, one has

$$f_{\mu,\sigma,\nu}(t) = \int_0^\infty \left\{ \left(\frac{r}{2\pi\sigma^2} \right)^{1/2} \exp\left[-\frac{r}{2\sigma^2}(t-\mu)^2 \right] \right\} \times \frac{\exp(-r\nu/2) r^{\nu/2-1} \nu^{\nu/2}}{2^{\nu/2}\Gamma(\nu/2)} \, dr. \tag{14}$$

We denote by r_{1i} and r_{2i} the mixing variables associated with the likelihoods corresponding to the x_i and the y_i respectively ($i = 1, \ldots, n$), where the r_{1i} and r_{2i} have gamma distributions as in (14). We shall write $r_1 = (r_{11}, \ldots, r_{1n})$, $r_2 = (r_{21}, \ldots, r_{2n})$, $x = (x_1, \ldots, x_n)$ and $y = (y_1, \ldots, y_n)$. Then, with the prior $\pi(\theta, \mu, \sigma) \propto \sigma^{-\alpha} |\mu|^a (1+\theta^2)^{-(1-a)/2}$, and the transformation $u = \theta\mu$, the joint posterior of u, μ, σ^2, r_1 and r_2 given the x and y is

$$\begin{aligned}
&\pi(u, \mu, \sigma^2, r_1, r_2 \mid x, y) \\
&\quad \propto (\sigma^2)^{-(2n+\alpha+1)/2} (\mu^2 + u^2)^{-(1-a)/2} \\
&\quad \times \exp\left[-\frac{1}{2\sigma^2} \sum_{i=1}^n r_{1i}(x_i - \mu)^2 - \frac{1}{2\sigma^2} \sum_{i=1}^n r_{2i}(y_i - u)^2 \right] \\
&\quad \times \prod_{i=1}^n \left[\exp\left(-\frac{1}{2} r_{1i}\nu \right) \exp\left(-\frac{1}{2} r_{2i}\nu \right) r_{1i}^{(\nu-1)/2} r_{2i}^{(\nu-1)/2} \right] (\mu^2 + u^2)^{-(1-a)/2}.
\end{aligned}$$

This leads to the full conditionals

$$\mu \mid u, \sigma^2, r_1, r_2, x, y \propto \exp\left[-\frac{1}{2\sigma^2}\sum_{i=1}^{n} r_{1i}(x_i-\mu)^2\right](\mu^2+u^2)^{-(1-a)/2};$$

$$u \mid \mu, \sigma^2, r_1, r_2, x, y \propto \exp\left[-\frac{1}{2\sigma^2}\sum_{i=1}^{n} r_{2i}(y_i-u)^2\right](\mu^2+u^2)^{-(1-a)/2};$$

$$\sigma^{-2} \mid \mu, u, r_1, r_2, x, y$$
$$\sim \text{Gamma}\left[\frac{\sum_{i=1}^{n} r_{1i}(x_i-\mu)^2+\sum_{i=1}^{n} r_{2i}(y_i-u)^2}{2}, \frac{2n+\alpha-1}{2}\right],$$

where a Gamma(b,c) pdf is proportional to $\exp(-bz)z^{c-1}$;

$$r_{1i} \mid u, \mu, \sigma^2, r_{1j}(j\neq i), r_2, x, y \sim \text{Gamma}\left[\nu+\frac{(x_i-\mu)^2}{2\sigma^2}, \frac{\nu+1}{2}\right];$$

$$r_{2i} \mid u, \mu, \sigma^2, r_1, r_{2j}(j\neq i), x, y \sim \text{Gamma}\left[\nu+\frac{(y_i-u)^2}{2\sigma^2}, \frac{\nu+1}{2}\right].$$

In actual implementation, we shall consider only the two cases $a = 0$ and $a = 1$ respectively. With α appropriately chosen, this corresponds to the reference prior, and the second order matching prior respectively.

The conditionals of μ or u given the rest being nonstandard, the Metropolis-Hastings algorithm is used to generate samples from these conditionals along the lines of Chib and Greenberg (1995).

In each case we computed the 0.05th and 0.95th percent posterior quantiles from a sample of size 10000 (discarding the first 5000) and repeated the iterations 1500 times to estimate the coverage probability. These sample sizes give an empirical variance of the order 10^{-4}. We ran these simulations for different values of (θ, μ) when $\sigma^2 = 1, n = 5, 10$ and $v = 1, 10, \infty$, where $\nu = 1$, and ∞ correspond respectively to the Cauchy and normal distributions. The results are given in Tables 1–6. The computing was done on Sun Sparc10 workstation using interface between C and Splus. All random numbers are generated by Splus. In the normal case, these results agree with the output obtained directly by numerical integration.

It is clear from the tables that π^m performs better than π^R in matching the target coverage probabilities. This is intuitively clear since π^m is a second order matching prior, but π^R is not. For both normal and Cauchy likelihoods, with the prior π^m, there is close matching except for small $|\mu|$. With π^R, matching is close except for small $|\mu|$ or for large $|\theta|$.

The matching via the nominal quantile points are similar across the normal, t_{10} and the Cauchy when the prior is π^R or π^m. This suggests the universality of the priors π^R or π^m for different f, with light, medium or heavy tails.

It appears also from our calculations that when $|\mu|$ is small, the percentile points $\theta^{(1-\alpha)}(\pi^m, x, y)$ or $\theta^{(1-\alpha)}(\pi^R, x, y)$ on an average (in the frequentist

TABLE 1. Frequentist coverage probabilities of 0.05(0.95) posterior quantiles of θ when $n = 5, \nu = \infty$(normal)

	θ	0.1		1		10		100	
		0.05	0.95	0.05	0.95	0.05	0.95	0.05	0.95
μ									
0.1	π^m	0.004	0.990	0.000	0.963	0.000	0.538	0.000	0.614
	π^R	0.004	0.997	0.001	0.980	0.000	0.449	0.015	0.523
1	π^m	0.025	0.964	0.015	0.948	0.001	0.949	0.005	0.944
	π^R	0.021	0.975	0.016	0.966	0.059	0.932	0.362	0.638
10	π^m	0.058	0.951	0.053	0.943	0.045	0.942	0.057	0.945
	π^R	0.037	0.949	0.046	0.949	0.067	0.932	0.274	0.700
100	π^m	0.049	0.950	0.047	0.939	0.043	0.957	0.041	0.946
	π^R	0.051	0.951	0.053	0.951	0.071	0.929	0.294	0.710

TABLE 2. Frequentist coverage probabilities of 0.05(0.95) posterior quantiles of θ when $n = 10, \nu = \infty$(normal)

	θ	0.1		1		10		100	
		0.05	0.95	0.05	0.95	0.05	0.95	0.05	0.95
μ									
0.1	π^m	0.003	0.990	0.000	0.956	0.000	0.759	0.000	0.815
	π^R	0.001	0.993	0.001	0.981	0.000	0.638	0.041	0.531
1	π^m	0.030	0.958	0.015	0.952	0.006	0.943	0.005	0.946
	π^R	0.035	0.970	0.041	0.959	0.071	0.923	0.267	0.672
10	π^m	0.056	0.955	0.048	0.953	0.052	0.941	0.059	0.961
	π^R	0.050	0.956	0.049	0.953	0.067	0.930	0.308	0.709
100	π^m	0.045	0.942	0.058	0.949	0.051	0.954	0.056	0.948
	π^R	0.053	0.945	0.047	0.953	0.072	0.932	0.281	0.709

sense) become small yielding thereby poor frequentist coverage probabilities for fixed θ. $\theta^{(1-\alpha)}(\pi^R, x, y)$ seems to have the same problem when the true θ is large, but $\theta^{(1-\alpha)}(\pi^m, x, y)$ seems to adjust itself in this case, and yields good frequentist coverage probabilities.

The poor performance of all the priors for certain regions of the parameter space is not very surprising. Gleser and Hwang (1987, Theorem 1) show that based on any sample of arbitrary but fixed size n, any confidence interval for θ of finite expected length has coverage probability (taking the infimum over all points in the parameter space) equal to zero. In the present case, this poor performance happens primarily in the neighborhood of $\mu = 0$. We want to emphasize, however, that our primary interest is the construction of Bayesian credible sets for θ. The coverage probabilities are then conditional on the data, and are not based on the infimum

TABLE 3. Frequentist coverage probabilities of 0.05(0.95) posterior quantiles of θ when $n = 5, \nu = 10$

μ	θ	0.1		1		10		100	
		0.05	0.95	0.05	0.95	0.05	0.95	0.05	0.95
0.1	π^m	0.006	0.992	0.000	0.955	0.000	0.547	0.000	0.615
	π^R	0.003	0.990	0.001	0.968	0.000	0.445	0.013	0.493
1	π^m	0.019	0.970	0.006	0.940	0.002	0.945	0.002	0.945
	π^R	0.018	0.977	0.011	0.963	0.066	0.932	0.231	0.633
10	π^m	0.041	0.951	0.060	0.937	0.051	0.961	0.059	0.947
	π^R	0.044	0.942	0.049	0.950	0.082	0.938	0.277	0.695
100	π^m	0.046	0.941	0.055	0.943	0.055	0.957	0.046	0.962
	π^R	0.046	0.942	0.061	0.943	0.061	0.918	0.267	0.678

TABLE 4. Frequentist coverage probabilities of 0.05(0.95) posterior quantiles of θ when $n = 10, \nu = 10$

μ	θ	0.1		1		10		100	
		0.05	0.95	0.05	0.95	0.05	0.95	0.05	0.95
0.1	π^m	0.005	0.993	0.000	0.955	0.000	0.715	0.000	0.789
	π^R	0.002	0.997	0.000	0.976	0.000	0.599	0.045	0.524
1	π^m	0.043	0.961	0.016	0.948	0.003	0.946	0.003	0.943
	π^R	0.027	0.965	0.038	0.956	0.075	0.923	0.262	0.673
10	π^m	0.049	0.955	0.055	0.948	0.049	0.955	0.051	0.951
	π^R	0.051	0.946	0.055	0.950	0.078	0.919	0.295	0.696
100	π^m	0.051	0.951	0.043	0.959	0.051	0.947	0.056	0.947
	π^R	0.051	0.943	0.047	0.945	0.077	0.922	0.283	0.716

TABLE 5. Frequentist coverage probabilities of 0.05(0.95) posterior quantiles of θ when $n = 5, \nu = 1$

μ	θ	0.1 0.05	0.1 0.95	1 0.05	1 0.95	10 0.05	10 0.95	100 0.05	100 0.95
0.1	π^m	0.003	0.992	0.001	0.938	0.000	0.421	0.000	0.341
	π^R	0.002	0.994	0.001	0.970	0.000	0.343	0.014	0.355
1	π^m	0.017	0.973	0.005	0.948	0.004	0.939	0.001	0.944
	π^R	0.016	0.985	0.007	0.967	0.044	0.865	0.216	0.616
10	π^m	0.044	0.951	0.050	0.952	0.055	0.953	0.047	0.945
	π^R	0.051	0.945	0.045	0.942	0.084	0.912	0.305	0.717
100	π^m	0.049	0.955	0.056	0.957	0.053	0.953	0.043	0.954
	π^R	0.056	0.969	0.057	0.958	0.075	0.912	0.294	0.709

TABLE 6. Frequentist coverage probabilities of 0.05(0.95) posterior quantiles of θ when $n = 10, \nu = 1$

μ	θ	0.1 0.05	0.1 0.95	1 0.05	1 0.95	10 0.05	10 0.95	100 0.05	100 0.95
0.1	π^m	0.000	0.994	0.000	0.953	0.000	0.562	0.000	0.591
	π^R	0.003	0.993	0.001	0.977	0.000	0.449	0.035	0.419
1	π^m	0.023	0.962	0.011	0.947	0.004	0.943	0.004	0.947
	π^R	0.019	0.979	0.021	0.945	0.083	0.909	0.245	0.640
10	π^m	0.045	0.958	0.051	0.957	0.047	0.952	0.057	0.957
	π^R	0.052	0.948	0.049	0.943	0.101	0.897	0.307	0.702
100	π^m	0.052	0.953	0.051	0.951	0.050	0.941	0.051	0.945
	π^R	0.051	0.956	0.049	0.943	0.088	0.915	0.277	0.697

of all points in the parameter space. Thus, Bayesian credible sets can be constructed to provide any required coverage probability. Moreover, the asymptotic matching of the coverage probability of Bayesian credible sets with the corresponding frequentist coverage probability does not contradict Gleser and Hwang (1987, Theorem 1). Indeed, these authors point out (p. 1361) that large sample approximate $100(1-\alpha)\%$ $(0 < \alpha < 1)$ confidence intervals of finite length exist almost surely for any function of θ.

6 Appendix

Proof of Theorem 1. The joint posterior of θ, μ and σ is given by

$$\pi_{a,\alpha}(\theta,\mu,\sigma \mid x,y) \propto \sigma^{-2n-\alpha} \times \exp\left[-\sigma^{-\delta}\left(\sum_{i=1}^{n}|x_i-\mu|^{\delta}+\sum_{i=1}^{n}|y_i-\theta\mu|^{\delta}\right)\right]|\mu|^{a}(1+\theta^2)^{-(1-a)/2}.$$

Integrating with respect to σ, the joint posterior of θ and μ is given by

$$\pi_{a,\alpha}(\theta,\mu \mid x,y) \propto \frac{1}{(\sum_{i=1}^{n}|x_i-\mu|^{\delta}+\sum_{i=1}^{n}|y_i-\theta\mu|^{\delta})^{(2n-1+\alpha)/\delta}(1+\theta^2)^{(1-a)/2}}.$$

Now, letting $u=\theta\mu$, the joint posterior of u and μ is given by

$$\pi_{a,\alpha}(u,\mu|x,y) \propto \frac{1}{(\sum_{i=1}^{n}|x_i-\mu|^{\delta}+\sum_{i=1}^{n}|y_i-u|^{\delta})^{(2n-1+\alpha)/\delta}(u^2+\mu^2)^{(1-a)/2}}.$$

For $|\mu| \le 3|\overline{x}|/2$, use the inequality $\sum_{i=1}^{n}|x_i-\mu|^{\delta} \ge \sum_{i=1}^{n}|x_i-x_{\text{med}}|^{\delta}$, where $x_{\text{med}} = \text{median}(x_1,\ldots,x_n)$, while for $|\mu| > 3|\overline{x}|/2$, use the inequality $\sum_{i=1}^{n}|x_i-\mu|^{\delta} \ge n|\overline{x}-\mu|^{\delta}$. Similarly, for $|u| \le 3|\overline{y}|/2$, use the inequality $\sum_{i=1}^{n}|y_i-u|^{\delta} \ge \sum_{i=1}^{n}|y_i-y_{\text{med}}|^{\delta}$, where $y_{\text{med}} = \text{median}(y_1,\ldots,y_n)$, while for $|u| > 3|\overline{y}|/2$, use the inequality $\sum_{i=1}^{n}|y_i-u|^{\delta} \ge n|\overline{y}-u|^{\delta}$. Now by a polar transformation of coordinates, one can verify that $(\mu^2+u^2)^{-(1-a)/2}$ is integrable when $|\mu| \le 3|\overline{x}|/2$ and $|u| \le 3|\overline{y}|/2$. Also for $|\mu| > 3|\overline{x}|/2$ and $|u| > 3|\overline{y}|/2$, $|\mu-\overline{x}| > 1/|\overline{x}|2$ and $|u-\overline{y}| > 1|\overline{y}|/2$, and hence

$$\int_{|\mu|>\frac{3}{2}|\overline{x}|}\int_{|u|>\frac{3}{2}|\overline{y}|}(|\mu-\overline{x}|^{\delta}+|u-\overline{y}|^{\delta})^{-(2n+\alpha-1)/\delta} \times (\mu^2+u^2)^{-(1-a)/2}\,d\mu\,du$$

$$\le \left[\frac{9}{4}(\overline{x}^2+\overline{y}^2)\right]^{-(1-a)/2}$$

$$\times \int_{|\mu|>3|\bar{x}|/2} \int_{|u|>3|\bar{y}|/2} \left(|\mu-\bar{x}|^{\delta}+|u-\bar{y}|^{\delta}\right)^{-(2n+\alpha-1)/\delta} d\mu\, du$$

$$\leq \left[\frac{9}{4}(\bar{x}^2+\bar{y}^2)\right]^{-(1-a)/2}$$

$$\times \int_{|\mu|>1/2|\bar{x}|} \int_{|u|>1/2|\bar{y}|} \left(|\mu|^{\delta}+|u|^{\delta}\right)^{-(2n+\alpha-1)/\delta} d\mu\, du$$

$$\leq \infty$$

if $2n-3+\alpha > 0$, which can be verified once again by a polar transformation of coordinates. Similar arguments can be used when $|\mu| \leq 3|\bar{x}|/2$, $|u| > 3|\bar{y}|/2$ or $|\mu| > 3|\bar{x}|/2$, $|u| \leq 3|\bar{y}|/2$. □

Proof of Theorem 2. For $\pi_{a,\alpha}(\theta,\mu,\sigma) \propto \sigma^{-\alpha}|\mu|^{a}(1+\theta)^{-(1-a)/2}$, the joint posterior of θ, μ and σ is given by

$$\pi_{a,\alpha}(\theta,\mu,\sigma \mid x,y) \propto \prod_{i=1}^{n}\left\{\sigma^{-2}\left[\left(1+\frac{(x_i-\mu)^2}{\sigma^2\nu}\right)\left(1+\frac{(y_i-\theta\mu)^2}{\sigma^2\nu}\right)\right]^{-(\nu+1)/2}\right\} \times \sigma^{-\alpha}|\mu|^{a}(1+\theta^2)^{-(1-a)/2}.$$

Letting $u=\theta\mu$, the joint posterior of u, μ and σ is given by

$$\pi_{a,\alpha}(u,\mu,\sigma \mid x,y) \propto \prod_{i=1}^{n}\left\{\sigma^{-2}\left[\left(1+\frac{(x_i-\mu)^2}{\sigma^2\nu}\right)\left(1+\frac{(y_i-u)^2}{\sigma^2\nu}\right)\right]^{-(\nu+1)/2}\right\} \times \sigma^{-\alpha}(u^2+\mu^2)^{-(1-a)/2}. \quad (15)$$

Let $\varepsilon=\min(\nu,1)$. Since

$$\prod_{i=1}^{n}\{[1+(x_i-\mu)^2\sigma^{-2}\nu^{-1}][1+(y_i-u)^2\sigma^{-2}\nu^{-1}]\}^{(\nu+1)/2}$$
$$\geq \{1+n[(\mu-\bar{x})^2+(u-\bar{y})^2]\sigma^{-2}\nu^{-1}\}^{(\nu+1)/2}$$
$$\geq \{1+n[(\mu-\bar{x})^2+(u-\bar{y})^2]\sigma^{-2}\nu^{-1}\}^{(\varepsilon+1)/2},$$

for $\sigma > 1$, the right hand side of (15) is

$$\leq \frac{1}{\sigma^{2n+\alpha-\varepsilon-1}} \frac{1}{[1+(\mu-\bar{x})^2\nu^{-1}+(u-\bar{y})^2\nu^{-1}]^{(\varepsilon+1)/2}(u^2+\mu^2)^{(1-a)/2}}.$$

Hence it is easy to see that $\int_1^{\infty}\int_{-\infty}^{\infty}\int_{-\infty}^{\infty}\pi_{a,\sigma}(u,\mu,\sigma)du\, d\mu\, d\sigma < \infty$, if $2n+\alpha>3$.

On the other hand, for

$$0 < \sigma \le 1, \quad \prod_{i=1}^{n} \left[\left(1 + (x_i - \mu)^2 \sigma^{-2} \nu^{-1}\right) \left(1 + (y_i - u)^2 \sigma^{-2} \nu^{-1}\right) \right]$$

$$\ge \prod_{i=1}^{n} \left[1 + (x_i - \mu)^2 (y_i - u)^2 \sigma^{-4} \nu^{-2} \right]$$

$$\ge 1 + 2\sigma^{-4} \nu^{-2} \sum_{i=1}^{n} \sum_{j=1}^{i-1} \prod_{k \ne i,j} (x_k - \mu)^2 (y_k - u)^2.$$

Hence the right hand side of (15)

$$\le \frac{1}{\sigma^{2n+\alpha}} \frac{1}{\{1 + 2\sum_{i=1}^{n} \sum_{j=1}^{i-1} \prod_{k \ne i,j} [(x_k - \mu)^2 (y_k - u)^2]/(\nu^2 \sigma^4)\}^{(\nu+1)/2}} \times \frac{1}{(u^2 + \mu^2)^{(1-a)/2}}$$

$$\le \frac{\sigma^{2(n-2)(\nu+1)-2n-\alpha}}{\{\sum_{i=1}^{n} \sum_{j=1}^{i-1} \prod_{k \ne i,j} [(x_k - \mu)^2 (y_k - u)^2]/(\nu^2)\}^{(\nu+1)/2}} \times \frac{1}{(u^2 + \mu^2)^{(1-a)/2}}.$$

We know that with probability 1, no two of the x_i or the y_i are equal. Let $D_{ij} = O\big[(x_i, y_j), \delta\big]$, which is a sphere centered at (x_i, y_j) with radius δ, δ being chosen small enough such that $D_{i'j'} \cap D_{ij} = \phi$, for all $i \ne i'$ or $j \ne j'$. It is easy to see that $\pi_{a,\alpha}(u, \mu, \sigma \mid x, y)$ is integrable under D_{ij}, $0 < \sigma \le 1$, for $i, j = 1, 2, \ldots, n$ and $\cap_{i,j=1}^{n} D_{ij}^{c}$, $0 < \sigma \le 1$, if $2(n-2)(\nu+1) - 2n - \alpha = 2(n-2)\nu - 4 - \alpha > -1$. □

Acknowledgments: Research partially supported by NSF Grant Number SBR-9423996 and SBR-9810968.

7 References

Barnard, G.A. and D.A. Sprott (1983). The generalised problem of the Nile: Robust confidence sets for parametric functions. *Ann. Statist. 11*, 104–113.

Barndorff-Nielsen, O.E. (1983). On a formula for the distribution of the maximum likelihood estimator. *Biometrika 70*, 343–365.

Berger, J.O. and J.M. Bernardo (1989). Estimating a product of means: Bayesian analysis with reference prior. *J. Amer. Statist. Assoc. 84*, 200–207.

Berger, J.O. and J.M. Bernardo (1992). On the development of reference priors (with discussion). In J. Bernardo, J. Berger, A. Dawid, and A. Smith (Eds.), *Bayesian Statistics IV*, pp. 35–60. Oxford University Press.

Berger, J.O., B. Liseo, and R. Wolpert (1999). Integrated likelihood methods for eliminating nuisance parameters. *Statist. Sci. 14*, 1–22.

Bernardo, J.M. (1977). Inferences about the ratio of normal means: A. Bayesian approach to the Fieller–Creasy problem. In *Recent Developments in Statistics, Proceedings of the 1976 European Meeting of Statisticians*, pp. 345–350.

Bernardo, J.M. (1979). Reference posterior distributions for Bayesian inference (with discussion). *J. Roy. Statist. Soc. Ser. B 41*, 113–147.

Carlin, B.P. and T.A. Louis (2000). *Bayes and Empirical Bayes Methods for Data Analysis.* London: Chapman and Hall.

Chib, S. and E. Greenberg (1995). Understanding the Metropolis-Hastings algorithm. *Amer. Statist. 49*, 327–335.

Cox, D.R. and N. Reid (1987). Parameter orthogonality and approximate conditional inference (with discussion). *J. Roy. Statist. Soc. Ser. B 49*, 1–39.

Creasy, M.A. (1954). Limits for the ratio of means (with discussion). *J. Roy. Statist. Soc. Ser. B 16*, 186–194.

Datta, G.S. and M. Ghosh (1995). Some remarks on noninformative priors. *J. Amer. Statist. Assoc. 90*, 1357–1363.

Datta, G.S. and M. Ghosh (1996). On the invariance of noninformative priors. *Ann. Statist. 24*, 141–159.

Fieller, E.C. (1954). Some problems in interval estimation (with discussion). *J. Roy. Statist. Soc. Ser. B 16*, 175–185.

Ghosh, M. and R. Mukerjee (1998). Recent developments on probability matching priors. In S. Ahmed, M. Ahsanullah, and B. Sinha (Eds.), *Applied Statistical Science III. Nonparametric Statistics and Related Topics*, pp. 227–252. Nova Science Publ., Inc.

Gleser, L.J. and J.T. Hwang (1987). The nonexistence of $100(1-\alpha)\%$ confidence sets of finite expected diameters in error-in-variable and related models. *Ann. Statist. 15*, 1351–1362.

Johnson, R.A. (1970). Asymptotic expansions associated with posterior distributions. *Ann. Math. Statist. 41*, 851–864.

Kappenman, R.F., S. Geisser, and C.F. Antle (1970). Bayesian and fiducial solutions to the Fieller-Creasy problem. *J. Roy. Statist. Soc. Ser. B 32*, 331–340.

Liseo, B. (1993). Elimination of nuisance parameters with reference priors. *Biometrika 80*, 295–304.

McCullagh, P. and R. Tibshirani (1990). A simple method for the adjustment of profile likelihood. *J. Roy. Statist. Soc. Ser. B 52*, 325–344.

Mendoza, M. (1996). A note on the confidence probabilities of reference priors for the calibration model. preprint.

Mukerjee, R. and D.K. Dey (1993). Frequentist validity of posterior quantiles in the presence of a nuisance parameter: Higher order asymptotics. *Biometrika 80*, 499–505.

Mukerjee, R. and M. Ghosh (1997). Second order probability matching priors. *Biometrika 84*, 970–975.

Phillipe, A. and C.P. Robert (1994). A note on the confidence properties of reference priors for the calibration model. preprint.

Reid, N. (1995). Likelihood and Bayesian approximation methods (with discussion). In J. Bernardo, J. Berger, A. Dawid, and A. Smith (Eds.), *Bayesian Statistics*, Volume 5, pp. 611–618. Oxford University Press.

Sendra, M. (1982). Reference posterior distribution for the Fieller-Creasy problem. *Test 33*, 55–72.

Stephens, D.A. and A.F.M. Smith (1992). Sampling-resampling techniques for the computation of posterior densities in normal means problems. *Test 1*, 1–18.

Tibshirani, R.J. (1989). Non-informative priors for one parameter of many. *Biometrika 76*, 604–608.

10

The Estimation of Ratios From Paired Data

D.A. Sprott

ABSTRACT Fisher many times expressed the view that in inductive inference comparatively slight differences in the mathematical specification of a problem may have logically important effects on the inferences possible, and that in complicated cases, such effects may be very puzzling. The problem of making inferences about ratios from paired data is examined from this viewpoint. The existence and role of likelihood function is emphasized.

1 Introduction

General considerations

Fisher (1991b, p. 142) expressed the importance in inductive inference of the mathematical specification of a problem as follows: "It is particularly to be noted in this example that the differences in logical form of the available inferences flow from quite simple differences in the mathematical specification of the problem. Not to formal principles only, but there is also needed attention to particular analytical details". Mathematical specification can be interpreted to mean the assumptions upon which the inferences rest. The importance of these assumptions is manifest by the extent to which they impinge upon the inferences. This is evidenced by the extent to which changes in the assumptions change the inferences. Obviously, as Fisher states, to examine this requires exhibiting the analytical details, and their numerical consequences, flowing from various different scientifically reasonable sets of assumptions. This is the problem of adaptive robustness, (Sprott, 1978, 1982).

The scientific inferences about a parameter are in the form of estimation statements, that is, quantitative statements of plausibility using all of the sample parametric information. These will be, where possible, in the form of a complete set of likelihood-confidence intervals. This is a set of intervals the end points of which, taken as a whole, reproduce the likelihood function, and which in repeated samples have the constant coverage frequency property which defines confidence intervals. The likelihood property ensures that the intervals are properly conditioned on the observed data, and so conveys all of the parametric information in the data. This requires the existence of a likelihood function. The confidence property requires the existence of an associated pivotal quantity in terms of which this likelihood function can be expressed.

The purpose in what follows is to illustrate these points on the estimation of a ratio β with n paired observations (x_i, y_i) from comparative experiments. Attention was drawn to this problem, and different, seemingly contradictory, methods of inference about β, even assuming normality, have been obtained by Creasy (1954), Fieller (1954), and by the ensuing discussion, in particular by Fisher (1954).

Pivotal quantity

A pivotal quantity is a known function $u(X;\beta)$ of the observations X and parameters β having a distribution $g(u)$ that does not involve any unknown parameters. Their use in inferences allows knowledge of densities g to be replaced by knowledge of the functional form of the pivotals u. The observations X are stochastically related to the parameters β by the functional form of the pivotal u, not by the density g. A typical example is the location-scale model in which this relationship is $x = \beta + \sigma u$. Then $u = (x-\beta)/\sigma$ is a pivotal whose distribution need not be specified exactly. In particular this frees the analysis from the assumption of normality, as illustrated in Example (1b), Section 5. This leads to the concept of robust likelihoods, Barnard and Sprott (1983).

The Fieller pivotal

Suppose that

$$u_i = \frac{y_i - \beta x_i}{\sigma\sqrt{1+\beta^2}}, \quad i = 1, \ldots, n, \tag{1}$$

are independent standard normal pivotal quantities. It then follows that

$$t = t(\beta) = \frac{\overline{u}\sqrt{n}}{s_u}, \tag{2}$$

is a Student t_{n-1} pivotal that can be used for inferences about β, where

$$\overline{u} = \sum u_i/n = (\overline{y} - \beta\overline{x})/\sigma\sqrt{1+\beta^2},$$

$$(n-1)s_u^2 = \sum (u_i - \overline{u})^2 = (S_{yy} - 2\beta S_{xy} + \beta^2 S_{xx})/\sigma^2(1+\beta^2),$$

the S's being sums of squared deviations and cross products as usual. The standard procedure for inferences about β is based on the pivotal (2), which is often called the Fieller pivotal, Fieller (1940, 1954).

An awkward feature of (2) is that there is always a confidence level beyond which the confidence intervals for β are $-\infty, +\infty$. If $\overline{x}$, $\overline{y}$ are small enough this will include most confidence levels of interest, indicating that in this case (2) contains little information about β. See Example 2, Section 5. A possible justification of this occurrence is that small values of $\overline{x}, \overline{y}$ imply that β cannot be precisely estimated, or is of the form 0/0 and so not well defined. This justification however implies that the information about β is contained in $\overline{x}, \overline{y}$.

Assumptions about the $\{x_i, y_i\}$

In setting up (2), nothing has been said about the behaviour of x_i, y_i. Fisher (1954), Fisher (1991b, p. 138) emphasized that seemingly slight differences in the mathematical specification of a problem may have important effects on the inferences. That assumptions about the x_i, y_i can be important is easily seen in the present case by considering the effect of assuming that the x_i's are constants. If the $\{x_i\}$ are fixed constants, then the problem is a normal regression, $y_i \sim N(\beta x_i, \sigma^2)$, for which $\sum x_i y_i / \sum x_i^2$ is a sufficient statistic estimating β with precision $1/\sqrt{\sum x_i^2}$, as indeed was noted by Creasy (1954, p. 222). The means $\overline{x}, \overline{y}$ are not involved. That is, although inferences (2) based on the normal distribution of $\overline{y} - \beta\overline{x}$ are mathematically correct, they will be irrelevant if they are improperly conditioned. And this will depend on assumptions about the x_i, y_i.

The purpose in what follows is to examine the effect on inferences about β of various assumptions concerning the x_i, y_i. The assumptions to be considered are:

(a) nothing is specified about $\{x_i, y_i\}$, Section 2;

(b) the distances $\left\{\sqrt{x_i^2 + y_i^2}\right\}$ of $\{x_i, y_i\}$ from the origin are regarded as fixed constants that determine the precision with which β can be estimated, but nothing is specified about the behaviour of $\{x_i, y_i\}$ individually, Section 3;

(c) $\{x_i, y_i\}$ are location-scale random variables with location parameters $\{\xi_i, \beta\xi_i\}$ respectively, Section 4, the linear functional relationship, Section 6.

The possible effects of these different assumptions will be illustrated by three data sets in Section 5.

2 The Standard Analysis; Assumption (a)

The use of (2) is widespread, e.g. Fieller (1954), Fisher (1991c, pp. 142–144), Fisher (1991a, pp. 194–195), Barnard (1994). In fact the use of (2) seems to be the standard procedure. Barnard (1994) emphasized its dependence on assumption (a), and that this is a weaker assumption than others used by Fisher, Creasy, Fieller and others. But Fisher (1954) stressed assumption (a).

The standard normality of (1) by itself leads more generally to

$$t(\beta) = \tilde{u}/s_{\tilde{u}} \sim t_{n-1}, \tag{3}$$

which is similarly a student t_{n-1} pivotal that can be used for inferences about β, where

$$\tilde{u} = \sum c_i u_i, \quad s_{\tilde{u}}^2 = \sum (u_i - c_i \tilde{u})^2/(n-1), \quad \text{and} \quad \sum c_i^2 = 1,$$

the $\{c_i\}$ being otherwise arbitrary fixed constants. Thus a further assumption is necessary to single out the special case $\{c_i\} = \{1/\sqrt{n}\}$ of (3) which is (2). Assumption (a) would seem to justify this. The problem is invariant under permutations of the u_i. Any other assumption about the x_i, y_i might violate this symmetry, as in Section 3 below.

In this formulation the only random variables are the u_i. These are not location-scale pivotals. The most important feature is perhaps that no likelihood function of β can be deduced from the density function of the u_i. Also (2) involves only $\overline{u}$, $s_{\overline{u}}$, and so ignores the information in the residuals $(u_i - \overline{u})/s_u$, which are functions of β. Barnard (1994) argued that for the $N(0,1)$ model these residuals have the spherical uniform distribution, and so contain little information about β. This seems to imply that the above method is restricted to approximate normal distributions, or more generally, distributions having this property.

To add flexibility to the analysis and allow a further assessment of assumptions, in what follows the quantity (1) will be generalized to

$$u_i = \frac{y_i - \beta x_i}{\sigma\sqrt{\lambda^2 + \beta^2}}. \tag{4}$$

The quantity λ is not a parameter to be estimated on the same logical level as β. It is a quantity that can be varied to ascertain its effect on the inferences about β. It is a model adjustment parameter that allows for a difference in precision of the x_i's relative to the y_i's. This is equivalent to replacing y_i by y_i/λ. If x_i and y_i are normal variates, λ^2 is the variance ratio. Varying λ will have no effect on (2) and the resulting analysis, since it cancels out. But it may have an effect when something is specified about the x_i's and y_i's.

3 An Approximate Conditional Location-Scale Model; Assumption (b)

The general case

To deal with assumptions (b) and (c) in the following sections a transformation to polar coordinates is convenient and suggestive,

$$x_i = r_i \cos \widehat{\tau}_i, \quad y_i = \lambda r_i \sin \widehat{\tau}_i, \quad \beta = \lambda \tan \tau,$$

so that

$$\widehat{\tau}_i = \tan^{-1}(y_i/\lambda x_i), \quad r_i = \mathrm{sgn}(x_i)\sqrt{x_i^2 + (y_i/\lambda)^2}, \quad \tau = \tan^{-1}\beta/\lambda, \tag{5}$$

where

$$-\frac{1}{2}\pi \le \tau \le \frac{1}{2}\pi, \quad -\frac{1}{2}\pi \le \widehat{\tau}_i \le \frac{1}{2}\pi, \qquad -\infty \le r_i \le \infty.$$

The quantity (4) can now be written as

$$u_i = \frac{y_i - \beta x_i}{\sigma\sqrt{\lambda^2 + \beta^2}} \equiv \frac{1}{\sigma} r_i \sin(\widehat{\tau}_i - \tau). \tag{6}$$

For values of τ such that $\sin(\widehat{\tau}_i - \tau) \approx \widehat{\tau}_i - \tau$, u_i can be approximated by

$$\begin{aligned} u_i &\approx r_i(\widehat{\tau}_i - \tau)/\sigma \\ &\approx r_i\left(\frac{\widehat{\tau}_i - \widehat{\tau}}{s} + \frac{\widehat{\tau} - \tau}{s}\right)\frac{s}{\sigma} = r_i(\widehat{u}_i + t)z. \end{aligned} \tag{7}$$

In (7) $\widehat{\tau}_i$ and τ occur together in the location model relationship $\widehat{\tau}_i - \tau$, while r_i occurs alone like a scale parameter. This suggests that the information about the value of τ is contained in the $\widehat{\tau}_i$'s. Large or small values of the $\widehat{\tau}_i$'s imply a correspondingly large or small value of τ. The r_i's by themselves imply nothing about the value of τ, but determine the amount of information about τ contained in the corresponding $\widehat{\tau}_i$. Therefore in the absence of any knowledge concerning the behaviour of the individual x_i, y_i, it seems reasonable to regard the r_i as fixed constants that, along with σ, determine the precision of $\widehat{\tau}_i$. The result is an approximate location-scale model with location parameter τ and scale parameters σ/r_i.

Moreover, (7) is purely an algebraic approximation to u_i. No distributional assumptions or approximations are involved. Thus (7) is equally applicable irrespective of the distribution. Using the observed r_i, the density of the pivotals u_i can be any function $g(u_i)$. This frees the analysis from the assumption of normality and allows changes of distributional form (adaptive robustness) to be assessed.

Equation (7) then implies the reduction of the $\{u_i\}$ to the reduced pivotals t, z, with distribution conditional on $\{\widehat{u}_i\}$,

$$f\left(t, z; \{r_i\} \mid \{\widehat{u}_i\}\right) \propto z^{n-1} \prod_{i=1}^{n} g\left[r_i(\widehat{u}_i + t)z\right], \tag{8}$$

Fisher (1934), Barnard and Sprott (1983). Note that there is considerable flexibility in the choice of $\widehat{\tau}$ and s. Provided they are used conditionally on the observed $\widehat{u}_i$, their choice will not affect inferences based on the likelihood function (likelihood-confidence intervals). Also, (8) depends on λ through (5).

This procedure is essentially a linearization of (4) with respect to $\tau = \tan^{-1}(\beta/\lambda)$. An indication of the accuracy of (7) is given by

$$\sum \left| r_i \sin(\widehat{\tau}_i - \widehat{\tau}) \right| \Big/ \sum \left| r_i(\widehat{\tau}_i - \widehat{\tau}) \right|. \tag{9}$$

A necessary condition for a good approximation is that (9) should be close to 1. A more detailed examination can be obtained by comparing individually $\{\sin(\widehat{\tau}_i - \tau)\}$ with $\{\widehat{\tau}_i - \tau\}$ for plausible values of τ.

Here, as in Section 2, the residuals $\widehat{u}_i$ have a distribution approximately independent of τ if the above linearization approximation is adequate. However, unlike the residuals in Section 2, the residuals $\widehat{u}_i$ are themselves parameter-free, and so are approximate ancillary pivotals. Inferences about τ are based on (8) with z integrated out.

Robust likelihood

Since (7) is an approximate location-scale $(\tau, \sigma/r_i)$ model defined by the location-scale pivotals $u_i = (\widehat{\tau}_i - \tau)r_i/\sigma$, an approximate pivotal likelihood function of τ can be obtained, based on the distribution of the linear pivotal t, as discussed by Chamberlin and Sprott (1989), Sprott (1990). This approximate likelihood is proportional to the density of t expressed as a function of τ, which is proportional to (8) with z integrated out. From (7), this can be written

$$L_p(\tau) \propto \int_{z=0}^{\infty} z^{n-1} \prod_{i=1}^{n} g\left[zr_i(\widehat{\tau}_i - \tau)/s\right] dz. \tag{10}$$

In terms of β, from (6) and (7) this is approximately

$$L_p(\beta) \propto \int_{z=0}^{\infty} z^{n-1} \prod_{1=1}^{n} g\left[z(y_i - \beta x_i)/s\sqrt{\lambda^2 + \beta^2}\right] dz, \tag{11}$$

obtained by reversing the approximation in (7), using $\widehat{\tau}_i - \tau \approx \sin(\widehat{\tau}_i - \tau)$.

This likelihood is a function of the model adjustment parameter λ. Also the density g is left unspecified, and so may be varied arbitrarily. These features allow the assessment of the sensitivity of the inferences to departures from an assumed model, usually the normal model discussed in the next section and illustrated in Section 5. Such likelihoods may therefore be called robust likelihoods.

The normal case

For the normal model, $u_i \sim N(0,1)$, the calculations are considerably simplified by taking $\widehat{\tau}$ and s in (7) to be the maximum likelihood estimate and its estimated standard error

$$\widehat{\tau} = \sum r_i^2 \widetilde{\tau}_i \Big/ \sum r_i^2, \quad s^2 = \sum r_i^2 (\widetilde{\tau}_i - \widehat{\tau})^2 \Big/ \Big[(n-1) \sum r_i^2\Big]. \qquad (12)$$

These give $\sum r_i^2 \widehat{u}_i^2 = (n-1) \sum r_i^2$ and $\sum r_i^2 \widehat{u}_i = 0$. Then integrating z out of the resulting density (8) gives the student t_{n-1} density of $t(\tau) = (\widehat{\tau} - \tau)/s$. This can be written in the form (3) with $c_i = r_i \,/\, \sqrt{\sum r_i^2}$.

The inferences take the classical form in τ, which can immediately be converted algebraically into corresponding inferences about β,

$$\tau = \widehat{\tau} \pm s t_{n-1}, \quad \beta = \tan(\widehat{\tau} \pm s t_{n-1}). \qquad (13)$$

These inferences depend on λ through (5).

For the normal model the approximate likelihood functions (10), (11) are

$$\begin{aligned} L_p(\tau) = L_p(\beta) &\propto \left[1 + t^2/(n-1)\right]^{-n/2} \propto \left[\sum r_i^2 (\widetilde{\tau}_i - \tau)^2\right]^{-n/2}, \\ &\dot{\propto} \left[\sum (y_i - \beta x_i)^2/(\lambda^2 + \beta^2)\right]^{-n/2}. \end{aligned} \qquad (14)$$

Thus the inferences (13) constitute a complete nested set of approximate likelihood-confidence intervals. This means that the nested set of intervals reproduces the likelihood function, and so the intervals are conditioned to be relevant to the observed data, and are also approximate confidence intervals. The corresponding intervals in β also have these properties.

In contrast the likelihood aspect of confidence intervals derived from the highly nonlinear pivotal (2) cannot be assessed under assumption (a).

The inferences (13) constitute a family of statements as λ varies from 0 to ∞, somewhat in the manner exemplified by Sprott and Farewell (1993) for the difference between two normal means. The two extreme boundaries $\lambda = 0$ and $\lambda = \infty$ yield respectively the regression of x_i on y_i with $r_i = y_i$ fixed and slope $1/\beta$, and the regression of y_i on x_i with $r_i = x_i$ fixed and slope β. The intermediate values of λ produce intermediate results, with likelihood functions lying between these two extreme regression likelihoods.

4 A Full Location-Scale Model; Assumption (c)

The general case

Under assumption (c) x_i and y_i are location-scale random variates with location parameters ξ_i and $\beta\xi_i$ and scale parameters σ and $\lambda\sigma$. Therefore at the outset there is a likelihood function $L(\beta, \sigma, \xi_i; x_i, y_i)$ proportional to the density of x_i, y_i. The difficulty is the elimination of the ξ_i's.

To this end, let $p_i = x_i - \xi_i$, $q_i = y_i - \beta\xi_i$ be the location pivotals with respect to ξ_i, with a distribution depending on the scale parameters σ and $\lambda\sigma$, and possibly also on β, but not on ξ_i. Thus their density is proportional to

$$(1/\sigma^2)g\big[(p_i/\sigma), (q_i/\lambda\sigma); \beta\big], \tag{15}$$

independent of ξ_i.

The $1-1$ transformation p_i, $q_i \leftrightarrow u_i^*$, v_i^*

$$\begin{aligned} u_i^* &= \frac{q_i - \beta p_i}{\sqrt{\lambda^2+\beta^2}} = \sigma u_i = \frac{y_i - \beta x_i}{\sqrt{\lambda^2+\beta^2}}, \\ v_i^* &= \frac{\beta q_i + \lambda^2 p_i}{\sqrt{\lambda^2+\beta^2}} = \frac{\beta y_i + \lambda^2 x_i}{\sqrt{\lambda^2+\beta^2}} - \xi_i\sqrt{\lambda^2+\beta^2}, \end{aligned} \tag{16}$$

has Jacobian 1 independently of all the parameters. The likelihood function L, which is proportional to (15) expressed as a function of the parameters, is therefore proportional to the joint density function of u_i^*, v_i^* expressed as a function of the parameters. The essential point is that this density will explicitly involve σ as a scale parameter, and possibly β, but not ξ_i. The parameter ξ_i enters only through v_i^* in (16).

Thus ξ_i can be eliminated by integrating v_i^* out of the joint density of u_i^*, v_i^*. The marginal or pivotal likelihood of β, σ is proportional to the marginal density of u_i^*

$$L_{mi}(\beta, \sigma) \propto \frac{1}{\sigma}h\left(\frac{u_i^*}{\sigma}; \beta, \lambda\right) = \frac{1}{\sigma}h\left(\frac{y_i - \beta x_i}{\sigma\sqrt{\lambda^2+\beta^2}}; \beta, \lambda\right). \tag{17}$$

This merely states that in h, σ occurs only in combination with u_i^* as u_i^*/σ, while β and λ can occur explicitly separately from u_i^*. The overall likelihood of β, σ is $\prod_i L_{mi}$.

Presumably the marginal density h can be assigned arbitrarily at the outset without considering the initial distribution (15). This yields a robust likelihood allowing an analysis similar to Section 3 with arbitrary distribution, not dependent on normality, as exemplified in Example (1b), Section 5.

The normal case

For the normal model the density function g in (15) is $(1/\sigma^2)\exp\big[-(p_i^2 + q_i^2/\lambda^2)/2\sigma^2\big]$. From (16) the density of u_i^*, v_i^* is $\exp\big[-(u_i^{*2}+v_i^{*2}/\lambda^2)/2\sigma^2\big]/\sigma^2$, so that the marginal density (17) is proportional to

$$\frac{1}{\sigma}\exp\left[-\frac{1}{2}\left(\frac{u_i^*}{\sigma}\right)^2\right] = \frac{1}{\sigma}\exp\left(-\frac{1}{2}u_i^2\right).$$

From (6) this gives the pivotal likelihood

$$L_{mi}(\tau,\sigma) = L_{mi}(\beta,\sigma) \propto \frac{1}{\sigma}\exp\left[-\frac{1}{2}\frac{r_i^2\sin^2(\widehat{\tau}_i-\tau)}{\sigma^2}\right]$$
$$\propto \frac{1}{\sigma}\exp\left[-\frac{1}{2}\frac{(y_i-\beta x_i)^2}{\sigma^2(\lambda^2+\beta^2)}\right].$$

A sample of n pairs (x_i, y_i) yields $L_m \propto (1/\sigma^n)\exp(-1/2\sum u_i^{*2}/\sigma^2)$. The maximum likelihood estimate of σ^2 based on this likelihood is proportional to $\sum u_i^{*2}$. This gives the maximized or profile pivotal likelihood of τ or β as

$$L_m(\tau) = L_m(\beta) \propto \left(\sum u_i^{*2}\right)^{-n/2}$$
$$= \left[\sum r_i^2 \sin^2(\widehat{\tau}_i - \tau)\right]^{-n/2}$$
$$= \left[\sum (y_i-\beta x_i)^2/(\lambda^2+\beta^2)\right]^{-n/2}. \tag{18}$$

This is the same form as the pivotal likelihood (14), but is logically different. In (14) the r_i's are assumed constant, leading to the approximate location scale analysis. Here they are precluded from being constant by the integration from $-\infty$ to ∞ with respect to $v_i^* \equiv \lambda\big[r_i\cos(\widehat{\tau}_i-\tau) - (\xi_i/\cos\tau)\big]$. This is a marginal model, so there is not an approximate location-scale structure. The analysis of Section 3 is not available.

The likelihood (18) is the same in form and structure as the marginal likelihood obtained by Kalbfleisch and Sprott (1970), except their exponent is $(n-1)/2$. It also was based on the model of this section. Their method is based on assumptions about volume elements, which appear to limit consideration to the normal model. The above makes no such assumption, and appears generalizable to non-normal models using (17).

A t approximation

The likelihood $L_m(\tau)$ of (18) is maximized at $\widehat{\tau}$ given by

$$\sum r_i^2 \sin 2(\widehat{\tau}_i - \widehat{\tau}) = 0. \tag{19}$$

The odd derivatives of $\log L_m$ at $\widehat{\tau}$ are zero indicating symmetry about $\widehat{\tau}$. The observed information and fourth standardized derivative, at $\widehat{\tau}$ are

$$I_\tau = \partial^2 \log L_m/\partial\widehat{\tau}^2 = n\sum r_i^2\cos 2(\widehat{\tau}_i-\widehat{\tau}) \Big/ \sum r_i^2\sin^2(\widehat{\tau}_i-\widehat{\tau}),$$
$$F_{4,\tau} = \left[\partial^4 \log L_m(\tau)/\partial\widehat{\tau}^4\right] I_\tau^{-2} = (6/n) + (4/I_\tau). \tag{20}$$

The approximating t_d likelihood with d degrees of freedom is given by

$$t_d = (\hat{\tau} - \tau)/s, \quad \text{where } d = (6/F_{4,\tau}) - 1, \quad s^2 = (d+1)/dI_\tau, \qquad (21)$$

Viveros and Sprott (1987).

5 Examples

Example 1. The Darwin data, (Fisher, 1991c, p. 30, pp. 194–195).These are the heights of fifteen matched pairs, crossed-versus self-fertilized, x_i, y_i, of maize plants in 1/8th inches:

$$\begin{array}{rclllllllll} \{x_i\} & = & 188 & 96 & 168 & 176 & 153 & 172 & 177 & 163 \\ \{y_i\} & = & 139 & 163 & 160 & 160 & 147 & 149 & 149 & 122 \\ \{x_i\} & = & 146 & 173 & 186 & 168 & 177 & 184 & 96 \\ \{y_i\} & = & 132 & 144 & 130 & 144 & 102 & 124 & 144. \end{array}$$

(a) Consider the method of Section 2, based on assumption (a). Setting (2) equal to 2.145, the 5% point of t_{14}, the resulting 95% confidence (according to Fisher, fiducial) interval is

$$.76209 \le \beta \le .999802,$$

(Fisher, 1991a, pp. 194–195). It does not appear possible to apply this procedure to a non-normal distribution.

(b) Consider the method of Section 3, based on assumption (b). With $\lambda = 1$ the quantities (12) are $\hat{\tau} = .71137$, $s = .0312$, so that the complete set of likelihood-confidence intervals (13) is

$$\tau = .71137 \pm .0312 t_{14}, \quad \beta = \tan(.71137 \pm .0312 t_{14}).$$

The quantity (9) is .992, indicative of a reasonably accurate linear approximation.

The 95% approximate likelihood-confidence interval is

$$.6444 \le \tau \le .7783, \quad .7515 \le \beta \le .9859$$

which is a 12% likelihood interval. This is reasonably close to the results of 1(a). Barnard (1994) uses the approximation $\beta = .87 \pm .055 t_{14}$, which gives (.752, .989). Curiously, this approximation is closer to Example 1(b) than to 1(a) of which it is supposed to be an approximation.

The similarity of the results of assumptions (a) and (b) is probably due to the approximate equality of the c_i's of (3), which vary between .206 and .284, and also to the distance of $\overline{x} = 140.6$, $\overline{y} = 161.5$ from the origin.

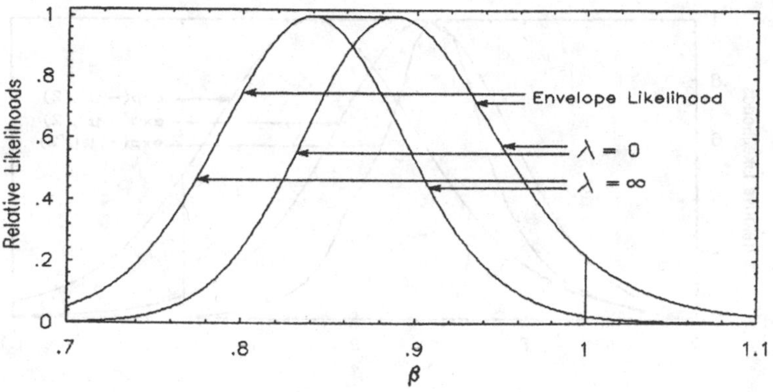

FIGURE 1. Relative likelihoods and envelope likelihood, Example 1(a).

Adaptive robustness

Figure 1 shows the likelihoods arising from $\lambda = 0$, $\lambda = \infty$ and the resulting envelope likelihood containing all the likelihoods produced as λ varies from 0 to ∞. For $\lambda = 0, \infty$ the 95% likelihood-confidence intervals are $(.781, 1.028)$ and $(.725, .954)$, respectively. Thus the inferences are in general not very sensitive to changes in λ. But the assessment of the specific value of scientific interest $\beta = 1$ is problematic, the relative likelihoods being $R(\beta = 1; \lambda = 0, 1, \infty) = 0.2201, 0.0745, 0.0242$, respectively.

As discussed in Sections 3, 3, using (8) the effect of the distributional form can also be assessed. Suppose, for example, the u_i's are independent double exponential variates. Then using the maximum likelihood estimates $\widehat{\tau}$ (the median) and $s = \sum r_i|\widehat{\tau}_i - \widehat{\tau}|/2n$, the approximate marginal distribution of $t = (\widehat{\tau} - \tau)/s$ is

$$f(t \mid \{\widehat{u}_i, r_i\}) \propto \left[\sum |r_i(\widehat{u}_i + t)|\right]^{-n}.$$

Using $\lambda = 1$,

$$\widehat{\tau} = .7086, \quad \widehat{\beta} = \tan\widehat{\tau} = .8571, \quad s = 8.9667.$$

Figure 2 shows the relative likelihood functions of β arising from assuming u has the normal, the double exponential and the $\exp(-1/2u^4)$ distribution.

Analyzing the ratios this way produces the same results as analyzing the differences, Sprott (1978), Sprott (1982). The relative likelihoods of $\beta = 1$ vary considerably from extreme implausibility 0.00417 (thick tailed distributions—Cauchy, double exponential), doubtful plausibility 0.0745 (normal distribution), to high plausibility 0.4765 (thin tailed distributions—$\exp(-1/2u^4)$, uniform).

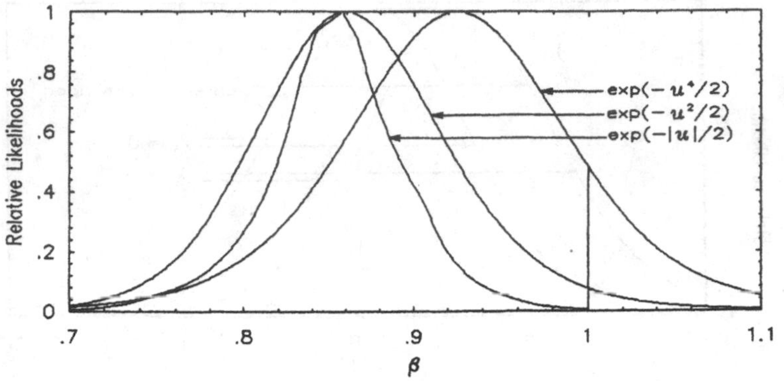

FIGURE 2. Comparison of relative likelihoods based on different distributional assumptions, Example 1(a).

(c) Consider the method of Section 4, based on assumption (c) and a normal model. From (19) the maximum likelihood estimate is $\widehat{\tau} = .71002$. From (20) and (21), $I_\tau = 1095.26$, $F_4 = .4037$, $d = 13.864$, $s = .03013$. The inferences then are $\tau = .71002 \pm .03097t_{14}$, giving the 95% likelihood-confidence interval

$$\tan .6436 = .749 \leq \beta \leq \tan .7771 = .984.$$

Thus procedures based on assumptions (a), (b), and (c), all yield essentially the same inferences about β for these data.

Example 2. The following artificial data illustrate an extreme case of the difficulty with (2) discussed in Section 1,

$$\begin{aligned} \{x_i\} &= 1.0, \quad -1.1, \quad -.9, \quad 1.2, \\ \{y_i\} &= 1.1, \quad -1.0, \quad -.91, \quad .95. \end{aligned}$$

(a) The Fieller pivotal quantity (2) is

$$(.035 - .05\beta)2/\sqrt{1.3119 - 2.7680\beta + 1.4833\beta^2},$$

which has a maximum absolute value of about .18. Thus the confidence intervals are $(-\infty, +\infty)$ for most confidence levels of interest.

(b) Using $\lambda = 1$, for these data the value of (9) is .999 indicating a reasonable linear approximation. From (5),

$$\begin{aligned} \{\widehat{\tau}_i\} &= .83298, \quad .73781, \quad .79092, \quad .66964, \\ \{r_i\} &= 1.4866, \quad -1.4866, \quad -1.2799, \quad 1.5305. \end{aligned}$$

From (12), $\widehat{\tau} = .7542$, $\widehat{\beta} = \tan\widehat{\tau} = .9394$, $s = .0363$. The resulting approximate likelihood-confidence intervals (13) are

$$\tau = \widehat{\tau} \pm st_{(3)} = .7542 \pm .0363t_{(3)}.$$

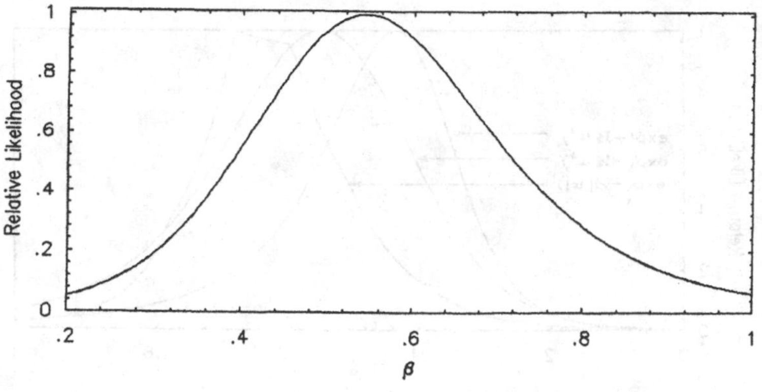

FIGURE 3. Relative likelihood —— and its t_8 approximation ······, Example 3c.

(c) From (19), (20), and (21), the relevant quantities are

$$\hat{\tau} = .7542, \quad I_\tau = 1004.13, \quad F_4 = 1.50398, \quad d = 2.989, \quad s = .0364,$$

giving the same numerical results as (b). But the reference set is different so the coverage frequency properties require further examination.

Thus with these data (b) and (c) yield similar results, but differ considerably from those of (a).

Example 3. The Cushney–Peebles data, (Fisher, 1991c, pp. 121, 142–144). These are the additional hours of sleep gained with ten patients using two supposedly soporific drugs A and B:

$$\begin{aligned} A\colon \{x_i\} &= +1.9\ +0.8\ +1.1\ +0.1\ -0.1\ +4.4\ +5.5\ +1.6\ +4.6\ +3.4 \\ B\colon \{y_i\} &= +0.7\ -1.6\ -0.2\ -1.2\ -0.1\ +3.4\ +3.7\ +0.8\ \ 0.0\ \ +2.0. \end{aligned}$$

(a) Setting (2) equal to 2.262, the 5% point of t_9, the resulting 95% interval is

$$-.4848 \le \beta \le +.6566,$$

Fisher (1991c, p. 144).

(b) These data produce values of $\{\hat{\tau}_i\}$ which are so variable that for no value of τ is the linear approximation adequate for all i, (9) being .797. The approximation is particularly bad for pairs $i = 2$, 4 and not so good for $i = 3$. Thus the method of Section 3 cannot be used on these data.

(c) From (19), (20), and (21), the relevant quantities are are

$$\hat{\tau} = .49763, \quad \sqrt{I_\tau} = 9.6611, \quad d \approx 8, \quad s = .1095.$$

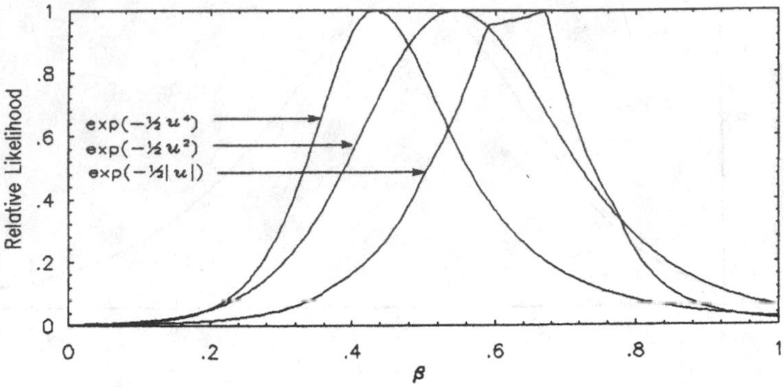

FIGURE 4. Relative likelihoods Example 3c.

The resulting estimation statements are

$$\tau = .49763 \pm .1095t_8.$$

If these can be interpreted as confidence intervals, then the .95 interval is

$$.245 \leq \tau \leq .750, \quad .250 \leq \beta \leq .932.$$

The pivotal profile likelihood function of β (18) and its t_8 approximation (21) are given in Figure 3. Inferences based on these likelihoods disagree markedly from inferences based on the Fieller pivotal, (2). In particular, negative values of β are essentially precluded by this likelihood, while $\beta = 1$ is more plausible than under assumption (a).

Figure 4 compares the pivotal profile relative likelihood functions of β derived from (17) assuming h is the normal, the double exponential, and the $\exp(-1/2u^4)$ distributions. Although the results are as variable as those of Example 1(a), the evidence against $\beta = 1$ is much less in doubt. Thus Examples 1 and 3 illustrate that adaptive robustness is a property not only of the assumptions, and of the data as emphasized by Barnard (1994), but also of the question being asked.

6 The Linear Functional Relationship

Suppose (4) is extended to $u_i = (y_i - \delta - \beta x_i)/\sigma\sqrt{\lambda^2 + \beta^2}$. The method of Section 2 cannot be applied here for inferences about β, since the pivotal (2) actually provides a test for $\delta = 0$ as a function of β. A 95% interval for β obtained from the Fieller pivotal can be interpreted as the interval containing all those values of β consistent with the hypothesis $\delta = 0$ at the 5% level of significance.

However the methods of Sections 3 and 4 can be applied. The model corresponding to Section 4 is $p_i = (x_i - \xi_i)/\sigma$, $q_i = (y_i - \delta - \beta x_i)/\lambda\sigma$. In this form it is referred to as the linear functional relationship.

Following Section 3, without specifying the distribution of (x_i, y_i), consider the pivotal quantity $u_i = (y_i - \delta - \beta x_i)/\sigma\sqrt{\lambda^2 + \beta^2}$. The results of Section 3 are modified as follows. The transformation (5) remains the same. Then

$$u_i = \frac{y_i - \delta - \beta x_i}{\sigma\sqrt{\lambda^2 + \beta^2}} = \frac{1}{\sigma}[r_i \sin(\widehat{\tau}_i - \tau) - \gamma], \quad \gamma = \frac{\delta}{\lambda}\cos\tau.$$

For values of τ such that $\sin(\widehat{\tau}_i - \tau) \approx \widehat{\tau}_i - \tau$, (7) becomes

$$\begin{aligned} u_i &\approx r_i(\widehat{\tau}_i - \tau - r_i^{-1}\gamma)/\sigma \\ &= r_i\left(\frac{\widehat{\tau}_i - \widehat{\tau} - r_i^{-1}\widehat{\gamma}}{s} + \frac{\widehat{\tau} - \tau}{s} + \frac{\widehat{\gamma} - \gamma}{r_i s}\right)\frac{s}{\sigma} = r_i\left(\widehat{u}_i + t_1 + \frac{t_2}{r_i}\right)z. \end{aligned}$$

This is an approximate regression model $\tau + r_i^{-1}\gamma, \sigma/r_i$. The resulting distribution of the reduced pivotals t_1, t_2, and z analogous to (8) is

$$f(t_1, t_2, z \mid \{\widehat{u}_i, r_i\}) \propto z^{n-1} \prod g[r_i(\widehat{u}_i + t_1 + r_i^{-1}t_2)z].$$

The relevant distribution of t_1 is obtained by integrating out z and t_2.

For the normal model, (12) is replaced by

$$\begin{aligned} \widehat{\tau} &= \sum r_i(r_i - \bar{r})\widehat{\tau}_i \Big/ \sum (r_i - \bar{r})^2, \\ \widehat{\gamma} &= \sum r_i(\widehat{\tau}_i - \widehat{\tau})/n, \\ s^2 &= \sum r_i^2(\widehat{\tau}_i - \widehat{\tau} - r_i^{-1}\widehat{\gamma})^2 \Big/ (n-2)\sum (r_i - \bar{r})^2, \end{aligned}$$

so that $\sum r_i^2\widehat{u}_i = \sum r_i\widehat{u}_i = 0$ and $\sum r_i^2\widehat{u}_i^2 = (n-2)\sum(r_i - \bar{r})^2$. The resulting density of t_1, t_2, z is the marginal distribution

$$f(t_1, t_2, z) \propto z^{n-1}\exp\left\{\frac{1}{2}z^2\left[(n-2)\sum(r_i - \bar{r})^2 + t_1^2\sum r_i^2 + nt_2^2 + 2nt_1t_2\bar{r}\right]\right\}.$$

It follows that

$$t_\tau = t_1 = \frac{\widehat{\tau} - \tau}{s}, \quad t_\gamma = \frac{\sqrt{n}t_2}{\sqrt{\sum r_i^2}} = \frac{\widehat{\gamma} - \gamma}{s\sqrt{\sum r_i^2/n}}$$

have approximate marginal t_{n-2} distributions, and $(n-2)\sum(r_i - \bar{r})^2 z^2 = \sum r_i^2(\widehat{\tau}_i - \widehat{\tau} - r_i^{-1}\widehat{\gamma})^2/\sigma^2$ has an approximate $\chi^2_{(n-2)}$ distribution.

Example 4. The following data are taken from (Fuller, 1987, p. 41), who analysed pairs of counts of two different kinds of cells using data from Cohen et al. (1977) who argued they could be assumed to be pairs of Poisson variates. Fuller used their square roots, which were assumed to be normally distributed with a constant variance $\sigma^2 = .25$, and with a linear relationship between their means as above, yielding

$$\begin{aligned} \{x_i\} &= \{18.358 \quad 11.874 \quad 13.304 \quad 10.770 \quad 9.381\} \\ \{y_i\} &= \{7.211 \quad 2.449 \quad 3.741 \quad 2.236 \quad 2.236\} \end{aligned}$$

The above results give

$$\widehat{\tau} = .5481, \quad \widehat{\gamma} = -3.6402, \quad s = .06367,$$

so that $\widehat{\beta} = .6105$, $\widehat{\delta} = -4.265$. The inferences are

$$\beta = \tan\tau = \tan(.5481 \pm .06367 t_{(3)}).$$

The approximate 95% likelihood-confidence interval is $.3455 \leq \tau \leq .7507$, or $.3599 \leq \beta \leq .9330$.

The estimated standard error of $\widehat{\gamma}$ is $s(\sum r_i^2/5)^{1/2} = .8735$. The inferences about γ are $\gamma = -3.640 \pm .8733 t_{(3)}$. The hypothesis $\gamma = \delta = 0$ gives $t_{(3)} = -3.640/.8733 = -4.168$ with a two-tailed P-value of .025.

The value of (9) is .985 indicating a reasonable linear approximation.

The estimate $s^2 \sum (r_i - \overline{r})^2 = .2496$ does not contradict the assumption $\sigma^2 = .25$. In fact, since the estimate is numerically so close to $\sigma^2 = .25$, assuming σ^2 is known to be .25 does not change the above analysis numerically, except that the approximate $t_{(3)}$ distribution of the t_τ and t_γ pivotals is replaced by the $N(0,1)$ distribution, resulting in much stronger inferences.

Schneeweiss et al. (1987), and Chamberlin and Sprott (1987), also discussed this example assuming that the $\{x_i, y_i\}$ are independently and normally distributed, using a similar polar transformation and conditioning argument. It is difficult to see how to generalize their approach to an arbitrary density $g(u_i)$.

7 Discussion

The problem of "paired ratios" is in sharp contrast to the corresponding problem of paired differences. The latter is based on the pivotal $u_i = (d_i - \delta)/\sigma$, $d_i = x_i - y_i$, which is a location-scale pivotal. A standard conditional analysis is available, Fisher (1934). Also here the standard normality of u_i implies the normality of the difference $x_i - y_i = d_i$, from which a likelihood function of δ can be deduced. Assumptions (a), (b), and (c), and possibly many others, make no difference.

The conditions under which (2) is the appropriate procedure are rather ambiguous in the literature. Fieller (1954) obtained (2) assuming bivariate normality of the x_i, y_i with means ξ, $\beta\xi$ and unknown covariance matrix. The same result can also be obtained under assumption (c) with the additional assumption that the ξ_i are independent $N(\xi, \sigma_\xi^2)$ variates. These are more stringent assumptions than (a). While the standard normality of u_i follows from the assumed normality of x_i, y_i, the converse is not true. In analysing the Cushney–Peebles data, (Fisher, 1991c, pp. 142–144), Example 3(a) above, Fisher (1954) stated "... no attempt is made to estimate any property of a hypothetical population from which the patients supplying the data might be supposed to have been drawn. Nobody has suggested that they were a random sample even of hospital patients.... No fiducial distribution is mentioned." However, in applying the same analysis to the Darwin data, Example 1(a), (Fisher, 1991a, p. 195) stated "... the data contradict, at the 5 per cent level of significance, any statement of the form 'The true average for self-fertilization is the fraction α [here β] of the true average for cross-fertilization', whenever α lies outside the limits 76.209 per cent and 99.980 per cent. The probability that α lies between these limits is 95 per cent." On the surface these statements seem contradictory.

In Section 3 the model itself is conditional. The distribution (8) with r_i fixed under assumption (b) is taken as primary. It is not the derived conditional distribution $f(t, z \mid \{r_i\}) = f(t, z, \{r_i\})/f(\{r_i\})$. Even if these two latter distributions exist, they cannot be deduced under assumption (b). In fact, it is the assertion that nothing is assumed about the x_i, y_i individually that justifies the model of Section 3. Any other assumption which allows the derivation of the marginal distribution of r_i would probably contradict the assertion that r_i contains no information about the value of τ, on which the analysis of Section 3 is based. It is this that gives the distributional flexibility in Section 3, where the procedure is not tied to the normal distribution.

The main difference between assumptions (a) and (b), (c) is the existence of likelihood functions in the latter but not the former. This allows the assessment of the likelihood properties of the confidence intervals. This is not possible under assumption (a). Thus, for example, under (a) there is no reason to obtain a 95% confidence interval by setting (2) equal to ± 1.96. The resulting interval has no property that would suggest it is any better than a multitude of other 95% confidence intervals.

(Fisher, 1991b, p. 138) expressed the view "It is a noteworthy peculiarity of inductive inference that comparatively slight differences in the mathematical specification of a problem may have logically important effects on the inferences possible. In complicated cases, such effects may be very puzzling...". It seems curious, even in the light of this, that such apparently innocuous changes in the assumptions can result in such large changes in the results as seen in Examples 3(a) and 3(c).

Acknowledgments: I should like to thank R.J. Cook and V.T. Farewell for reading previous drafts of this paper and for helpful suggestions.

8 References

Barnard, G.A. (1994). Pivotal inference illustrated on the darwin maize data. In A. Smith and P. Freeman (Eds.), *Aspects of Uncertainty.* New York: Wiley.

Barnard, G.A. and D.A. Sprott (1983). The generalised problem of the Nile: Robust confidence sets for parametric functions. *Ann. Statist. 11*, 104–113.

Chamberlin, S.R. and D.A. Sprott (1987). Some logical aspects of the linear functional relationship. *Statist. Papers 28*, 291–299.

Chamberlin, S.R. and D.A. Sprott (1989). Linear systems of pivotals and associated pivotal likelihoods with applications. *Biometrika 76*, 685–691.

Cohen, J.E., P. D'Eustachio, and G.M. Edelman (1977). The specific antigen-binding cell populations of individual fetal mouse spleens: repertoire composition, size and genetic control. *J. Exper. Med. 146*, 394–411.

Creasy, M.A. (1954). Limits for the ratio of means (with discussion). *J. Roy. Statist. Soc. Ser. B 16*, 186–194.

Fieller, E.C. (1940). The biological standardization of insulin. *J. Roy. Statist. Soc. Suppl. 7.*

Fieller, E.C. (1954). Some problems in interval estimation (with discussion). *J. Roy. Statist. Soc. Ser. B 16*, 175–185.

Fisher, R.A. (1934). Two new properties of mathematical likelihood. *Proc. Roy. Statist. Soc. Ser. A 144*, 285–307.

Fisher, R.A. (1954). Contribution to the symposium on interval estimation. *J. Roy. Statist. Soc. Ser. B 16*, 212–213.

Fisher, R.A. (1991c). *Statistical Methods for Research Workers* (14th ed.). Statistical Methods, Experimental Design, and Scientific Inference. Oxford: Oxford University Press.

Fisher, R.A. (1991a). *The Design of Experiments* (8th ed.). Statistical Methods, Experimental Design, and Scientific Inference. Oxford: Oxford University Press.

Fisher, R.A. (1991b). *Statistical Methods and Scientific Inference* (3rd ed.). Statistical Methods, Experimental Design, and Scientific Inference. Oxford: Oxford University Press.

Fuller, W.A. (1987). *Measurement Error Models*. New York: John Wiley and Sons.

Kalbfleisch, J.D. and D.A. Sprott (1970). Application of likelihood methods to models involving large numbers of parameters (with discussion). *J. Roy. Statist. Soc. Ser. B 32*, 175–208.

Schneeweiss, H., Sprott, D.A., and R. Viveros (1987). An approximate conditional analysis of the linear functional relationship. *Statist. Papers 28*, 183–202.

Sprott, D.A. (1978). Robustness and nonparametric procedures are not the only or safe alternatives to normality. *Canad. J. Psychology 32*, 180–185.

Sprott, D.A. (1982). Robustness and maximum likelihood. *Communications in Statist. Theory and Method 11*, 2513–2529.

Sprott, D.A. (1990). Inferential estimation, likelihood, and linear pivotals (with discussion). *Canad. J. Statist. 18*, 1–15.

Sprott, D.A. and V.T. Farewell (1993). The difference between two normal means. *Amer. Statistician 47*, 126–128.

Viveros, R. and D.A. Sprott (1987). Allowance for skewness in maximum likelihood estimation with application to the location-scale model. *Canad. J. Statist. 15*, 349–361.

11

Meta-Analysis: Conceptual Issues of Addressing Apparent Failure of Individual Study Replication or "Inexplicable" Heterogeneity

K. O'Rourke

1 Introduction

This paper is about issues of applying statistics to a particular area: meta-analysis (MA) of randomized clinical trials (RCTs) and possibly also observational clinical trials. Applying statistics is "messy". As no theory or model is ever correct, when applying statistics we can only attempt to find the "least wrong" model or approach. I will identify some current approaches to MA as *not* being "least wrong". This does not mean that they could not be "least wrong" for other applications. Discussing the applying of statistics—unfortunately for me at least—requires "wordy" explanations and I hope the reader will bear with me.

MA *should* be an attempt to thoroughly investigate the replication of results in essentially similar studies—the goal being both the summarization of what has been learned and the summarization of the "real" uncertainty that remains. Emphatically—MA is *not* just getting a pooled estimate of a treatment effect plus or minus some hypothetical standard error. Now, the theme of this paper concerns the question—is some sort of random effects modeling approach such as Empirical Bayes (EB) "least wrong" for MA? My claim is no "plus or minus some explanation". The more general question of when is it advantageous to treat inexplicable variation as being random underlies this theme and there is some discussion of this question in Cox (1958).

Treating "inexplicable variation as being random" is somewhat vague so some immediate discussion of why and how seems in order. As for why, my primary concern is to get confidence intervals that have correct coverage for "unconfounded" treatment effects. Without delving too much into material on causal inference, I will simply act as if confidence intervals from ideally conducted RCTs will have the correct confidence interval coverage (at least individually, by trial) where as those from flawed RCTs or observational studies will not (Rubin, 1974, 1978, 1990, 1991). A secondary concern is that the confidence interval coverage be approximately symmetric—missing above and below the true treatment effect size roughly (very roughly) equally often. Currently I have no concern about confidence intervals for a *particular* study's treatment effect estimate—often referred to as "borrowing strength"—as I have yet to find this of important interest in a MA, per se. (The reasons for this will hopefully become clear later.)

As for how the "random" inexplicable variation needs to be taken into account in order to get correct coverage, there are some apparently different approaches:

1. A compound distribution is postulated where the effect size is thought of as being random. The effect size is first sampled from a "level two distribution" and then given this effect size the observations come from a "level one distribution". Likelihood or other analyses are then based on this compound distribution. This will be referred to as "compound likelihood" (CL). (One of the reviewers brought a paper by Lee and Nelder (1996) to my attention. Initially in this paper Lee and Nelder used the term *h*-likelihood for what I have referred to as the "compound likelihood"—but in their reply to discussants they argued that it is *the* likelihood. The essential novelty being that the random variable from the level two distribution is fixed but unknown whereas with standard likelihood all the random variables are known.)

2. With the same mathematical specification as in 1, the level two distribution is thought of as being a "prior" distribution for treatment effect size and the parameters of this prior are empirically estimated. This empirically estimated prior is then updated by the likelihood from the level one distribution or, perhaps more intuitively put for MA, the level one likelihoods are combined using this estimated prior as in Efron (1996). This will be referred to as empirical Bayes (EB).

3. Further to 1 or 2, yet another distribution—a third level distribution—is added being considered a prior distribution on the level two distribution. Under 1, this prior is updated by the compound likelihood. Under 2, this third level distribution —often called a hyper-prior—"mixes" the prior resulting in a compound prior which is then updating by a simple "noncompound" likelihood. This will be referred to as hierarchical Bayes (HB).

Note that in this third variant the prior distribution (third level) could place enough probability on a "constant effect size being sampled" so that the second level distribution has no effect and we are back to a standard Bayesian "fixed effect" analysis. More refined and rigorous definitions of other and similar variations can be found in Morris (1983) and also Efron (1996).

When being critical of how various approaches or models may or may not apply to empirical research it is perhaps useful to be clear about what is being criticized. It is not the implications of a model as these are necessarily true. It is rather the appropriateness of the assumed model from which the implications follow. For example, if one assumes a particular distribution for the level two distribution, say $N(\mu, \sigma^2)$, that implies a certain weighting for combining study estimates which although correct under the assumed model may be at odds with background information. The weights are not wrong, the assumed model is. The implications of assuming a particular level two distribution will be more fully discussed later.

2 MA and RCT Background

Getting back to MA, there is an opportunity to investigate replication quantitatively. We have more than just an "internal assessment" of uncertainty but can empirically get at the "real" uncertainty (Mosteller and Tukey, 1977). With a single trial, even if multi-centered, there is but one protocol and its effect can not be empirically assessed. Now, there are many problems with how RCTs are actually planned, carried out and reported on. It is extremely hard to do RCTs properly and only a minority are completed without some potentially serious flaw (Schulz et al., 1995, 1994; Schulz, 1995). This is very different from many other areas such as for instance lab tests where only a small minority of tests are likely to be flawed and it is perhaps "least wrong" here to approach the problem as one of "robustness" or identifying outliers (Guttman et al., 1993).

The problems with RCTs do not arise because someone attempts to do a MA—i.e. quantitatively investigate replication of results amongst published trials—but the problems do become much more noticeable. Early criticisms of MA as being "unable" to correct for problems such as publication bias were misdirected against those doing MA rather than at those doing the original individual trials. For arguments that MA may be most needed when trials are largely flawed see O'Rourke and Detsky (1989). Also, drawing on comment made by Nelder in Chatfield (1995) that science actually requires "significant sameness" it would seem to follow that it is just as important to display lack of sameness as sameness itself.

A list of a few common RCT flaws (for empirical evidence see Schulz et al. (1995)):

1. lack of randomization concealment;

2. loss of blinding;

3. confounded "as treated" analysis (analyzing only the patients who complied) presented as if it were an "intent to treat" analysis (analyzing all the patients who were randomized);

4. informative drop out and noncompliance of patients;

5. faulty data entry.

But even apart from flaws, there has been a neglect of MA in the statistical literature, perhaps so much so that Chatfield coined the phrase "Statistical myth of a single study" (Chatfield, 1995). On the other hand, Pena (1997) has recently argued for the value of analyzing classical estimation problems from the point of view of meta-analysis (or how information is combined) on the basis of providing useful insights about the properties of estimates and their robustness. See also Cox (1982).

Now there are a number of questions a MA may attempt to answer:

1. Should another study be done?

2. If yes to 1, how should it be done?

3. If no to 1, should the treatment be adopted—does the estimated benefit outweigh the estimated costs given some as of now conventionally accepted benefit?

Now required properties of statistical methods for 2 and 3 may be different from 1 but 1 is arguably the most important question where tradition seems to call for correct confidence interval coverage of the unconfounded treatment effect. For 3 it may be minimum mean squared error of "benefit/cost" and for 2 it may be "minimax" in the less technical sense of "conservatively" estimating the required sample size for a definitive RCT.

Before getting to the statistics of MA let me characterize a "small area" sample survey as a MA. First there would usually be less than 12 areas sampled, although these would a priori be expected to be very similar if not identical for practical purposes. Now Professor Jon Rao would take a careful random sample of say area 1 and achieve an almost complete response. The other 11 areas each would be surveyed by a different individual some using random samples but with considerable nonresponse and others using convenience samples again with considerable nonresponse. Now how should we weight Professor Rao's estimates with the other 11 estimates to get a confidence interval for an unconfounded estimate of the average of the 12 areas? Should we adjust the specific estimate of area 1 by borrowing "weakness" from the other 11 "small area" samples?

3 History Overview

Fisher (1937) considered the case of analyzing the results from 10 "representative" farms and asked the question "Is the interaction between treatment and place the appropriate error term for testing treatment?". In answering yes to this question was he not proposing an EB or CL approach? (In his example there were equal n's per experiment and the within experimental error was assumed equal amongst the 10 experiments). He commented that the only value of the observed "within study errors" was to provide a check that the study was properly conducted (a quality threshold, below which a study was to be weighted by "0"?). He also commented that the effective samples were small (10 observations to estimate treatment by place interaction or the variance of the treatment effect) and that this would require very accurate methods indeed.

He also discussed what he called "categories of variation", one being indefinite where there is no means of reproducing the conditions at will and the other being definite where the conditions could be controlled or at least measured. It is with reference to these categories that considerations regarding study specific versus population averaged inferences can perhaps be made more explicit. Biological indefinite variation sets the requirement for population averaged inference where as definite variation or methodological indefinite variation do not. One may wonder which of these two categories Fisher would place the aforementioned "small area" sample survey or more generally MA. It may be that in agricultural experiments it is more likely that variation in observed treatment effect is due to actual variation in treatment effect (due to climate and soil variation) and less likely to be due to variation in the quality of experiments. In RCTs I believe it is the opposite, but many writers on the debate regarding the appropriateness of CL or EB in meta-analysis argue on the basis of the observed treatment variation being likely due to actual treatment variation. They then argue about whether one should be making inferences about future populations versus the actual population in which the experiments were done (Peto, 1987; Meier, 1987; Bailey, 1987; Laird and Mosteller, 1990; Normand, 1995).

Cochran (1937) proposed a "most general hypothesis" (for MA) that (the treatment effect size) μ_i is drawn from $N(\mu, \sigma^2)$—a level two distribution—and then (the observed treatment effects) x_i are drawn from $N(\mu_i, \sigma_i^2)$—a level one distribution. From this compound distribution he derived the likelihood function and equations of estimation for μ, σ^2 and σ_i^2 (what is being called CL or EB in this paper). Now the equations of estimation for μ of the level two distribution $N(\mu, \sigma^2)$ indicated that a kind of semi-weighted mean of x_is would be appropriate—the required weights being relative to $(\sigma_i^2 + \sigma^2)^{-1}$. Here the potential arbitrariness of CL or EB (or even HB) becomes readily apparent. The study weights may be quite arbitrary if $N(\mu, \sigma^2)$ is not close to the true level 2 distribution or if the estimated weights $(\widehat{\sigma_i^2} + \widehat{\sigma^2})^{-1}$ are poorly estimated. An instance of the first would be when there is more than one σ^2 i.e. a mix of "good" and "poor" studies and an instance of the second would when there is 12 or less studies/observations available to estimate σ^2 - the usual case in MA. For an insightful parallel between the arbitrary weighting of likelihoods based on quality scores (a measure of how well studies were thought to be conducted) and the resulting "weighting" of the likelihoods from assuming "poorly" motivated and almost impossible to verify level 2 distributions, see Goodman (1989).

4 Current Likelihood Based Methods for MA

A "menu" of likelihood based methods for MA can be used to describe most if not all of the current statistical methods used in MA:

1. Select a likelihood for treatment effect—for example profile, conditional, or partial

2. Choose a statistic based on 1—for example score, Wald, likelihood ratio, or some approximation

3. Decide on a "combined" versus "component analysis"—(Cox, 1982) and below.

4. Consider including a likelihood term that models inexplicable study by treatment variation.

5. Instead of 2, consider multiplying by a prior either the component or combined likelihoods.

6. Construct confidence intervals from 2 or 5.

7. Use convenient approximations.

A combined versus component analysis can easily be demonstrated using some convenient choices from the above menu. For instance using conditional loglikelihoods from each study the combined analysis would simply add these up, calculate say a score statistic and then get a confidence interval from the approximate distribution of this score statistic. A component analysis would calculate the score statistic separately from each study's conditional loglikelihood and convolve (usually sum) these individual score statistics into one statistic and then get a confidence interval from the distribution of the convolution. These will be exactly the same if a particular weighted sum is used (Greenland and Salvan, 1990) in which case it is simply two different ways of getting the same estimate/confidence interval. Otherwise they would be different but usually just slightly different and therefore from a practical perspective it is usually just a matter of taste.

An example though of a (slightly?) more "drastic" component analysis may be enlightening. Rather than individual score statistics take the p-values of the score statistics (under the null hypothesis) and either multiply them or sum them. For the trivial case of two studies their product—$P = p_1 \times p_2$ would have the probability distribution function $P - P \times \log P$

where as their sum—$P = p_1 + p_2$ would have the triangular probability distribution function (Fraser, 1976, p. 106). Adding up loglikelihoods as in the combined approach is justified simply by the independence of outcomes from independent studies whereas it is not obvious how to justify various different ways of (arbitrarily?) convolving components (Cox and Hinkley, 1974, p. 86).

As for a likelihood term to somehow model inexplicable study by treatment variation there is even more latitude. Any level two distribution could in principle be used to model the inexplicable study by treatment variation as random. Furthermore, as the "effect" of including level two distribution can conceptually be viewed as a "weighting" of the level 1 distribution likelihoods—we could consider other weighting schemes such as for instance "relevance weights" discussed in this volume by Hu and Zidek. Additionally, items 1 and 4 from the "likelihood based methods menu" have to be considered together, since the inexplicable study by treatment variation could depend on a nuisance parameter and choices in 1 such as the use of conditional likelihoods could be inappropriate for some choices in 4.

On the other hand, rather than struggling with the above mentioned menu, we could just do a sign test—i.e. ask if the percent of studies where the treatment group was observed to do better will simply allow us to reject a null hypothesis of 50%. But given the discussion so far we might also want to consider that a null hypothesis of 60 or 70% might be more appropriate. For if the treatment had no effect but the studies were flawed to the degree apparent then what percent of time should we expect the treatment group to do better than the control group?

5 Examples of EB, CL and HB Approaches

There are some more or less well known "selections" from the above "menu" in the literature that will illustrate the variations that need to be considered. Rubin (1992) has argued for response surface modeling where the treatment effect estimate and variance are both functions of both how flawed the studies were as well as how studies systematically differed from each other. Because of concerns with small effective sample sizes O'Rourke et al. (1991) and Detsky et al. (1992) suggested a graphical display to facilitate the identification of a quality threshold. If such an indication appeared clear, all studies below the threshold would be discarded. Largely it was an effort to prevent highly flawed studies from having an impact. In some sense EB, CL or HB can be viewed as evasions or less ambitious substitutes of these approaches—the modeling of inexplicable heterogeneity is abandoned and random effects are used to make allowances for this "modeling failure". It may be hard to put this better than (Cox and Snell, 1981, p. 46) where they referred to random effects modeling (EB, CL and HB) as being "conceptually the least satisfactory, because it provides merely a description of the fact that... (an estimate of interest) varies in an unpredictable way".

Now there are many variations of EB, CL and HB. DerSimonian and Laird (1986) provide an approximate noniterative estimate of σ^2 under an EB framework which is then used to weight study level estimates—a component analysis where individual study estimates are then summed. Morris and Normand (1992) provide a formal framework for both EB and HB and draw attention to the possible interest in a particular study's estimate and the possibility of "borrowing strength" from other studies. Efron (1996) conceptualizes the EB approach in MA as providing a way of combining likelihoods—i.e. a combined analysis. In this, he allows for more than just a simple symmetric level two distribution as well as for the possibility of a mixture of two level two distributions (by allowing for an option to opt out). DuMouchel et al. (1997) provide a framework for HB including software. Smith et al. (1996) implement HB with MCMC methods and software. Emerson et al. (1996) provide robust estimators of μ and σ^2 in a EB framework.

It is implied in all the above mentioned approaches that it is advantageous to treat inexplicable treatment effect variation as being random. Whether this is true or not, it is very unclear how the form of the level 2 distribution is to be chosen and then given some choice how the parameters should be estimated given very small effective sample sizes. Most approaches assume $N(\mu, \sigma^2)$ for the level two distribution or at least some fairly well behaved symmetric distribution with a single dispersion parameter σ^2. Now a single dispersion parameter is certainly not appropriate given the kinds of flaws that are experienced in RCTs. Unfortunately this ends up being very critical because it is likely that the least flawed studies will also be the larger studies and most EB or CL approaches with a single σ^2 will decreases the relative weights of larger less flawed studies.

As for the estimation of σ^2 since there is likely to be usually only 12 or fewer RCTs estimation will be poor and highly dependent on the assumed form of the level two distribution. And since there is likely to be only one or two out of the usual 12 or fewer RCTs without large flaws "robust methods" seem totally misdirected. If in these cases robust methods are used to estimate σ^2, it will be underestimated and overly narrow confidence intervals will be put about a confounded treatment effect estimate. If robust methods are also used to estimate μ, the treatment effect estimate will be even more confounded as the influence of any one study, in particular the one best study will be—if important—lessened. For an alternative view see Emerson et al. (1996).

Interestingly there is a quasi-likelihood approach that "purposefully" keeps the study weights fixed—that is takes a "nonweighted" sum of the individual study loglikellihoods—but then "down weights" the combined—loglikellihood (or at least in practice multiplies the standard errors of or divides observed statistics based on the combined loglikellihood by a factor greater than one) to reflect heterogeneity as a random effect (Firth, 1990). Because it avoids the two above-mentioned problems (incorrectly changing or limiting individual weights) I would like to call this a "least harmful" EB or CL for MA. Further discussion of the issues involved in estimating σ^2, often referred to as over-dispersion or anomalous dispersion, can be found in Firth (1990); Desmond and Godambe (1999) and Cox and Snell (1989).

Many applications in MA fit quite "naturally" into the framework of quasi-likelihood generalized linear models which in our "likelihood based methods menu" comes from picking a combined analysis and using profile likelihood. The Wald statistic for treatment effect is calculated and to allow for inexplicable heterogeneity its standard error estimate is multiplied by the square root of Chi-square for heterogeneity. The likelihood ratio statistic for treatment effect may also be calculated and allowance for inexplicable heterogeneity made by dividing it by the Chi-square for heterogeneity by analogy to normal-theory analysis of variance where the scale parameter needs to be estimated. In both cases this multiplication or division is only done if the Chi-square for heterogeneity is greater than one. To allow for uncertainty in estimating σ^2 use of t-and F-distributions to form tests and confidence intervals again by analogy to normal-theory analysis of variance has been recommended (Firth, 1990, p. 68). (For a recent detailed and critical review of these of quasi-likelihood methods see Tjur (1998).) The distribution theory is approximate at best and Monte Carlo simulation may easily show the tests and confidence intervals are far from correct in particular cases. Bootstrap techniques may also be used to form nonparametric tests and confidence intervals but except in simple straight forward situations the results may be quite misleading (McCullagh, 2000).

Another analogy may be helpful. This comes from viewing what Fisher did (Fisher, 1937) as dividing the loglikelihood from a normal-theory analysis of variance by the normalized deviance—this replaces the "known or essentially estimated without error" σ^2 in the loglikelihood with an estimate of the study treatment effect variance or interaction. The loglikelihood is

$$\text{loglik}(\mu, \mathbf{y}) = -\sum_{j=1}^{m} (y_j - \mu)^2/2\sigma^2,$$

y_j being the mean of study j and σ^2 being taken as known or essentially estimated without error and the deviance is

$$\text{dev}(\mathbf{y}, \mu) = \sum_{j=1}^{m} (y_j - \widehat{\mu})^2/\sigma^2,$$

$\widehat{\mu}$ being the pooled estimate and the loglikelihood divided by the normalized deviance is

$$\frac{\text{loglik}(\mu, \mathbf{y})}{\text{dev}(\mathbf{y}, \mu)/(m-1)} = \frac{-\sum_{j=1}^{m}(y_j - \mu)^2}{2\sum_{j=1}^{m}(y_j - \widehat{\mu})^2/(m-1)}.$$

If we now by analogy do this with profile loglikelihoods from generalized linear models we get the similar adjustments to the Wald, likelihood ratio and score. I feel it is unlikely that this or any other analogy will lead to a more rigorous justification for these adjustments as they are motivated not to correct for model failure but simply to make some *ad hoc* allowance for it.

Three published MA examples from the above mentioned papers (specifically Efron (1996); Smith et al. (1996) and Berkey et al. (1995)) were re-analyzed as quasi-likelihood generalized linear models using S-Plus (1997) and the first two are presented in the appendix. Some points that arose from this re-analysis exercise:

1. The simple quasi-likelihood generalized linear modeling method produced almost exactly the same confidence interval as Efron's EB combined conditional likelihood approach.

2. The single level two distribution used in Smith et al. (1996) down-weighted the likely better RCTs (as they were larger) likely resulting in a poorer confidence interval for the unconfounded treatment effect.

3. The studies in Emerson et al. (1996) were a subset of the studies used in an earlier paper on "nonrobust" EB regression methods (Berkey et al., 1995). Between the publication of the two papers the authors became aware that not all of the studies were RCTs as they had believed and stated in the first paper.

6 Future Directions

At this point it may seem tempting to try to get the level 2 distribution "correct" as well as possibly use informative or noninformative level 3 distributions. Perhaps the first correction to the level 2 distribution would be to allow more than one σ^2 such as a mixture $p \times N(\mu, \sigma_1^2) + (1-p) \times N(\mu, \sigma_2^2)$ or even more "adventurous" $N\left[\mu, f(\sigma^2)\right]$ where $f(\sigma^2)$ is some function of how flawed an RCT was. Rao (1988) stated that "the problem in such a case [MAs with inexplicable treatment effect variation] is the estimation of the prior distribution of θ $\left[\mu \text{ and } p,\, b,\, \sigma^2,\, f(\sigma^2)\right]$, which provides all the information". The main point is that there are two sample sizes in a meta-analysis—one for inexplicable heterogeneity or between study effects and one for the common treatment effect given no heterogeneity (or given appropriate allowance for it). The first would be the number of studies (usually very small) and the second would be the total number of subjects (usually very large).

Perhaps it would be better to stop thinking about the flaws in the RCTs as something to be treated as random but as something that confounds the treatment effect estimates by various amounts and directions. Such a conceptualization is clearly and succinctly provided in Cox (1982).

From this perspective we could think of there being unknown shifts of the individual study likelihoods, say b_i and we would need to be concerned with the cumulative bias of the combined likelihoods say b_c. This b_c would be a function of the b_is and the influence or "weights" of the individual likelihoods on the combined likelihood. If we knew the b_is we would simply re-shift the likelihoods so that $b_c = 0$ (or we could figure out the b_c and re-shift the combined likelihood). If we knew the $|b_i|$s we could figure out the largest possible $-b_c$ and $+b_c$ and shift the combined likelihood both ways and get Sprott's "envelope" around both likelihoods and extract a confidence interval from that. Being a bit smarter we could vary the contributions of the likelihoods so the confidence interval from the envelope likelihood was narrowest (Cox, 1982). The important issue here is that with knowledge of b_i or $|b_i|$ we can narrow the confidence interval based solely on a unconfounded study by combining with it confounded studies and by making an appropriate adjustment not lose the correct confidence interval coverage. Now, can we do this with empirical estimates of b_is with or without an informative prior for the b_is?

Some simple cases to consider are:

1. one apparently unflawed RCT and one apparently flawed RCT;
2. one apparently unflawed RCT and one observational study;
3. a 2×2 cross-over trial with insufficient washout or "for sure" carry-over.

In case 1 there is some uncertainty as to which study is confounded (first, second, both or neither) but in case 2 it is almost a given that the second is confounded (or at least more confounded) and in case 3 if we think of an extremely well conducted cross-over study with the only flaw being an insufficient wash-out period, the comparison of treatment groups in period one will be unconfounded whereas the comparison of treatment groups in period two will be confounded because of the carry-over effect.

The case of the 2×2 trial may be very informative, for Grizzle (1965) suggested that when there was a concern about carry-over effects (in our terminology pooling unconfounded and confounded estimates of treatment effect) that one first test for carry over and if the test is significant use only the first period (known to be unconfounded) estimate. Otherwise just use the usual between subject treatment estimates—which pools or combines the unconfounded and possibly confounded treatment estimates. See Ahmed (2000) in this volume for discussion of pre-test estimators and their relation to Stein's estimate. Many authors drew exception to Grizzle's advice, drawing attention to the potential loss of confidence interval coverage. Later study concluded that the 2×2 cross-over trial should not be used unless the possibility of carry-over can a priori be completely ruled out (Jones and Lewis, 1995, p. 1032).

Cox (1982, p. 47) suggested something apparently very similar for MA: that suggests we start with (but always keep separate) an unconfounded estimate by pooling only those studies known to be unconfounded (in our context RCTs without flaws) and then get a second estimate by pooling into the unconfounded estimate those studies that may be confounded (in our context flawed RCTs and or observational studies) as long as these appear consistent with our unconfounded estimate. It is somewhat different in that a formal test was not suggested as in Grizzle (1965) and in the insistence in Cox (1982) that both unconfounded and the "pooled unconfounded with possibly confounded" estimates be presented separately in the final statement of conclusions.

The following extreme but simple example identifies the problem that arises when pooling apparently similar confounded estimates with unconfounded estimates. Let $Y_{\textbf{unconfounded}} \sim N(\mu, 1)$ and $Y_{\textbf{confounded}} \sim N(\mu + 2, .01)$. The unconfounded estimate is most likely to equal the confounded estimate when it is close to $\mu + 2$ (i.e. $Y_{\textbf{unconfounded}}$ 2 standard deviations away from μ rather than $Y_{\textbf{confounded}}$ 20 standard deviations away from $\mu + 2$). This is definitely not when we want to shrink the confidence interval for μ initially based just on $Y_{\textbf{unconfounded}}$. In fact, without knowing by how much and in which direction to shift the interval the only possible "improvement" would come from widening the confidence interval for μ based just on $Y_{\textbf{unconfounded}}$!

That the confidence interval coverage depends importantly on the observed difference between the unconfounded and confounded estimates is an instance of an inference problem referred to as relevant or recognizable subsets. As Casella (1992) put it "There should not be a subset of the sample space (a *recognizable* subset) on which the inference from a procedure can be substantially altered. If such subsets exist, then inference from the procedure is suspect.... but if this recognizable set is not a meaningful reference set, then the criticism may be vacuous." The observed difference between the unconfounded and confounded estimates is such a set and arguably a very meaningful one. (I have recently become aware of a paper by Sen (1996) that addressed the same issue in the 2×2 cross-over study, clearly identifying practical issues of analysis.)

7 Initial Conclusions

Given this recognizable subset "glitch" it would seem that rather than ranking RCTs by the apparent lack of flaws and sequentially pooling them while they appear similar (for instance see (O'Rourke et al., 1991) and Detsky et al. (1992)) we should just take the least flawed study and throw the rest away. At least that is for question 1—do we need to do another study? As for question 2 (say for estimating the sample size required given another study needs to be done) "minimax" may be better and for question 3 (estimating the benefit in order to do a cost/benefit study of the treatment given some as of now conventionally accepted benefit) minimum mean squared error is arguably more appropriate. But for question 1, the most important question in MA, the "throwing away" of "apparently more flawed" RCTs is likely to be unpopular. To this there are two immediate replies—firstly, RCTs are in themselves extreme hardships undertaken at "almost any cost" to avoid confounding and secondly, pleading hardship in general is no defense in science.

8 A Bayesian Afterthought

As an afterthought I would like to sketch out a fully Bayes approach with an informative level 3 distribution. To avoid complications about whether yesterday's posterior need be today's prior I will consider the following, perhaps not that unlikely situation. A genetic discovery is made that disease X is caused by the lack of a gene that allows the body to produce protein Xalpha123. Now the genetic researcher recalls that a medical supplement Beta275 is actually a close analogue of protein Xalpha123 although they are quite unaware of what uses Beta275 may have been put to in the past. But with the assistance of a local Bayesian statistician they are able to explicate a prior distribution of the supplement's effect on 5 year mortality in patients just recently diagnosed with disease X so they design an RCT and apply for funding. The funding agency remembers funding a study on medical supplement Beta275 for the very same disease and suggests that perhaps a meta-analysis of this and other studies (perhaps funded from elsewhere) might assist them in determining if they should fund this new study. Unfortunately the genetic researcher is at first unable to find a statistician to collaborate on such a meta-analytic exercise. But when discussing the lack of statisticians willing to collaborate on meta-analyses with the Bayesian statistician who helped them initially explicate their prior distribution (whom they did not think to approach about the meta-analytic exercise) they are informed that it is just Bayes. We simply need to take your prior and multiply it by the likelihoods from the RCTs already conducted (that you had no knowledge of when we explicated your prior) and use the posterior to answer the funding agencies question about the need to fund another study. They get lucky initially in the MA because it turns out that medical supplement Beta275 is only manufactured by one company and they know that only two studies were ever conducted. One was an almost flawless RCT conducted by a group that have conducted many RCTs and the other it turns out was not randomized but patients were chosen for treatment or control by the principal investigator. Now the Bayesian statistician is somewhat concerned about the "confounded" likelihood from the second study. They first multiply the prior by the likelihood from the almost flawless RCT to get a posterior. Now should they use this posterior or should they treat it as "today's" prior and multiply by some possibly adjusted likelihood from the observational study? Drawing on an example given by Rubin (1978, p. 46) where confounding is conceptualized as preferential selection of patients into the treatment groups, although it would depend on the specifics of the trial and the prior on the selection preferences, the correct posterior interval given both studies will likely be wider than just the posterior from first study. Quoting from Rubin "In real data problems with moderate sample sizes and ... [confounding] , the consideration of realistic models ... may easily lead to such a variety of inferences under the different models [priors on selection preferences], or such a large posterior variance when a specific prior distribution is placed on the models, that the Bayesian may consider his data worthless for inference for causal effect [correct posterior coverage of unconfounded treatment effect]".

9 Final Conclusions

In empirical research we have to be pragmatic. It is all well to encourage flawless RCTs but foolhardy to expect such to (usually) be the case. In fact, RCTs may even become unethical or simply unachievable after some given point in time. A meta-analysis that for good reasons uses the result of only one study (or even none) arguably is a better meta-analysis than one that for the wrong reasons or in the wrong way uses the results of all studies. This talk attempted to provide quantitative justifications for using only the least flawed studies. For qualitative arguments see Shapiro (1997). A qualitative "fruit salad" metaphor may though best convey my conclusions. From the frequentist (Fisherian) point of view the existence of recognizable subsets is a disquieting warning that there may well be "inferential monsters" lurking in any "apple sauce" made from sweet (unconfounded) and sour (confounded) apples. From the Bayesian point of view (given Rubin's explicit use of a prior for confounding) it may be that there is not enough knowledge about the direction and size of confounding in flawed RCTs in general (empirical Bayes) or a priori (full Bayes) to adequately sweeten the sour apples.

Acknowledgments: The author is grateful for early and continued advice on MA from Professor D.F. Andrews and encouragement and key references on MA from Professor Sir D.R. Cox.

10 Re-Analysis of Examples (available at www.LRI.CA/PROGRAMS/ASA)

Efron's Meta-analysis Example (Efron, 1996, p. 543).

The reported MLE of the log odds using the quadratic density as the level two distribution and excluding study "s40" was -1.22 with a jackknifed standard error estimate of .26.

```
Generalized Linear Model with all studies.

> glm1_glm(cbind(cs,cf) ~ study + rx,
 + family=binomial, data= xx)
> ss_summary(glm1,dispersion=0)
> log.odds_ss$coef[42,1:2]
> log.odds
     Value Std. Error
 -1.220301  0.2350947

Generalized Linear Model without study ''s40''.

> glm1_glm(cbind(cs,cf) ~ study + rx,
 + family=binomial, data= xx[xx$study != ''s40'',])
> ss_summary(glm1,dispersion=0)
> log.odds_ss$coef[41,]
> log.odds
     Value Std. Error
 -1.043364  0.2024657

Data Set (referred to as ''xx'' in SPlus glm program above)

   study cs cf rx
 1    s1  7  8  1
 2    s2  8 11  1
 3    s3  5 29  1
 4    s4  7 29  1
 5    s5  3  9  1
 6    s6  4  3  1
 7    s7  4 13  1
 8    s8  1 15  1
 9    s9  3 11  1
10   s10  2 36  1
11   s11  6  6  1
12   s12  2  5  1
13   s13  9 12  1
```

14	s14	7	14	1
15	s15	3	22	1
16	s16	4	7	1
17	s17	2	8	1
18	s18	1	30	1
19	s19	4	24	1
20	s20	7	36	1
21	s21	6	34	1
22	s22	4	14	1
23	s23	14	54	1
24	s24	6	15	1
25	s25	0	6	1
26	s26	1	9	1
27	s27	5	12	1
28	s28	0	10	1
29	s29	0	22	1
30	s30	2	16	1
31	s31	1	14	1
32	s32	8	16	1
33	s33	6	6	1
34	s34	0	20	1
35	s35	4	13	1
36	s36	10	30	1
37	s37	3	13	1
38	s38	4	30	1
39	s39	7	31	1
40	s40	0	34	1
41	s41	0	9	1
42	s1	11	2	0
43	s2	8	8	0
44	s3	4	35	0
45	s4	4	27	0
46	s5	0	12	0
47	s6	4	0	0
48	s7	13	11	0
49	s8	13	3	0
50	s9	7	15	0
51	s10	12	20	0
52	s11	8	0	0
53	s12	7	2	0
54	s13	7	17	0
55	s14	5	20	0
56	s15	11	21	0
57	s16	6	4	0
58	s17	8	2	0

```
59    s18  4 23  0
60    s19 15 16  0
61    s20 16 27  0
62    s21 13  8  0
63    s22  5 34  0
64    s23 13 61  0
65    s24  8 13  0
66    s25  6  0  0
67    s26  5 10  0
68    s27  5 10  0
69    s28 12  2  0
70    s29  8 16  0
71    s30 10 11  0
72    s31  7  6  0
73    s32 15 12  0
74    s33  7  2  0
75    s34  5 18  0
76    s35  2 14  0
77    s36 12  8  0
78    s37  2 14  0
79    s38  5 14  0
80    s39 15 22  0
81    s40 34  0  0
82    s41  0 16  0
```

Smith et al's Meta-analysis Example (Smith et al., 1996, p. 425).

Using MCMC the hierarchical Bayes estimate of the odds ratio was .25 with a 95% probability interval of (.16,.36).

```
> glm1_glm(cbind(cx,cn - cx) ~ study+rx,
 + family=binomial,data= xx)
> ss_summary(glm1,dispersion=0)
> odds.ratio_exp(-ss$coef[23,1])
> odds.ratio.ci_exp(c(-(ss$coef[23,1] + 2*ss$coef[23,2]),
+ -(ss$coef[23,1] - 2*ss$coef[23,2])))
> odds.ratio
[1] 0.3459918
> odds.ratio.ci
[1] 0.2495078 0.4797858
```

```
Data Set (referred to as ''xx'' in SPlus glm program above)

   study cx  cn rx
 1    s1 25  54  1
 2    s2 24  41  1
```

3	s3	37	95	1
4	s4	11	17	1
5	s5	26	49	1
6	s6	13	84	1
7	s7	38	170	1
8	s8	29	60	1
9	s9	9	20	1
10	s10	44	47	1
11	s11	30	160	1
12	s12	40	185	1
13	s13	10	41	1
14	s14	40	185	1
15	s15	4	46	1
16	s16	60	140	1
17	s17	12	75	1
18	s18	42	225	1
19	s19	26	57	1
20	s20	17	92	1
21	s21	23	23	1
22	s22	6	68	1
23	s1	7	47	0
24	s2	4	38	0
25	s3	20	96	0
26	s4	1	14	0
27	s5	10	48	0
28	s6	2	101	0
29	s7	12	161	0
30	s8	1	28	0
31	s9	1	19	0
32	s10	22	49	0
33	s11	25	162	0
34	s12	31	200	0
35	s13	9	39	0
36	s14	22	193	0
37	s15	0	45	0
38	s16	31	131	0
39	s17	4	75	0
40	s18	31	220	0
41	s19	7	55	0
42	s20	3	91	0
43	s21	14	25	0
44	s22	3	65	0

11 What is a MA and When Should It Be Done

MA is first and foremost a scientific endeavor and therefore must be done in a manner that is explicit (both in what are considered to be the facts and the proper inferences between facts) and fully replicable by other parties (Gardin, 1981). More specifically (L'Abbe et al., 1987) there should be adequate compliance with the following requirements:

1. An explicit and detailed working protocol.

2. A literature search strategy that can be replicated.

3. Inclusion and exclusion criteria for research reports and a list of exclusions with reasons given.

4. Verification of independence of published studies (use of separate patients in different studies.)

5. A careful exploration of differences in treatment effect estimates with the aim of explaining them on the basis of relevant clinical differences (Fisher's definite biological variability), differences in quality of research (Fisher's definite methodological variability) or simply sampling variability with appropriate combination of treatment effect estimates *if and where* indicated. In order that this analysis can be easily replicated, a listing of individual study results or inputs along with a listing of what is believed to be the most relevant clinical and methodological differences between studies will be required.

6. A careful consideration of the potential effects of publication bias along with an explicit request for any leads regarding studies that were not included or listed as exclusions.

7. A set of conclusions which includes a summary of what was believed to be done adequately and what was done inadequately along with how this is reflected in estimates of uncertainty.

8. A set of suggestions and directions for future research—either for the particular questions or areas of similar research.

That most researchers realize the need for stating their results in the context of previous studies is clear from the inclusion of literature review section in almost all scientific journals. MA is a further development and refinement of this approach offering a more rigorous and coherent treatment of past research work. It is tempting to propose (O'Rourke and Detsky, 1989) that no experimental results should be published without the inclusion of an appropriate MA in place of the *ad hoc* literature review section.

12 References

Bailey, K.R (1987). Inter-study differences: How should they influence the interpretations and analysis of results? *Statist. Med. 6*, 351–358.

Berkey, C.S., D.C Hoaglin, F. Mosteller, and G.A Colditz (1995). A random-effects regression model for meta-analysis. *Statist. Med. 14*, 395–411.

Casella, G. (1992). Conditional inference from confidence sets. In M. Ghosh and P. K. Pathak (Eds.), *Current Issues in Statistical Inference: Essays in honor of D. Basu.* Institute of Mathematical Statistics.

Chatfield, C. (1995). Uncertainty, data mining and inference. *J. Roy. Statist. Soc. Ser. A 158*, 418–466.

Cochran, W.G. (1937). Problems arising in the analysis of a series of similar experiments. *J. Roy. Statist. Soc. Suppl. 4*, 102–118.

Cox, D.R. (1958). The interpretation of the effects of nonadditivity in the Latin square. *Biometrika 45*, 69–73.

Cox, D.R. (1982). Combination of data. In S. Kotz and N. Johnson (Eds.), *Encyclopedia Statisti. Sci. 2.* New York: Wiley.

Cox, D.R. and D.V. Hinkley (1974). *Theoretical Statistics.* London: Chapman and Hall.

Cox, D.R. and E.J. Snell (1981). *Applied Statistics.* London: Chapman and Hall.

Cox, D.R. and E.J. Snell (1989). *Analysis of Binary Data: Second edition.* London: Chapman and Hall.

Data Analysis Product Division, MathSoft (1997). *S-Plus 4 guide to statistics.* Seattle: Data Analysis Product Division, MathSoft.

DerSimonian, R. and N. Laird (1986). Meta-analysis in clinical trials. *Cont. Clin. Trials 7*, 177–188.

Desmond, A.F. and V.P. Godambe (1999). Estimating functions. In P. Armitage and T. Colton (Eds.), *Encyclopedia of Biostatistics.* New York: Wiley.

Detsky, A.S., C.D. Naylor, K. O'Rourke, A. McGeer, and K.A. L'Abbe (1992). Incorporating variations in the quality of individual randomized trials into meta-analysis. *J. Clin. Epid. 45*, 255–265.

DuMouchel, W., D. Fram, Z. Jin, S.L. Normand, B. Snow, S. Taylor, and R. Tweedie (1997). *MetaGraphs: Software for Exploration and Modeling of Meta-analyses.* Belmont Research Incorporated.

Efron, B. (1996). Empirical Bayes methods for combining likelihoods (with discussion). *J. Amer. Statist. Assoc. 91*, 538–565.

Emerson, J.D., D.C. Hoaglin, and F. Mosteller (1996). Simple robust procedures for combining risk differences in sets of 2x2 tables. *Statist. Med. 15*, 1465–1488.

Firth, D. (1990). Generalized linear models. In D. Hinkley, N. Reid, and E. Snell (Eds.), *Statistical Theory and Modelling.* London: Chapman and Hall.

Fisher, R.A. (1937). *The Design of Experiments.* Edinburgh: Oliver and Boyd.

Fraser, D.A.S. (1976). *Probability and Statistics: Theory and Application.* North Scituate: Duxbury Press.

Gardin, J.C. (1981). *La Logic du Plausible: Essais d'Epistémologie Pratique.* Ann Arbor, Michigan: University Microfilms International.

Goodman, S.N. (1989). Meta-analysis and evidence. *Cont. Clin. Trials 10*, 188–204.

Greenland, S. and A. Salvan (1990). Bias in the one-step method for pooling study results. *Statist. Med. 9*, 247–252.

Grizzle, J.E. (1965). The two-period change over design and its use in clinical trials. *Biometrics 21*, 467–480.

Guttman, I., I. Olkin, and R. Philips (1993). Estimating the number of aberrant laboratories. Technical Report 9301, Department of Statistics, The University of Toronto.

Jones, B. and J. Lewis (1995). The case for cross-over trials in phase III. *Statist. Med. 14*, 1025–1038.

L'Abbe, K.A., A.S. Detsky, and K. O'Rourke (1987). Meta-analysis in clinical research. *Ann. Internal Med. 107*, 224–33.

Laird, N.M. and F. Mosteller (1990). Some statistical methods for combining experimental results. *Internat. J. Techn. Assessment in Health Care 6*, 5–30.

Lee, Y. and J.A. Nelder (1996). Hierarchical generalized linear models. *J. Roy. Statist. Soc. Ser. B 58*, 619–678.

McCullagh, P. (2000). Re-sampling and exchangeable arrays. *Bernouilli 6*, 303–322.

Meier, P. (1987). Commentary. *Statist. Med. 6*, 329–331.

Morris, C.N. (1983). Parametric empirical Bayes inference: Theory and applications. *J. Amer. Statist. Assoc. 78*, 47–65.

Morris, C.N. and S.L. Normand (1992). Hierarchical models for combining information and for meta-analysis. In J. Bernardo, J. Berger, A. Dawid, and A. Smith (Eds.), *Bayesian Statistics 4*. Oxford: Oxford University Press.

Mosteller, F. and J.W. Tukey (1977). *Data Analysis and Regression*. New York: Addison-Wesley.

Normand, S.L. (1995). Meta-analysis software: A comparative review. *Amer. Statist. 49*, 298–309.

O'Rourke, K. and A.S. Detsky (1989). Meta-analysis in medical research: Strong encouragement for higher quality in individual research efforts. *J. Clin. Epid. 42*, 1021–1024.

O'Rourke, K., A. McGeer, C.D. Naylor, K.A. L'Abbe, and A.S. Detsky (1991). Incorporating quality appraisals into meta-analyses. Technical Report 9103, Department of Statistics, The University of Toronto.

Pena, D. (1997). Combining information in statistical modeling. *Amer. Statist. 51*, 326–332.

Peto, R. (1987). Discussion. *Statist. Med. 6*, 242.

Rao, C.R. (1988). Comment on Iyengar and Greenhouse's "Selection models and the file drawer problem". *Statist. Sci. 3*, 131.

Rubin, D.B. (1974). Estimating causal effects of treatments in randomized and nonrandomized studies. *Edu. Psychology 66*, 688–701.

Rubin, D.B. (1978). Bayesian inference for causal effects: The role of randomization. *Annals of Statistics 6*, 34–58.

Rubin, D.B. (1990). Neyman (1923) and causal inference in experiments and observational studies. *Statist. Sci. 5*, 472–480,.

Rubin, D.B. (1991). Practical implications of modes of statistical inference for causal effects and the critical role of the assignment mechanism. *Biometrics 47*, 1213–1234.

Rubin, D.B. (1992). Meta-analysis: Literature synthesis or effect-size surface estimation? *J. Edu. Statist. 17*, 363–374.

Schulz, K.F. (1995). Unbiased research and the human spirit: The challenges of randomized controlled trials. *Canad. Med. Assoc. J. 153*, 783–786.

Schulz, K.F., I. Chalmers, D.A. Grimes, and D.G. Altman (1994). Assessing the quality of randomization from reports of controlled trials published in obstetrics and gynecology journals. *J. Amer. Med. Assoc. 272*, 125–128.

Schulz, K.F., I. Chalmers, R.J. Hayes, and D.G. Altman (1995). Empirical evidence of bias: Dimensions of methodological quality associated with estimates of treatment effects in controlled trials. *J. Amer. Med. Assoc. 273*, 408–412.

Sen, S (1996). The AB/BA cross-over: How to perform the two stage analysis if you can't be persuaded that you shouldn't. In H. de Ridder (Ed.), *Liber Amicorum Roel Van Strik*, Erasmus University.

Shapiro, S. (1997). Is meta-analysis a valid approach to the evaluation of small effects in observational studies? *J. Clin. Epid. 50*, 223–229.

Smith, T.C., D.J. Spiegelhalter, and M.H.K. Parmar (1996). Bayesian meta-analysis of randomized trials using graphical models and BUGS. In D. Berry and D. Stangl (Eds.), *Bayesian Biostatistics*. New York: Marcel Dekker.

Tjur, T. (1998). Nonlinear regression, quasi likelihood, and over dispersion in generalized linear models. *Amer. Statist. 52*, 222–227.

12

Ancillary Information for Statistical Inference

D.A.S. Fraser and N. Reid

ABSTRACT Inference for a scalar interest parameter in the presence of nuisance parameters is obtained in two steps: an initial ancillary reduction to a variable having the dimension of the full parameter and a subsequent marginalization to a scalar pivotal quantity for the component parameter of interest. Recent asymptotic likelihood theory has provided highly accurate third order approximations for the second marginalization to a pivotal quantity, but no general procedure for the initial ancillary conditioning. We develop a second order location type ancillary generalizing Fraser and Reid (1995) and a first order affine type ancillary generalizing Barndorff-Nielsen (1980). The second order ancillary leads to third order p-values and the first order ancillary leads to second order p-values. For an n dimensional variable with p dimensional parameter, we also show that the only information needed concerning the ancillary is an array $V = (v_1 \dots v_p)$ of p linearly independent vectors tangent to the ancillary surface at the observed data. For the two types of ancillarity simple expressions for the array V are given; these are obtained from a full dimensional pivotal quantity. A brief summary describes the use of V to obtain second and third order p-values and some examples are discussed.

1 Introduction

Recent asymptotics indicates that statistical inference for a scalar parameter component is obtained in two distinct steps: a conditioning reduction from some initial data dimension n to an effective variable having the same dimension p as the parameter; a subsequent marginalization to a pivotal quantity for the component parameter of interest.

For the second step with variable and parameter of the same fixed dimension p, the procedures for higher order inference are straightforward and summarized in Section 2. The theory shows that third order p-values for any component parameter, say $\psi(\theta)$, can be calculated using only two items of information concerning the statistical model and the observed data. The two items are the observed likelihood function,

$$\ell^0(\theta) = \ell(\theta; y^0) = a + \log f(y^0; \theta), \tag{1}$$

which as usual is left indeterminate to an additive constant a, and the observed likelihood gradient,

$$\varphi'(\theta) = \ell_{;y}(\theta; y^0) = \frac{\partial}{\partial y'} \log f(y;\theta)\big|_{y^0}. \tag{2}$$

The observed likelihood gradient is used as a nominal reparametrization that mimics the canonical parametrization of an exponential model. Any affinely equivalent version $\widetilde{\varphi}(\theta) = b + C\varphi(\theta)$ with C nonsingular $p \times p$ will work equally, where C could correspond to a change of y coordinates.

The computational procedure uses $\ell^0(\theta)$, $\varphi(\theta)$ together with the interest parameter value $\psi(\theta) = \psi$ to be assessed, and produces a left tail p-value $p(\psi)$, which is an approximation to $P(\widehat{\psi} \leq \widehat{\psi}^0; \psi)$. The summary in Section 2 records essential details for the computation.

This paper focuses on the reduction from an initial data variable y of dimension n to a reduced variable of dimension p. In full exponential models this can be achieved by a sufficiency reduction, but in general is obtained by conditioning on an approximate ancillary $a(y)$. If we have such an ancillary of dimension $n - p$ then the model factorizes as

$$f(y;\theta) = g\big(x \mid a;\theta\big)h(a),$$

and the likelihood function for inference is $\log g\big(x \mid a;\theta\big)$. As stated above we need to compute its observed value and its observed gradient. These can be computed from the full likelihood. First, the observed likelihood in the conditional model is equal to the observed likelihood from the full model:

$$\ell\big(\theta; x^0 \mid a^0\big) = \ell(\theta; y^0).$$

To compute the observed likelihood gradient, let $V = (v_1 \dots v_p)$ be p linearly independent vectors tangent to the ancillary surface at the data y^0. These provide a basis for local coordinates in the conditional model and thus the reparametrization $\varphi(\theta)$ can be obtained by differentiating the full likelihood in the directions V,

$$\begin{aligned}\varphi'(\theta) &= \frac{\partial}{\partial V}\ell(\theta;y)\big|_{y^0} = \frac{\partial}{\partial y'}\ell(\theta;y)\big|_{y^0}V = \ell_{;y}(\theta;y^0)V \\ &= \left[\sum_1^n \frac{\partial}{\partial y_i}\ell(\theta;y^0)v_{i1}, \dots, \sum_1^n \frac{\partial}{\partial y_i}\ell(\theta;y^0)v_{ip}\right],\end{aligned} \tag{3}$$

where the notation $(\partial/\partial V)\ell(\theta;y)$ denotes the vector of directional derivatives $[(\partial/\partial v_i)\ell(\theta;y), \ \dots, (\partial/\partial v_p)\ell(\theta;y)]$ and we have $(\partial/\partial v)\ell(\theta;y) = (\partial/\partial t)\ell(\theta;y+tv)\big|_{t=0}$. The final equality holds under the assumption of independence of the n coordinates of y.

For approximating the p-value $p(\psi)$, it suffices to have V tangent to a first order ancillary to obtain second order accuracy and V tangent to a second order ancillary to obtain third order accuracy (Skovgaard, 1986; Fraser and Reid, 1995).

In this paper we develop simple formulas for a matrix V_2 corresponding to a second order ancillary and thus to third order inference, and for a matrix V_1 corresponding to a first order ancillary and thus to second order inference. The formulas, developed in detail below, are:

$$V_2 = \left.\frac{\partial y}{\partial \theta'}\right|_{(y^0,\hat{\theta}^0)}, \quad V_1 = \left.\frac{\partial E(y;\theta)}{\partial \theta'}\right|_{\hat{\theta}^0}. \tag{4}$$

The formula for V_2 giving third order accuracy requires coordinate-by-coordinate pivotal variables $z_i = z_i(y_i, \theta)$ which for interpretation are taken to describe how a variable y_i measures θ; the partial differentiation is then for fixed pivotal z. The formula for V_1 requires each component variable y_i to in fact be the score variable at the observed maximum likelihood value; the θ differentiation in the formula then gives an expected information matrix for that vector coordinate. The first formula with components recorded in (17) is developed in Sections 3 to 6. The second formula with details given in (42) is derived in Section 7 using properties of the first formula but taking account of the use of score variables.

The second order ancillary needed for third order inference is constructed using local coordinates at the observed data. For a nearby point with the same value of the ancillary a question arises whether the construction would give the same ancillary value. We view this as a foundational issue that will be addressed elsewhere, but the ancillary is the same to the required order.

2 Third Order Statistical Inference

Likelihood asymptotics provides computationally accessible formulas for deriving p-values $p(\psi)$ for assessing a value ψ for a scalar interest parameter $\psi(\theta)$; these formulas however tend in practice to be limited to the case with a variable y having the same dimension p as the parameter θ, that is, assuming that standard sufficiency and conditioning reductions have already been applied. Also in most cases there is an assumption of a continuous variable and parameter and of reasonable regularity.

Most of the formulas use specialized first order quantities $r = r(y;\psi)$, $q = q(y;\psi)$ and combine these to obtain $p(\psi)$ by one or other of the expressions

$$\begin{aligned} \Phi_1(r,q) &= \Phi(r) + (r^{-1} - q^{-1})\phi(r), \\ \Phi_2(r,q) &= \Phi\left[r - r^{-1}\log\left(\frac{r}{q}\right)\right], \end{aligned} \tag{5}$$

developed for particular contexts respectively by Lugannani and Rice (1980) and by Barndorff-Nielsen (1986). In (5) Φ and ϕ are the standard normal distribution and density functions.

The usual definition for r is the signed likelihood root obtained from the observed likelihood function

$$r = \operatorname{sgn}(\widehat{\psi} - \psi) \cdot \{2[\ell(\widehat{\theta}; y) - \ell(\widehat{\theta}_\psi; y)]\}^{1/2}, \tag{6}$$

where $\widehat{\theta}_\psi$ is the maximum likelihood value under the constraint $\psi(\theta) = \psi$.

For q, various formulas are available depending on the model type and the approach (Lugannani and Rice, 1980; Barndorff-Nielsen, 1986, 1991; Fraser and Reid, 1993, 1995; Fraser et al., 1999). For present purposes we follow the tangent exponential model approach (Fraser, 1988; Fraser and Reid, 1993, 1995) and use the locally defined canonical parameter φ of (2) or (3). The interest parameter $\psi(\theta)$ is replaced by a linear function of the $\varphi(\theta)$ coordinates

$$\chi(\theta) = \frac{\psi_\varphi(\widehat{\theta}^0_\psi)}{|\psi_\varphi(\widehat{\theta}^0_\psi)|}\varphi(\theta), \tag{7}$$

where the subscript to ψ denotes differentiation and the numerical coefficients of the linear function are based on the observed constrained maximum likelihood value $\widehat{\theta}^0_\psi$, with the superscript to emphasize the dependence on the observed values. The function q is given as the standardized maximum likelihood departure

$$q = \operatorname{sgn}(\widehat{\psi} - \psi) \cdot |\chi(\widehat{\theta}) - \widehat{\chi}(\widehat{\theta}_\psi)| \cdot \left[\frac{|\jmath_{(\theta\theta)}(\widehat{\theta})|}{|\jmath_{(\lambda\lambda)}(\widehat{\theta}_\psi)|}\right]^{1/2}, \tag{8}$$

where $|\jmath_{(\theta\theta)}(\widehat{\theta})|$ and $|\jmath_{(\lambda\lambda)}(\widehat{\theta}_\psi)|$ are the full and nuisance information determinants recalibrated in the φ parametrization,

$$\begin{aligned} |\jmath_{(\theta\theta)}(\widehat{\theta})| &= |-\ell_{\theta\theta}(\widehat{\theta})|\,|\varphi_\theta(\widehat{\theta})|^{-2}, \\ |\jmath_{(\lambda\lambda)}(\widehat{\theta}_\psi)| &= |-\ell_{\lambda\lambda}(\widehat{\theta}_\psi)|\,|\varphi'_\lambda(\widehat{\theta}_\psi)\varphi_\lambda(\widehat{\theta}_\psi)|^{-1}. \end{aligned} \tag{9}$$

A generalized expression for q which does not require an explicit nuisance parametrization and is applicable to the Bayesian context is developed in Fraser et al. (1999). The same paper also establishes the equivalence of the above q and the u used by Barndorff-Nielsen (1986), for the cases where the dimensions n and p are equal or where the dimension n is greater than p and the directions V correspond to the ancillary used for the calculation of u.

Formula (8) can be rewritten in an alternative form following Pierce and Peters (1992):

$$q = \operatorname{sgn}(\widehat{\psi} - \psi)|\widehat{\chi} - \widehat{\chi}_\psi|\,|\widehat{\jmath}_{(p)}|^{1/2}\rho^{-1}(\widehat{\theta}, \widehat{\theta}_\psi), \tag{10}$$

where $\jmath_{(p)}$ is the recalibrated information from the profile likelihood for ψ and $\rho^2(\widehat{\theta}, \widehat{\theta}_\psi)$ measures the effect of integrating out a conditional distribution concerning the nuisance parameter λ,

$$\rho^2(\widehat{\theta},\widehat{\theta}_\psi) = \frac{|j_{(\lambda\lambda)}(\widehat{\theta}_\psi)|}{|j_{(\lambda\lambda)}(\widehat{\theta})|}.$$

3 First Derivative Ancillary

A first derivative ancillary at θ_0 is a variable $a(y)$ whose distribution as given by its density g is free of θ to first derivative at θ_0,

$$g_\theta(a;\theta_0) = 0, \tag{11}$$

where the subscript θ denotes differentiation. This type of ancillary was proposed in Fraser (1964) for local inference and adapted to asymptotic inference in Fraser and Reid (1995).

For the case of a scalar parameter and n independent scalar components y_i, an $n-1$ dimensional first derivative ancillary was developed in Fraser and Reid (1995) from location model theory and then adjusted to be a second order ancillary at the observed maximum likelihood value $\theta = \widehat{\theta}^0$; the adjustment does not alter the tangent direction of the ancillary curve at the data point. The tangent direction for a second order ancillary is thus given as the tangent direction for the particular first derivative ancillary.

The ancillary is derived in Fraser and Reid (1995) as follows. Let $F^i(y_i;\theta)$ be the distribution function for the ith coordinate, and define

$$x_i = \int^{y_i} -\frac{F^i_y(y;\theta_0)}{F^i_{;\theta}(y;\theta_0)}\,dy, \tag{12}$$

which is in location relation to θ at θ_0; the subscripts denote differentiation. Let $g^i(x_i;\theta)$ be the density for x_i; then $g^i\left[x_i - (\theta-\theta_0);\theta_0\right]$ and $g^i(x_i;\theta)$ coincide to first derivative at θ_0, and the ancillary $(x_i - \bar{x},\ldots,x_n - \bar{x})$ for the location model is a first derivative ancillary for the original model. In terms of the x coordinates the ancillary direction is the one vector and in terms of the original coordinates the ancillary direction is

$$v' = \left[-\frac{F^1_{;\theta}(y_1;\theta_0)}{F^1_y(y_1;\theta_0)},\ldots,-\frac{F^n_{;\theta}(y_n;\theta_0)}{F^n_y(y_n;\theta_0)}\right]. \tag{13}$$

This vector is described by $\partial y/\partial\theta\big|_{(y^0,\theta_0)}$ where differentiation is for fixed value of $\left[F^1(y_1,\theta),\ldots,F^n(y_n;\theta)\right]$ or for fixed value of an equivalent pivotal variable.

As a simple example suppose y_i has density $f^i(y_i - a_i\beta)$. Then $y - a\beta$ is pivotal and $V = a$. The corresponding canonical parameter is

$$\varphi(\theta) = \sum a_i \ell^i_{;y_i}(y^0_i - a_i\beta) = -\ell_\beta(y^0 - a\beta),$$

where $\ell^i = \log f^i$, and for this location model is equivalent to the score function at the data point. Section 2 then gives a simple expression for the p-value $p(\beta)$ for assessing β; this leads to and agrees with earlier expressions in DiCiccio et al. (1990).

We now generalize this first derivative ancillary to the vector parameter context but at this point consider only directional change $\theta = \theta_0 + \Delta u$ for given unit direction u. In the next section we then adjust the ancillary to be second order ancillary at θ_0 but in doing this do not alter its direction at the data point.

Initially let $z(y;\theta)$ be a p dimensional pivotal quantity based on variable y and parameter θ each of dimension p. We assume that the pivotal quantity in some manner expresses how the variable measures the parameter; we view this as an important supplemental concept for statistical inference. The choice of the pivotal must be sensible in the context of the particular application.

Now we consider the directional change $\theta = \theta_0 + \Delta u$, where u is a unit vector giving a fixed direction and Δ designates a differential increment for that direction. We have $d\theta = \Delta u$ in θ at θ_0, so for given pivotal value the change in y is

$$\frac{\partial y}{\partial \theta}\Delta u = -z_y^{-1}(y;\theta_0)z_{;\theta}(y;\theta_0)\Delta u = v(y)\Delta.$$

This gives a vector field $[v(y)]$ on R^p which in wide generality integrates to give curves $y = y(a, x)$ where x is a scalar in location relation with Δ and a is a $p-1$ dimensional variable indexing the curves.

Let $f(a, x; \theta)$ be the distribution of (a, x). Then the location model

$$f(a, x - \Delta; \theta_0) = h(a;\theta_0)g(x - \Delta \mid a; \theta_0), \tag{14}$$

coincides with the original model $f(a, x; \theta_0 + \Delta u)$ to first derivative at $\Delta = 0$. In the new coordinates (a, x), the location model (14) has a marginal density h for a and a conditional density g for x; we then note that the locally developed location model (14) has location trajectories that are lines parallel to the pth axis. We refer to (14) as the tangent location model at θ_0 relative to the parameter change direction u.

Now consider n independent coordinates y_i where each coordinate has pivotal quantity $z^i(y_i;\theta)$ as just discussed. We also consider again a scalar directional change $d\theta = \Delta u$ at θ_0 and let (a_i, x_i) be the corresponding new coordinates. Then the composite location model

$$\Pi f^i(a_i, x_i - \Delta; \theta_0) = \Pi h^i(a_i;\theta_0)\Pi g^i(x_i - \Delta \mid a_i; \theta_0),$$

coincides with the given model $\Pi f^i(a_i, x_i; \theta)$ to first derivative at θ_0. The ancillary for this location model can be given as $(a_1, \ldots, a_n, x_1 - \overline{x}, \ldots, x_n - \overline{x})$, with tangent direction $(0, \ldots, 0, 1, \ldots, 1)$; $\hat{\Delta}$ can replace $\overline{x}$. Let $(b_1, \ldots,$

b_{n-1}) be orthonormal coordinates in R^n orthogonal to the one dimensional linear space $\mathcal{L}(1)$ defined by the one-vector and let t be a complementing coordinate as coefficient of the one vector in the modified space, all for given $a_1, \ldots, a_n$. Then the ancillary can be given as $(a_1, \ldots, a_n, b_1, \ldots, b_{n-1})$ with ancillary direction $(0, \ldots, 0, 0, \ldots, 0, 1)$. In the modified coordinates the tangent location model has the form

$$\begin{aligned}\Pi h^i(a_i;\theta_0)\Pi g^i(x_i - \Delta \mid a_i;\theta_0) &= h(a;\theta_0)h_0(b \mid a;\theta_0)g(t - \Delta \mid a, b;\theta_0) \\ &= h(A;\theta_0)g(t - \Delta \mid A;\theta_0),\end{aligned} \tag{15}$$

where $A = (a, b)$, the ancillary direction is $(0, \ldots, 0, 1)$, and t becomes a location variable, say $\hat{\Delta}$.

In terms of the original coordinates y_i the ancillary direction is

$$v_u(y^0) = \left.\frac{\partial y}{\partial \theta}\right|_{(y^0,\theta_0)} u = V(y^0)u, \tag{16}$$

where the $np \times p$ matrix $\partial y / \partial \theta$ at the data involves partial differentiation for fixed pivotal and has ith $p \times p$ block given as

$$\begin{aligned}V^i(y^0) = \left.\frac{\partial y_i}{\partial \theta}\right|_{(y^0,\theta_0)} &= -\left[\frac{\partial z^i(y_i;\theta)}{\partial y_i'}\right]^{-1} \left.\frac{\partial z^i(y_i;\theta)}{\partial \theta'}\right|_{(y^0,\theta_0)} \\ &= -\left[z^i_{y_i}(y^0, \theta_0)\right]^{-1} z^i_{;\theta}(y^0;\theta_0);\end{aligned} \tag{17}$$

also we take $V(y^0)$ to be the stacked array obtained from the $V^i(y^0)$.

Now consider various directional changes, in particular those in the coordinate axis directions at θ_0; we then obtain the array of p vectors

$$V(y^0) = \left.\frac{\partial y}{\partial \theta}\right|_{(y^0,\theta_0)}. \tag{18}$$

We do note that the individual location models do not in general generate a joint location model due to the general non integrability of the composite vector field. We shall show however that $V(y^0)$ has appropriate properties for inference, and for this we need some asymptotic properties and some local second order coordinates developed in the next two sections. The integrability of the $V(y)$ to the required order will be examined elsewhere.

As a simple example consider $(y_1, \ldots, y_n)$ where each coordinate y_i has the density $f^i(y_i;\beta,\gamma) = c_i^{-1}\gamma^{-1}h^i[c_i^{-1}\gamma^{-1}(y_i - a_i\beta)]$. If we examine a coordinate pair we obtain a natural pivotal

$$z(y_1, y_2;\beta,\gamma) = \begin{bmatrix} c_1^{-1}\gamma^{-1}(y_1 - a_1\beta) \\ c_2^{-1}\gamma^{-1}(y_2 - a_2\beta)\end{bmatrix},$$

with derivatives

$$z_y = \begin{pmatrix} c_1^{-1}\gamma^{-1} & 0 \\ 0 & c_2^{-1}\gamma^{-1}\end{pmatrix}, \quad z_{;\theta} = \begin{bmatrix} -a_1c_1^{-1}\gamma^{-1} & -c_1^{-1}\gamma^{-2}(y_1 - a_1\beta) \\ -a_2c_2^{-1}\gamma^{-1} & -c_2^{-1}\gamma^{-2}(y_2 - a_2\beta)\end{bmatrix},$$

giving

$$\frac{\partial y}{\partial(\beta,\gamma)} = -z_y^{-1} z_{;\theta} = \begin{bmatrix} a_1 & \gamma^{-1}(y_1 - a_1\beta) \\ a_2 & \gamma^{-1}(y_2 - a_2\beta) \end{bmatrix},$$

for the coordinate pair and giving

$$\frac{\partial y}{\partial(\beta,\gamma)} = \begin{bmatrix} a & \gamma^{-1}(y - a\beta) \end{bmatrix},$$

for the full model. An affinely equivalent pair of vectors in R^n is (a, y). Then in the composite notation of the full model we have $V = (a, y^0)$. The corresponding canonical parameter pair is

$$\varphi'(\theta) = \Big\{ \sum a_i c_i^{-1} \gamma^{-1} m_z^i [c_i^{-1}\gamma^{-1}(y_i^0 - a_i\beta)], \\ \sum y_i^0 c_i^{-1} \gamma^{-1} m_z^i [c_i^{-1}\gamma^{-1}(y_i^0 - a_i\beta)] \Big\},$$

where $m^i(z) = \log h^i(z)$. Section 2 then gives simple expressions for any p-values $p(\psi)$ for testing an interest parameter $\psi(\beta,\gamma) = \psi$; but the proof for this requires results of the next three sections. Related formulas are found in DiCiccio et al. (1990).

4 Bending and Tilting

Various types of ancillary can be developed by simple modifications to some initial ancillary such as the first derivative ancillary. Two methods used for these modifications, referred to as bending and tilting, are discussed in this section.

Consider a pair of variables x, y with a standardized asymptotic distribution as some parameter n becomes large. We assume that to the first order x is normal $(0, 1)$ and $y \mid x$ is normal $(\delta, 1)$. Let $\ell_1(x)$ and $\ell_2(y - \delta \mid x)$ be the corresponding log density functions involving first order standardized normal components plus cubic terms divided by $6n^{1/2}$; these are examined on a bounded region and the dependence on n is assessed.

Bending

Suppose we change the conditioning from $y \mid x$ to $y \mid X$ where $X = x - cy^2/(2n^{1/2})$ held fixed describes a curve bent to the right $(c > 0)$ but remaining parallel at $y = 0$ to the original y axis. The joint log density for (x, y) is

$$\begin{aligned} \ell_1(x) + \ell_2(y - \delta \mid x) &= \ell_1\left(X + \frac{cy^2}{2n^{1/2}}\right)\ell_2\left(y - \delta \,\middle|\, X + \frac{cy^2}{2n^{1/2}}\right) \\ &= \ell_1(X) - \frac{cXy^2}{2n^{1/2}} + \ell_2(y - \delta \mid X), \end{aligned}$$

to order $O(n^{-1})$; the middle term comes from the cross term in $-(1/2)\left[X+cy^2/(2n^{1/2})\right]^2$ and $\ell_2(t \mid x) = \ell_2(t \mid X)$. We integrate the joint density with respect to y for fixed X obtaining

$$\ell_1(X) - \frac{cX(\delta^2+1)}{2n^{1/2}}.$$

It follows that the bending adds a log-likelihood component $-cX\delta^2/(2n^{1/2})$ to the marginal likelihood for X. A version of bending was used in Fraser and Reid (1995).

Suppose in a parallel way that the trajectory of the mean of the distribution of (x,y) is changed from $(0,\delta)$ with free δ to $\left[c\delta^2/(2n^{1/2}),\delta\right]$, this being the same bending to the right ($c>0$) as above. Then by similar calculations we obtain a log-likelihood component $cx\delta^2/(2n^{1/2})$ which cancels that from the previous bending. This has use in some Bayesian analyses.

Tilting

Suppose we change the conditioning from $y \mid x$ to $y \mid X$ where $X = x - cy/n^{1/2}$ held fixed describes a line tilted to the right ($c > 0$). The joint log density for (x,y) is

$$\begin{aligned}\ell_1(x) + \ell_2(y-\delta \mid x) &= \ell_1\left(X + \frac{cy}{n^{1/2}}\right) + \ell_2\left(y-\delta \,\middle|\, X + \frac{cy}{n^{1/2}}\right)\\ &= \ell_1(X) - \frac{cXy}{n^{1/2}} + \ell_2(y-\delta \mid X),\end{aligned}$$

to order $O(n^{-1})$. Then integrating with respect to y for fixed X gives

$$\ell_1(X) - cX\delta(n^{1/2}),$$

to order $O(n^{-1})$. It follows that the tilting adds a log likelihood component $-cX\delta/n^{1/2}$ to the marginal likelihood for X.

In summary we note that linear and quadratic likelihood contributions in δ and δ^2 of order $n^{-1/2}$ can be added to the marginal distribution by tilting and bending. And if both are done the effect is additive, independent of the order.

We now apply bending to the pivotal context described in Section 3 with independent coordinates y_i, p dimensional effective pivotals $z^i\ (y_i;\theta)$, and scalar parameter change $d\theta = \Delta u$ at θ_0; tilting will be used in Section 6.

First consider a particular component and in terms of the coordinates y_i taken now to be say (a_i, x_i), we expand the logarithm of the density relative to the tangent location model

$$f^i(a_i,x_i;\theta) = f^i(a_i,x_i-\Delta;\theta_0)\exp\left[r_i(y_i)\frac{\Delta^2}{2} + s_i(y_i)\frac{\Delta^3}{6} + O(\Delta^4)\right],$$

where $E[r_i(y_i);\theta_0] = 0$ from the norming. For first order departures from θ_0 we use (15) and write $\Delta = \delta/n^{1/2}$, $t = z/n^{1/2}$ with z used temporarily here and not to be confused with the pivotal quantity; we then obtain

$$\begin{aligned} \Pi f^i(a_i, x_i; \theta) &= h(A;\theta_0)g[(z-\delta)n^{-1/2} \mid A;\theta_0] \\ &\quad \left[1 + n^{-1}\sum r_i(y_i)\frac{\delta^2}{2} + n^{-3/2}\sum s_i(y_i)\frac{\delta^3}{6}\cdots\right] \\ &= h(A;\theta_0)g[(z-\delta)n^{-1/2} \mid A;\theta_0] \\ &\quad \left(1 + c_1 w\frac{\delta^2}{2n^{1/2}} + c_2 z\frac{\delta^2}{2n^{1/2}} + c_3\frac{\delta^3}{6n^{1/2}}\cdots\right), \end{aligned} \tag{19}$$

where under θ_0 we have that $\widetilde{w} = n^{-1/2}\sum r_i(y_i) = c_1 w + c_2 z$ is first order normal with mean zero, $c_1 w$ is $\widetilde{w}$ evaluated with z replaced by zero and is first order normal with mean zero and $\widetilde{w} - c_1 w = c_2 z$ has been expanded to the first order in z.

The conditional log density for $z \mid A$ can be normed giving

$$g[(z-\delta)n^{-1/2} \mid A;\theta_0]n^{-1/2}\left[1 + c_2(z-\delta)\frac{\delta^2}{2n^{1/2}}\right], \tag{20}$$

leaving a marginal density

$$h(A;\theta_0)\exp\left(c_1 w\frac{\delta^2}{2n^{1/2}}\right), \tag{21}$$

with log likelihood component $c_1 w\delta^2/(2n^{1/2})$. Results from Section 3 in Fraser and Reid (1995) give an ancillary for the distribution (21) of w on the maximum likelihood surface $z = 0$, allowing the examination of w conditionally given that ancillary; for this, note that the observed maximum likelihood surface is cross-sectional to the orbits of the ancillary and the distribution of the ancillary can thus be examined as projected to the maximum likelihood surface. Then following Section 3 we bend the trajectories for x to remove the likelihood component in (21): $W = w - c_1 z^2/(2n^{1/2})$. It follows that the new conditioning variable replacing A is ancillary to the second order. The ancillary direction at the data $(A^0, 0)$ is $(0, \ldots, 0, 1)$ and in terms of the original coordinates is

$$\begin{aligned} v_u(y^0) &= -\left[\frac{\partial z(y;\theta)}{\partial y}\right]^{-1}\frac{\partial z(y;\theta)}{\partial \theta}\bigg|_{(y^0,\widehat{\theta}^0)} u \\ &= -\frac{\partial y}{\partial \theta}\bigg|_{(y^0,\widehat{\theta}^0)} u, \end{aligned}$$

where the inverse matrix is block diagonal and $\partial y/\partial\theta$ is calculated for fixed pivotal value.

The above analysis generalizes that in Section 5.1 of Fraser and Reid (1995).

5 Second Order Coordinates for a Data Component

We consider a p-dimensional data component y_i with p-dimensional pivotal quantity $z^i\ (y_i;\theta)$ and examine how changes in θ near some θ_0 produce data effects through the pivotal quantity. For notational simplicity we omit the designation i, and examine changes in θ in general directions as opposed to the fixed direction u used with (16); then

$$V(y) = -\left[z_y(y;\theta_0)\right]^{-1} z_{;\theta}(y;\theta_0), \tag{22}$$

records p vectors for the component p dimensional variable which describe the effect on y caused by change $(d\theta_1, \ldots, d\theta_p)$ at θ_0.

For second order coordinates we examine displacements of y of magnitude $n^{-1/2}$ near some initial y^0:

$$\begin{aligned} V(y) &= \left.\frac{dy}{d\theta}\right|_{\theta_0} \\ &= V + (y_1 - y_1^0)V^1 + \cdots + (y_p - y_p^0)V^p, \end{aligned}$$

where y_α here designates the αth coordinate of the data component, $V = V(y^0)$, and $V^\alpha = V_{y_\alpha}(y^0) = (\partial/\partial y_\alpha)V(y)\big|_{y^0}$ is the derivative of $V(y)$ with respect to y_α at y^0.

Now in the pattern of Section 3 let x designate the change induced by parameter change $(d\theta_1, \ldots, d\theta_p)$ at θ_p; we take x to be a linear transformation of $y - y^0$ chosen so that V is the identity matrix, and obtain

$$\begin{aligned} V(x) &= V + x_1V^1 + \cdots + x_pV^p \\ &= (v^1 \cdots v^p) + x_1(v^{11} \cdots v^{1p}) + \cdots + x_p(v^{p1} \cdots v^{pp}), \end{aligned} \tag{23}$$

and $v^1 \ldots v^p$ are unit coordinate vectors.

We will also be interested in a general change of variables $\widetilde{x} = \widetilde{x}(x)$ which for second order coordinates takes the form

$$\begin{aligned} \widetilde{x}_1 &= x_1 + \frac{(a_1^{11}x_1^2 + 2a_1^{12}x_1x_2 + \cdots)}{2}, \\ \widetilde{x}_p &= x_p + \frac{(a_p^{11}x_1^2 + 2a_p^{12}x_1x_2 + \cdots)}{2}, \end{aligned} \tag{24}$$

giving

$$\begin{aligned} \frac{d\widetilde{x}}{dx} &= I + x_1(a^{11} \cdots a^{1p}) + \cdots + x_p(a^{p1} \cdots a^{pp}) \\ &= I + x_1A^1 + \cdots + x_pA^p, \end{aligned} \tag{25}$$

where the column vectors A^α satisfy the symmetry relation $a^{\alpha\beta} = a^{\beta\alpha}$.

If we combine (23) and (24) we obtain

$$\left.\frac{d\widetilde{x}}{d\theta}\right|_{\theta_0} = V + x_1(A^1 + V^1) + \cdots + x_p(A^p + V^p). \tag{26}$$

Note that the arrays V^α can be arbitrary but the arrays A^α have the symmetry mentioned above; this relates to the general nonintegrability of the p-dimensional vector field $[V(x)]$.

First consider a general directional increment $t(\Delta_1, \ldots, \Delta_p)$ from θ_0. The equation (23) can be integrated to the second order giving

$$x = v^\alpha \Delta_\alpha + \frac{v^{\alpha\beta} \Delta_\alpha \Delta_\beta}{2}, \tag{27}$$

where summation notation on α, β is over indices $1, \ldots, p$: for the integration note that $dx/dt = b + ct$ has solution $x = bt + ct^2/2$.

We now single out a particular direction from θ_0 and for simplicity take this to be pth coordinate change. We then explore how the family of pth coordinate trajectories differ from radial trajectories from the initial y^0.

First consider an increment $t(\Delta_1, \ldots, \Delta_{p-1}, 0)$ from θ_0. In a similar way the equations integrate to give

$$x = v^\ell \Delta_\ell + \frac{v^{\ell m} \Delta_\ell \Delta_m}{2}, \tag{28}$$

where ℓ, m range over $1, \ldots, p-1$. Next consider a change $(0, \ldots, 0, \Delta_p)$ from some initial x^0. We obtain

$$x = x^0 + \left(v^p \Delta_p + \frac{v^{pp} \Delta_p \Delta_p}{2} \right) + x^0_\alpha v^{\alpha p} \Delta_p. \tag{29}$$

If $x^0 = 0$ the path is given by the expression in parentheses; otherwise the path has an offset term given by the final term.

We then take x^0 to be the expression (28) and from (29) obtain

$$x = \left(v^\ell \Delta_\ell + \frac{v^{\ell m} \Delta_\ell \Delta_m}{2} \right) + \left(v^p \Delta_p + \frac{v^{pp} \Delta_p \Delta_p}{2} \right) + v^{\ell p} \Delta_\ell \Delta_p; \tag{30}$$

the first parentheses contains the initial point, the second contains the standard displacement if the $\Delta_\ell = 0$, and the last term is an offset term due to non parallel action. By choosing $a^{\ell p} = -v^{\ell p}$ in (24) we can eliminate the offset term in the new coordinates and by choosing $a^{pp} = -v^{pp}$ we eliminate the quadratic term in the second parentheses. Equation (30) then becomes

$$x = \left(v^\ell \Delta_\ell + \frac{v^{\ell m} \Delta_\ell \Delta_m}{2} \right) + v^p \Delta_p. \tag{31}$$

This reflects in a local coordinate notation the general rectification results indicated by (15).

Now consider the general directional increment $(\Delta_1, \ldots, \Delta_p)$; this gives the trajectory (27) which differs from (30) by the displacement

$$d = (v^{p\ell} - v^{\ell p}) \frac{\Delta_\ell \Delta_p}{2}. \tag{32}$$

An adjustment (24) of coordinates cannot in general remove this term, which gives a measure of the nonintegrability, the Frobenius conditions. We use this displacement in the next section, to verify second order ancillarity.

6 Second Order Ancillary Directions

Consider a composite data variable y with n independent coordinates y_i having a common parameter θ of dimension p. We assume as in Section 3 that each y_i is p dimensional with an effective p dimensional pivotal quantity. Also we note that in wide generality a pivotal can be obtained by the successive coordinate-by-coordinate conditional distribution functions; with independent scalar variables this typically coincides with most reasonable pivotal quantities.

In this section we use the results from Sections 3, 4 and 5 to construct an $np - p$ dimensional second order ancillary that has tangent vectors $V = (v_1 \ldots v_p)$, at the data point y^0 as given by the first formula in (4) or by (18).

We begin by considering a directional change $d\theta = \Delta u$ for some initial direction u from $\theta_0 = \widehat{\theta}^0$ and use the first derivative ancillary results of Section 3 to develop a tentative $np - p$ dimensional variable that has certain ancillary properties. As a second step we develop a modified ancillary type variable that is first derivative first order ancillary under arbitrary directional change. Then as a third step we further modify the ancillary variable to be second order at $\widehat{\theta}^0$. The significant feature of these steps is that the tangent directions at the observed data point remain unchanged; it follows that the second order ancillary directions are given by the first derivative argument available from (18).

For the first step we apply the notation from Section 5 and let y_0 be some point adjacent to y^0 but with the same maximum likelihood value as y^0 and x_i be the departure for the ith coordinate for that adjacent y_0. Correspondingly let v_i^α and $v_i^{\alpha\beta}$ be the associated vectors as in (27) and (30). Then for the composite variable we let x, v^α, $v^{\alpha\beta}$ be the stacked vectors from the x_i, v_i^α, $v_i^{\alpha\beta}$; these are np dimensional vectors. We thus have an initial point y_0 on the observed maximum likelihood surface and an adjustment x giving a general point $y = y_0 + x$. Various definitions for $x = x\ (\cdot, y_0)$ will define various candidate ancillary-type surfaces through y_0.

First consider a radial change $(\Delta_1, \ldots, \Delta_{p-1}, 0)$ for the first $p-1$ parameter coordinates followed by a change Δ_p in the pth coordinate. From (30) we obtain the surface described by

$$x = v^\alpha t_\alpha + v^{\ell m} t_\ell t_m/2 + v^{pp} t_p^2/2 + v^{\ell p} t_\ell t_p, \tag{33}$$

using coordinates $(t_1, \ldots, t_p)$ on the surface. As noted with (31) a change in x coordinates can give the simpler expression,

$$x = v^{\alpha} t_{\alpha} + v^{\ell m} t_{\ell} t_m / 2, \tag{34}$$

based on having $v^{\alpha p} = 0$, and in these x coordinates the Δ_p trajectories are parallel and linear in accord with Section 3 results. For the investigation of asymptotic properties, we will use $\Delta_{\alpha} = \delta_{\alpha}/n^{1/2}$ and correspondingly $t_{\alpha} = z_{\alpha}/n^{1/2}$, and then rewrite (33) in terms of the standardized departures as

$$n^{1/2} x = v^{\alpha} z_{\alpha} + v^{\ell m} z_{\ell} z_m/(2n^{1/2}) + v^{pp} z_p^2/(2n^{1/2}) + v^{\ell p} z_{\ell} z_p / n^{1/2}, \tag{35}$$

to order $O(n^{-1})$; note that z is again used temporarily and is not to be confused with the pivotal quantity. Thus the first derivative ancillary curves corresponding to change in the u_p direction are

$$\begin{aligned} T_{u_p}(z_1, \ldots, z_{p-1}, y_0) = \{ y_0 + n^{-1/2} [v^{\ell} z_{\ell} + v^{\ell m} z_{\ell} z_m/(2n^{1/2}) + v^{p} z_p \\ + v^{pp} z_p^2/(2n^{1/2}) + v^{\ell p} z_{\ell} z_p / n^{1/2}]; z_p \text{ in } R \}; \end{aligned} \tag{36}$$

these are straight lines in the modified notation corresponding to (34). Then let S_{u_p} be the surface generated as the union of the preceding trajectories,

$$\begin{aligned} S_{u_p}(y_0) = \{ y_0 + n^{-1/2} [v^{\alpha} z_{\alpha} + v^{\ell m} z_{\ell} z_m/(2n^{1/2}) + v^{pp} z_p^2/(2n^{1/2}) \\ + v^{\ell p} z_{\ell} z_p / n^{1/2}] : z_{\alpha} \text{ in } R \}. \end{aligned} \tag{37}$$

Of course $S_{u_p}(y_0)$ is also first derivative ancillary.

The distribution of z_p given $z_1, \ldots, z_{p-1}$ is location with respect to δ and is asymptotically normal by Brenner et al. (1982). Also the log joint conditional density for $(z_1, \ldots, z_p)$ is $O(n)$ and thus asymptotically normal by the vector version (Fraser et al., 1997) of the preceding. By applying an appropriate linear transformation to $\Delta = (\theta - \theta_0)$ based on $\hat{\jmath}^0$ we obtain an identity covariance matrix for the limiting normal distributions of the z's. Then in an obvious extension of (15) the full density with parameter change Δu_p can be written

$$h(y_0) g(z_1, \ldots, z_{p-1} \mid y_0) g(z_p - \delta_p \mid z_1, \ldots, z_{p-1}, y_0), \tag{38}$$

where the two g densities are first order standardized normal and the dependence on θ_0 is implicit.

As a second step we modify the preceding ancillary type surface based on u_p to obtain first order independence of the choice of direction u for the parameter change. The modified surface is obtained as the union of the trajectories through y_0: the u trajectory through y_0 is obtained from (27),

$$T_u(y_0) = \{ y_0 + n^{-1/2} [v^{\alpha} t u_{\alpha} + v^{\alpha\beta} t^2 u_{\alpha} u_{\beta}/(2n^{1/2})] : t \text{ in } R \}; \tag{39}$$

and the surface formed from these is

$$S(y_0) = \{y_0 + n^{-1/2}[v^\alpha z_\alpha + v^{\alpha\beta} z_\alpha z_\beta/(2n^{1/2})] : z_\alpha \text{ in } R, \text{ free}\alpha\}. \quad (40)$$

A point on $S(y_0)$ differs from a point with the same coordinates on $S_{u_p}(y_0)$ by the displacement

$$d(z_1, \ldots, z_{p-1}, z_p; y^0) = (v^{p\ell} - v^{\ell p}) z_\ell z_p / (2n^{1/2}),$$

as calculated from (32). Let $w^{p\ell}$ be $v^{p\ell} - v^{\ell p}$ orthogonalized to the orthonormal vectors v^α. Then to order $O(n^{-1})$ the mapping $\tilde{x} = x + w^{p\ell} z_\ell z_p/(2n^{1/2})$ carries $S_{u_p}(y_0)$ to $S(y_0)$ and amounts to an $n^{-1/2}$ twisting of the surface.

Now consider the distributional effect of the mapping just described under parameter change $\delta u_p/n^{1/2}$. If we integrate out z_p we are integrating along a tilted trajectory (Section 4) and the marginal log density for $(z_1, \ldots, z_{p-1})$ is modified by a term $b^{p\ell} z_\ell \delta/(2n^{1/2})$. Then when we integrate over $z_1, \ldots, z_{p-1}$ we are integrating an odd function and obtain that the integral of the modifying term is zero. Thus the surface $S(y_0)$ has first order first derivative ancillarity with respect to u_p; but the surface $S(y_0)$ is not dependent on the direction u_p and is thus first order first derivative ancillary at θ_0 with respect to change in an arbitrary direction.

We now examine the quadratic effect expressed in terms of δ^2. The methods in Fraser and Reid (1995, Appendix) extend to the more general context discussed in Section 3 and show that the marginal log density of $z_1, \ldots, z_{p-1}$ acquires a log likelihood term $-cx\delta^2/(2n^{1/2})$ where x is a rescaled variable having a first order marginal and conditional normal distribution; of course c and x depend on the direction u_p.

Now consider a general departure $d\theta = \delta^\alpha u_\alpha/n^{1/2}$ where the $u_1, \ldots, u_p$ are the unit coordinate vectors. By straightforward extension of the preceding paragraph we have that the marginal log likelihood of y_0 acquires a term

$$\frac{-w_{\alpha\beta}\delta^\alpha\delta^\beta}{2n^{1/2}}, \quad (41)$$

where the $w_{\alpha\beta} = w_{\beta\alpha}$ are rescaled normal variables to the first order, and with the same accuracy near the observed ancillary surface can be replaced by their observed values $w^0_{\alpha\beta}$. We now apply a rotation to $(\delta^1, \ldots, \delta^p)$ to diagonalize the quadratic expression (41):

$$\frac{-\tilde{w}_{\alpha\beta}\tilde{\delta}^\alpha\tilde{\delta}^\alpha}{2n^{1/2}}.$$

The corresponding $\tilde{z}_\alpha$ are independent with a normal $(\tilde{\delta}^\alpha, 1)$ distribution to the first order. Then from the bending results at (19) we can bend the $\tilde{z}_\alpha$ axis appropriately and remove the marginal component $-\tilde{w}_{\alpha\alpha}\tilde{\delta}^\alpha\tilde{\delta}^\alpha/(2n^{1/2})$, and do this for $\alpha = 1, \ldots, p$. It follows that this third modified surface is second order ancillary and has unchanged tangent directions at the data point y^0. The ancillary matrix V in the original coordinates is given by (18).

Our procedure gives a second order ancillary surface at the observed data and the only information used concerning that ancillary is its tangent direction array at the data point. This apparent data dependence raises a question that seems primarily foundational at this stage of the development. If we go to some other point on the contour of the ancillary and repeat the procedure there we would generally get a different ancillary contour. However its tangent directions at the original data point would be second order equivalent to the original V, and would thus lead to the same inference; this is developed elsewhere.

7 First Order Ancillary Directions

An affine ancillary was developed by Barndorff-Nielsen (1980) for a k-dimensional exponential model with p-dimensional parameter; the model was assumed to have asymptotic properties as some initial sample size n became large. The affine ancillary is first order and reduces the dimension of the model from the initial k to the dimension p of the parameter. It is based on the asymptotic normality of the score variables in the saturated or full (k, k) exponential model, and its form near the maximum likelihood surface $\widehat{\theta} = \widehat{\theta}^{\circ}$ is to use scores from the full or saturated (k, k) embedding model and condition on those that correspond to nuisance parameters relative to θ at $\widehat{\theta}^{\circ}$.

A somewhat different affine ancillary was developed in Barndorff-Nielsen (1986) for a p-dimensional asymptotic model with p-dimensional parameter and r dimensional interest parameter $\psi(\theta)$. The ancillary is first order and corresponds to a fixed value of the interest parameter $\psi(\theta) = \psi$; its form near the maximum likelihood surface is to condition on the score for a nuisance parameter λ that complements $\psi(\theta)$. The marginal distribution of the ancillary is used to assess the value $\psi(\theta) = \psi$ of the interest parameter.

The two first order affine ancillaries have quite different objectives: for the first, to obtain a conditional assessment of the interest parameter; for the second, to obtain a conditioning variable so as to integrate out nuisance parameter effects. These ancillary techniques however are not focused on the reduction from some original dimension n to a parameter dimension such as p.

In this section we generalize the affine ancillary to handle the first order reduction from an initial n dimensional model to a model having dimension equal to that of the parameter. This leads to second order inference for an interest parameter using the procedures outlined in Section 2. The generalized affine ancillary is examined first for the exponential model and then for the general model.

Consider independent exponential models $f^i(y_i;\theta)$ with p dimensional canonical variables y_i and common p dimensional parameter θ; also let y designate the full variable obtained by stacking the y_i. We do not invoke sufficiency because we seek coordinate by coordinate properties to compare with the second order ancillary results. Then let $\theta_0 = \widehat{\theta}(y^0) = \widehat{\theta}^0$ be the observed maximum likelihood value for the full y. For the mean vector let $\mu^i(\theta) = E(y_i;\theta)$ be the mean of the component y_i and $\mu^i_\theta = \mu^i_\theta(\theta_0) = (\partial/\partial\theta)\mu^i(\theta)\big|_{\theta_0}$ be its gradient with respect to θ; and also let $\mu(\theta)$ and μ_θ be the corresponding stacked np-vector and $np \times p$ array for the full response y.

Let A be a matrix of $np - p$ linearly independent row vectors orthogonal to the p column vectors forming μ_θ. Then Ay has mean $A\mu(\theta)$, and at θ_0 has gradient equal to zero: $A\mu_\theta = 0$. We call Ay first derivative ancillary in mean at θ_0. An interesting and important feature of Ay is that its tangent directions at the data point are given by the vector μ_θ; we write $V_1 = \mu_\theta$.

Rather than checking directly the distributional ancillarity associated with the vectors V_1, we compare the accuracy of the corresponding canonical parameter with that obtained from the second order methods in Section 6. This is a comparison of the q departure measures as the r departure measure (6) is the same. For this it is appropriate to adjust for the summation over the sample coordinate elements and use the modified canonical parameters

$$\varphi_1' = \ell_{;y}(\theta; y^0)V_1 n^{-1} = n^{-1}\sum_{i=1}^{n} \ell_{;y_i}(\theta; y_i^0)V_1^i,$$

$$\varphi_2' = \ell_{;y}(\theta; y^0)V_2 n^{-1} = n^{-1}\sum_{i=1}^{n} \ell_{;y_i}(\theta; y_i^0)V_2^i,$$

where $V_1^i = \mu^i_\theta$ and $V_2^i = \partial y_i/\partial\theta\big|_{(y^0,\theta_0)}$ as developed for (18).

First we standardize the variables in several ways. Let θ be relocated to measure departure from the initial θ_0; then the new $\theta_0 = 0$. Next we reparametrize θ so that $\theta = \varphi_2$. Finally we rescale each y_i so that the canonical parameter $\varphi^i(\theta)$ of its exponential family model $f^i(y_i;\theta)$ has $\partial\varphi^i(\theta)/\partial\theta$ equal to the $p \times p$ identity I at θ_0, and then relocate so $\varphi^i(\theta_0) = 0$. We can then write

$$\ell(\theta; y_i) = \left[\theta + b^i(\theta)\right]' y_i + c^i(\theta) + a,$$

where an individual canonical parameter deviates from θ by $b^i(\theta)$ where $b^i(0) = 0$, $b^i_\theta(0) = 0$ and each coordinate $b^i_\alpha(\theta)$ of $b^i(\theta)$ then has the form $b^i_\alpha(\theta) = \theta' B^i_\alpha \theta/2 + \ldots$ with symmetric $p \times p$ second derivative matrix B^i_α.

Now let z_i be a pivotal variable for the Section 6 development and let $g^i(z_i;\theta) = y_i$ be the corresponding response presentation from the pivotal quantity. Then

$$\begin{aligned} y_i &= g^i(z_i;\theta) = g^i(z_i;0) + g^i_{\theta'}(z_i;0)\theta + \cdots, \\ \mu^i(\theta) &= E[g^i(z_i;0)] + E[g^i_{\theta'}(z_i;0)]\theta + \cdots = \mu^i + c^i\theta + \cdots, \\ y_i - E(y_i;\theta) &= g^i(z_i;\theta) - \mu^i + [g^i_{\theta'}(z_i;\theta) - c^i]\theta + \cdots, \\ V_1^i - V_2^i &= [c^i - g^i_{\theta'}(z_i;0)] = D^i(z_i), \end{aligned}$$

where the $p \times p$ matrix $D^i(z_i)$ has mean value zero. It follows then that $\varphi_1' - \varphi_2' = \varphi_1' - \theta'$ can be expressed as

$$\begin{aligned} \varphi_1' - \theta' &= n^{-1} \sum_{i=1}^{n} \ell_{;y_i}(\theta; y_i^0)(V_1^i - V_2^i) \\ &= n^{-1} \sum_{i=1}^{n} [\theta + b^i(\theta)]' D^i(z_i). \end{aligned}$$

Now consider the parameter difference in terms of the standardized departures $\varphi_2 = \theta = \delta/n^{1/2}$ and $\varphi_1 = \delta_1/n^{1/2}$

$$\delta_1' - \delta' = \delta' \cdot \frac{1}{n} \sum_{i=1}^{n} D^i(z_i) + \frac{1}{n^{1/2}} \cdot \frac{1}{n} \sum_{i=1}^{n} n b^i(\delta/n^{1/2}) D^i(z_i).$$

A coefficient of a coordinate of δ is an average of variables with means zero and thus has the form $0 + z/n^{1/2}$; this gives a contribution that is of order $O(n^{-1/2})$ but to this order is just a linear equivalent of the canonical parameter. A coefficient of a $\delta_\alpha\delta_\beta/n^{1/2}$ is also an average of the form as just described; this gives a quadratic contribution that is of order $O(n^{-1})$. We thus have that φ_1 and φ_2 are equivalent to the second order and it follows that V_1 leads to second order inference. The steps in the analysis suggest however that the convergence may be slow.

Now more generally consider independent continuous models $f^i(y_i;\theta)$ with p dimensional variables y_i and common p dimensional parameter θ. As before let $\theta_0 = \widehat{\theta}(y^0) = \widehat{\theta}^0$ be the observed maximum likelihood value from the full data y^0.

We approximate each component model by an exponential model centered at θ_0. Let t_i be the score variable for the ith component

$$t_i = \frac{\partial}{\partial\theta} \log f^i(y_i;\theta)\big|_{\theta_0},$$

$c_i(\theta)$ be its cumulant generating function, and $k_i(y_i;\theta_0)$ be the Jacobian between variables

$$k_i(y_i;\theta_0) = \frac{\partial t_i}{\partial y_i} = \frac{\partial}{\partial y_i}\frac{\partial}{\partial\theta} \log f^i(y_i;\theta)\big|_{\theta_0}.$$

Then the following exponential model coincides with the given model to first derivative at θ_0:

$$g^i(t_i;\theta)k_i(y_i;\theta_0) = \exp\left[\theta' t_i - c_i(\theta)\right] f^i(y_i;\theta_0).$$

It follows that first derivative properties at θ_0 of the two model are identical. By this we mean that any calculation using θ_0 or $(d/d\theta)\big|_{\theta_0}$ is the same for the two models; the higher order properties are then examined through the differences between the models. We apply the earlier results in this section to the composite model for the t_i and then adjust for the change of variable.

The usual calculations with score variables gives the information as gradient of the mean

$$\mu_\theta^i = \frac{\partial}{\partial\theta} E(t_i;\theta)\big|_{\theta_0} = i^i(\theta_0),$$

this can be calculated from the variance of the score in the original component model,

$$i^i(\theta_0) = \text{Var}(t_i;\theta_0).$$

The ancillary directions array V_1 for the t variables is thus

$$\widetilde{V}_1 = \begin{bmatrix} i^1(\theta_0) \\ \vdots \\ i^n(\theta_0) \end{bmatrix},$$

and for the original y variables is

$$V_1 = \begin{bmatrix} k_1^{-1}(y_1;\theta_0)i^1(\theta_0) \\ \vdots \\ k_n^{-1}(y_n;\theta_0)i^n(\theta_0) \end{bmatrix}. \tag{42}$$

For second order inference we can then use the array V_1 in (42) to calculate the nominal reparametrization $\varphi(\theta)$ in (3) which together with the observed likelihood $\ell^0(\theta)$ in (1) provides the second order inference for a component parameter using (5) with (6) and (8) in Section 2. Other second order methods may be found in Barndorff-Nielsen and Chamberlin (1994) and DiCiccio and Martin (1993). More recently Skovgaard (1996) developed an explicit test that avoids the direct determination of ancillary characteristics; the approach is more general in that large deviation properties are addressed but restricted in that sampling is in effect from a fixed dimension curved exponential model and the reduction to that dimension is by means of sufficiency. The approach here allows different general models for different components and the asymptotic properties correspond to an increasing number of components designated by the n used throughout.

8 Examples

For an asymptotic model with variable and parameter of the same dimension p, third order highly accurate p-values $p(\psi)$ are available (Section 3) for testing values of a scalar component parameter $\psi(\theta) = \psi$. The information needed for this is available from the observed likelihood $\ell^0(\theta) = \ell(\theta; y^0)$ and the observed likelihood gradient $\varphi(\theta) = \ell_{;y}(\theta; y^0) = (\partial/\partial y)\ell(\theta; y)\big|_{y^0}$.

In this paper we have developed techniques for extending these approximation methods to the case of an initial model of dimension larger than the dimension p of the parameter. All that is needed for this extension is an array $V = (v_1 \ldots v_p)$ of vectors tangent to an approximate ancillary. We then use the likelihood $\ell^0(y)$ and the observed likelihood gradient

$$\varphi(\theta) = \ell_{;y}(\theta; y^0)V = \ell_{;V}(\theta; y^0),$$

calculated in the directions V. For third and second order inference we calculate V_2 or V_1 by formulas (4) and (42) with the main details in Sections 6 and 7.

We examine the vector arrays V_2 and V_1 for several examples and cite the literature where the approximations have been examined and assessed. We do note that third order inference using V_2 needs only simple calculations with an appropriate pivotal quantity plus the standard likelihood ratio and maximum likelihood calculations; the particular case of the exponential model (Example 3) needs a somewhat new looking function which however is easily calculated. For second order inference the gradient of the mean score introduces coordinate by coordinate expected informations and thus by use of the expectation operator requires more than first derivative information at the data point.

Example 1. Consider the regression model $Y = \eta(\beta) + e$ where component error variables e_i have known distributions with mean zero. We have immediately that $e = y - \eta(\beta)$ is pivotal; thus

$$V_2 = \left.\frac{\partial y}{\partial \beta}\right|_{(y^0,\widehat{\theta}^0)} = \left.\frac{\partial \eta}{\partial \beta}\right|_{(y^0,\widehat{\theta}^0)} = X^0,$$

$$V_1 = \left.\frac{\partial E(y)}{\partial \beta}\right|_{\widehat{\theta}^0} = \left.\frac{\partial \eta}{\partial \beta}\right|_{\widehat{\theta}^0} = X^0,$$

where X^0 can be called the local design matrix. In the case of independent coordinates with density $f(z)$, we have then

$$\varphi' = \sum \ell'[y_i^0 - \eta_i(\beta)]\, X_i^0,$$

where $\ell'(z) = (d/dz)\log f(z)$ and X_i^0 is the ith row in X^0. For details and assessment, see Abebe et al. (1996).

Example 2. Consider the nonlinear regression model $y = \eta(\beta) + \sigma e$ where the errors e have a known distribution $f(e)$ with mean zero. We have then that $e = \sigma^{-1}\left[y - \eta(\theta)\right]$ is pivotal and V_2 and V_1 are given as

$$V_2 = (\hat{e}^0, X^0),$$

$$V_1 = \left.\frac{\partial E(y;\theta)}{\partial(\sigma,\beta)}\right|_{\hat{\theta}^0} = (0, X^0),$$

where X^0 is as given in Example 1 and $\hat{e}^0$ is the observed standardized residual $\left[y^0 - \eta(\hat{\beta}^0)\right]/\hat{\sigma}^0$. Details and simulations based on V_2 are discussed in Fraser et al. (1999). Note the small change from V_2 to V_1 that takes third order inference to second order inference.

Example 3. Generalized Linear Models. Consider independent coordinates y_i each with an exponential family model with a canonical parameter θ_i that is a given function $\theta_i = h^i(\beta)$ of a main parameter vector β of dimension p. For developing the second order ancillary we need for each coordinate variable y a function $d(y;\theta)$ that describes how the parameter θ forces the variable y

$$d(y;\theta) = -\frac{F_{;\theta}(y;\theta)}{F_y(y;\theta)},$$

where

$$F_y(y;\theta) = \exp\left[\theta y - c(\theta)\right]h(y) = f(y;\theta),$$

$$F_{;\theta}(y;\theta) = \int_{-\infty}^{y} \left[y - c'(\theta)\right] f(y;\theta)\, dy.$$

This is a well defined function of y and θ that depends on the exponential model and thus on $c(\theta)$, and is easily computed for many models. Indications are that a difference quotient works almost as well.

Let $F_i(y;\theta)$ be the model for y_i with forcing function $d^i(y;\theta)$ and cumulant function $c^i(\theta)$: then

$$\mu^i(\theta) = E(y_i;\theta) = \frac{\partial}{\partial\theta}c^i(\theta) = c^i_\theta(\theta),$$

$$\mu^i_\theta(\theta) = \mathrm{var}(y_i;\theta) = i^i(\theta) = \frac{\partial^2}{\partial\theta^2}c^i(\theta) = c^i_{\theta\theta}(\theta),$$

where $i^i(\theta)$ is the information for the ith coordinate y_i. Also let $(\partial/\partial\beta)h^i(\theta)$ be designated $h^i_\beta(\theta)$. Then

$$V_2 = \begin{pmatrix} d^1(y^0_1, \hat{\theta}^0_1)h^1_{\beta'}(\hat{\beta}^0) \\ \vdots \\ d^n(y^0_n, \hat{\theta}^0_n)h^n_{\beta'}(\hat{\beta}^0) \end{pmatrix}, \quad V_1 = \begin{pmatrix} i^1(\hat{\theta}^0)c^1_\theta(\hat{\theta}^0_1)h^1_{\beta'}(\hat{\beta}^0) \\ \vdots \\ i^n(\hat{\theta}^0)c^n_\theta(\hat{\theta}^0_n)h^n_{\beta'}(\hat{\beta}^0) \end{pmatrix}.$$

For a numerical example consider Example U from Cox and Snell (1981) with preliminary analyses in Fraser et al. (1994). The response is y the lifetime in weeks of leukemia patients and the concomitant variable is x the logarithm of the initial white blood cell count (Feigl and Zelen, 1965). The assumed model is ordinary exponential

$$f_i(y_i;\theta_i) = \exp(-y_i\theta_i + \log\theta_i),$$

and log mean life is taken to be linear in x. To maintain a monotone connection among parameter versions it is convenient to take the parameter to be the particular canonical parameter

$$\phi_i = -\theta_i = -\exp\bigl[-\alpha - \beta(x_i - \overline{x})\bigr];$$

the expectation parameter and its gradient are then

$$\mu^i = \theta_i^{-1} = \exp\bigl[\alpha + \beta(x_i - \overline{x})\bigr],$$
$$\mu^i_{-\theta} = \theta_i^{-2} = \exp\bigl\{2\bigl[\alpha + \beta(x_i - \overline{x})\bigr]\bigr\}.$$

For this example the forcing function $d(y;-\theta)$ is easily calculated from the pivotal quantities $z_i = \log(y_i\theta_i) = \log y_i - \alpha - \beta(x_i - \overline{x})$, each having the extreme value distribution $\exp(z - e^z)$. The forcing function and mean forcing function are

$$d^i(y_i^0, -\widehat{\theta}_i^0) = \frac{\partial y_i}{\partial \theta_i}\bigg|_{(y^0,\widehat{\theta}^0)} = \frac{y_i^0}{\widehat{\theta}_i^0}, \quad \mu_\theta^i(\widehat{\theta}_i^0) = \frac{\widehat{y}_i^0}{\widehat{\theta}_i^0},$$

and

$$V_2 = \begin{pmatrix} y_1^0 & y_1^0(x_1 - \overline{x}) \\ \vdots & \vdots \\ y_n^0 & y_n^0(x_n - \overline{n}) \end{pmatrix}, \quad V_1 = \begin{pmatrix} \widehat{y}_1^0 & \widehat{y}_1^0(x_1 - \overline{x}) \\ \vdots & \vdots \\ \widehat{y}_n^0 & \widehat{y}_n^0(x_n - \overline{x}) \end{pmatrix}.$$

where $\widehat{y}^0$ designates the estimated mean vector calculated from the estimated α and β.

For illustration the example has a computational advantage in being a location model as indicated by the pivotal above; exact p values for assessing the rate parameter are then available by numerical integration. We calculate the confidence percentage points by inverting the significance function $p(\beta)$ for various values of the significance $1-\gamma$ obtaining $\hat{\beta}_\gamma = p^{-1}(1-\gamma)$ for one-sided confidence levels with $\gamma = 0.5\%, 2.5\%, 50\%, 97.5\%, 99.5\%$ and doing this for the various methods discussed: $p_0(\beta) = \Phi\bigl[(\beta - \hat{\beta})(j^{\beta\beta})^{-1/2}\bigr]$ from the standardized maximum likelihood estimate, $p_1(\beta) = \Phi(r)$ from the signed likelihood root (6), the second order $p_2(\beta)$ from V_1, the third order $p_3(\beta)$ from V_2, and the exact $p(\beta)$ obtained by numerical integration. The values are recorded in Table 1.

The sample of 17 in the present context gives rather precise information and the likelihood values p_1 are already close to the exact p; and the

TABLE 1. One sided confidence percentage points for the rate parameter β with the Feigl and Zelen data

γ	0.5%	2.5%	50%	97.5%	99.5%
$p_0^{-1}(1-\gamma)$	−2.1747	−1.9199	−1.1092	−0.2987	−0.0040
$p_1^{-1}(1-\gamma)$	−2.1705	−1.9153	−1.1092	−0.2797	−0.0161
$p_2^{-1}(1-\gamma)$	−2.1815	−1.9243	−1.1050	−0.2847	−0.0167
$p_3^{-1}(1-\gamma)$	−2.1719	−1.9147	−1.1036	−0.2745	−0.0065
$p^{-1}(1-\gamma)$	−2.1724	−1.9152	−1.1041	−0.2750	−0.0069

TABLE 2. One sided confidence percentage points for the rate parameter β with the Feigl and Zelen data: subsample of n=5

γ	0.5%	2.5%	50%	97.5%	99.5
$p_0^{-1}(1-\gamma)$	−4.1760	−3.4983	−1.3414	0.8154	1.4934
$p_1^{-1}(1-\gamma)$	−4.6547	−3.8064	−1.3414	1.3703	2.4033
$p_2^{-1}(1-\gamma)$	−4.6794	−3.8069	−1.2562	1.5718	2.6467
$p_3^{-1}(1-\gamma)$	−4.7325	−3.8409	−1.2761	1.5699	2.6515
$p^{-1}(1-\gamma)$	−4.7254	−3.8474	−1.2772	1.5610	2.6484

third order values p_3 are very close to the exact p. An interesting point to note is that the second order values p_2 tend to correct in the wrong direction from the likelihood values; an explanation can be found in the fact (Barndorff-Nielsen and Chamberlin, 1994) that the likelihood for this example is already an adjusted and a modified likelihood and thus in itself produces unusually close values.

To obtain a picture of the small sample behaviour we drew a random subsample obtaining $(x, y) = [(-0.368, 16), (0.392, 22), (-0.178, 56), (0.572, 5), (-0.418, 39)]$. The various confidence percentage points are recorded in Table 2.

9 References

Abebe, F., D.A.S. Fraser, N. Reid, and A. Wong (1996). Nonlinear regression: third order significance. *Utilitas Mathematica 47*, 1–17.

Barndorff-Nielsen, O.E. (1980). Conditionality resolutions. *Biometrika 67*, 293–310.

Barndorff-Nielsen, O.E. (1986). Inference on full or partial parameters based on the standardized, signed log likelihood ratio. *Biometrika 73*, 307–322.

Barndorff-Nielsen, O.E. (1991). Modified signed log likelihood ratio. *Biometrika 78*, 557–563.

Barndorff-Nielsen, O.E. and S.R. Chamberlin (1994). Stable and invariant adjusted directed likelihoods. *Biometrika 81*, 485–499.

Brenner, D., D.A.S. Fraser, and P. McDunnough (1982). On asymptotic normality of likelihood and conditional analysis. *Canad. J. Statist. 10*, 163–172.

Cox, D.R. and E.J. Snell (1981). *Applied Statistics*. London: Chapman and Hall.

DiCiccio, T.J., C.A. Field, and D.A.S. Fraser (1990). Approximation of marginal tail probabilities and inference for scalar parameters. *Biometrika 77*, 77–95.

DiCiccio, T.J. and M.A. Martin (1993). Simple modifications for signed likelihood ratio statistics. *J. Roy. Statist. Soc. Ser. B 55*, 305–316.

Feigl, P. and M. Zelen (1965). Estimation of exponential survival probability with concomitant information. *Biometrika 21*, 826–838.

Fraser, D.A.S. (1964). Local conditional sufficiency. *J. Roy. Statist. Soc. Ser. B 26*, 52–62.

Fraser, D.A.S. (1988). Normed likelihood as saddlepoint approximation. *J. Mult. Anal. 27*, 181–193.

Fraser, D.A.S., P. McDunnough, and N.A. Taback (1997). Improper priors, posterior asymptotic normality, and conditional inference. In N. Johnson and N. Balakrishnan (Eds.), *Advances in the Theory and Practice of Statistics*, pp. 563–569. New York: Wiley.

Fraser, D.A.S., G. Monette, K.W. Ng, and A. Wong (1994). Higher order approximations with generalized linear models. In T. Anderson, K. Fang, and I. Olkin (Eds.), *Multivariate Analysis and Its Applications*, Volume 24 of *Inst. Math. Statist. Lect. Notes and Monograph Series*, Hayward, pp. 253–262.

Fraser, D.A.S. and N. Reid (1993). Simple asymptotic connections between densities and cumulant generating function leading to accurate approximations for distribution functions. *Statist. Sinica 3*, 67–82.

Fraser, D.A.S. and N. Reid (1995). Ancillaries and third order significance. *Utilitas Mathematica 47*, 33–53.

Fraser, D.A.S., N. Reid, and J. Wu (1999). A simple general formula for tail probabilities for frequentist and Bayesian inference. *Biometrika 86*, 249–264.

Fraser, D.A.S., A. Wong, and J. Wu (1999). Regression analysis, nonlinear or nonnormal: simple and accurate p-values from likelihood analysis. *J. Amer. Statist. Assoc. 94*, 1286–1295.

Lugannani, R. and S.O Rice (1980). Saddlepoint approximation for the distribution of the sums of independent random variables. *Adv. Appl. Prob. 12*, 475–490.

Pierce, D.A. and D. Peters (1992). Practical use of higher order asymptotics for multiparameter exponential families (with discussion). *J. Roy. Statist. Soc. Ser. B 54*, 701–738.

Skovgaard, I.M. (1986). Successive improvements of the order of ancillarity. *Biometrika 73*, 516–519.

13

The Relevance Weighted Likelihood With Applications

F. Hu and J.V. Zidek

ABSTRACT In this article we describe an extension based on relevance weighting of Wald's classical likelihood theory. The extension allows bias to be traded for precision in the likelihood setting like bias is traded for variance in nonparametric regression. All relevant sample information can thereby be used while bias is filtered out. We describe and demonstrate the use of both the nonparametric and parametric likelihoods. The latter is used to develop a method for forecasting goals in ice-hockey.

1 Introduction

In this paper we describe an extension of Fisher's classical likelihood that Hu (1994) introduces as the "Relevance Weighted Likelihood (REWL)". It generalizes the local likelihood, defined in the context of nonparametric regression by Tibshirani and Hastie (1987) that was extended as a local likelihood by Staniswalis (1989) and as a quasi-local-likelihood by Fan et al. (1995).

In contrast to the local likelihood, the REWL can be global as demonstrated by one of the applications in Hu and Zidek (1997) where the celebrated James–Stein estimator is found to be a maximum (relevance weighted) likelihood estimator (with relevance weights estimated from the data).

The relevance weights allow bias to be traded for precision in the likelihood setting, as bias is traded for variance in the nonparametric regression setting. The need for such a theory has become increasingly important as the scale of modern experimental science has grown in its space-time scales thanks to demand (for example in environmental science) combined with feasibility (for example through information technology). On these scales, replicating an experiment may be completely infeasible thus leading to the need for a theory that embraces bias without sacrificing the goal of efficiency so fundamental to modern statistical thinking.

The theory described here enables the bias-precision trade off to be made without relying on the Bayesian approach (Berger, 1985). The latter permits the bias-variance trade off to be made in a conceptually straightforward manner. Reliance on empirical Bayes methods softens the demands for realistic prior modeling in complex problems. Efron (1996) illustrates the empirical Bayes approach in such problems and uses the term "relevance" in a manner similar to that of Hu (1994).

Our theory offers a simple alternative to the empirical Bayesian approach for complex problems. At the same time it links within a single formal framework, a diverse collection of statistical domains such as weighted least squares, nonparametric regression, meta-analysis and shrinkage estimation. That synergistically suggests new methods and problems in these domains as we will attempt to show. Meanwhile the REWL comes with an underlying general theory that extends Wald's theory for the maximum likelihood estimator (Hu, 1997). We will try to demonstrate its advantages in this article.

In the next section, we define the REWL in both the parametric and nonparametric cases. In the latter case we include large sample theory obtained by borrowing results from the well-developed theory of empirical processes. Moreover we describe its counterpart for the parametric case. Then in Section 3 we show through simulated experiments how the nonparametric REWL can be applied. In Section 4 we turn to the parametric REWL and complete an analysis begun in Hu and Zidek (1997). In particular we show how the REWL may be used to develop a method for forecasting goals in ice-hockey using the Poisson model. We conclude with a brief discussion in Section 5.

2 The Relevance Weighted Likelihood

In this section we describe the REWL in both the nonparametric and parametric cases. To do so we assume Y_i, $i = 1, \ldots, n$ are independently distributed random variables or vectors, each having an associated population distribution with probability density and cumulative distribution (pdf and cdf, respectively) f_i and F_i. Let $\mathbf{Y} = (Y_1, \ldots, Y_n)$ be the vector or matrix of these measurable attributes. From each population i, $n_i \geq 0$ items are randomly and independently sampled, yielding $\mathbf{Y}_i = (Y_{i1}, \ldots, Y_{in_i})$, Y_{ij} representing the Y_i measured on the jth item sampled from the ith population $j = 1, \ldots, n_i$, $i = 1, \ldots, n$ (the null vector when $n_i = 0$). Assume the Y_{ij}, $j = 1, \ldots, n_i$ are independent and identically distributed, each having its associated population distribution. Denote the realization of $\mathbf{Y}_i$ by $\mathbf{y}_i$, $i = 1, \ldots, n$.

Inferential interest concerns attributes of a population distribution which may be:

(1) the joint population distribution of the Y_i i.e. of $\mathbf{Y}$;

(2) exactly one of these distributions; or

(3) another distribution thought to resemble these n populations.

For definiteness we will present the theory for case (2). (Hu and Zidek (1997) treat the general case.) Thus we will describe the REWL that enables data from all populations to assist in inference about population i.

Starting from the Akaike entropy maximization principle (Akaike, 1973, 1977, 1978, 1982, 1983, 1985), Hu and Zidek (1997) derive the REWL in the nonparametric and parametric cases (the NP-REWL and P-REWL, respectively). As a function of the unknown population probability density function (pdf) g, the NP-REWL is

$$g \to \prod_{j=1}^{n} \prod_{l=1}^{n_j} g^{\lambda_{ij}/n_j}(y_{jl}). \tag{1}$$

Its parametric counterpart is

$$\theta \to \prod_{j=1}^{n} \prod_{l=1}^{n_j} f_i^{\lambda_{ij}/n_j}(y_{jl} \mid \theta_i), \tag{2}$$

where $\theta = (\theta_1, \ldots, \theta_n)$. In both cases $\lambda_{ij} \geq 0$ and we take $\lambda_{ij}/n_j = 0$ when $n_j = 0$ for all i and j.

The relevance weights $\{\lambda_{ij}\}$ enable the investigator to trade off bias for precision in estimating the likelihood for population i using the data from the remaining populations. Although Hu and Zidek (1997) suggest a general method for selecting the weights, they note that the choice of these weights is best made within the context of the particular problem being considered. In particular, low dimensional parametric models for these weights may suggest themselves. We use a conjunction of modeling and the proposed Hu–Zidek approach in Section 4 to demonstrate how these weights can be selected in practice. We now proceed with the assumption that the relevance weights have been determined.

The maximum REWL estimator (MREWLE) of the population pdf using (1) in the nonparametric case is easily shown to be:

$$\hat{f}_i = \sum_{j=1}^{n} \omega_{ij} f_j^{\text{emp}}, \tag{3}$$

where f_j^{emp} denotes the discrete empirical density function for the jth population obtained from the empirical cumulative distribution function (cdf) F_j^{emp} while $\omega_{ij} \propto \lambda_{ij}$ satisfy $\sum_{j=1}^{n} \omega_{ij} = 1$. From (3) we obtain the MREWLE of the cdf for population i

$$\widehat{F}_i = \sum_{j=1}^{n} \omega_{ij} F_j^{\text{emp}}. \tag{4}$$

We call this the Relevance Weighted Empirical Distribution (REWED) after Hu and Zidek (1993). It is a special weighted empirical process (Shorack and Wellner, 1986).

The MREWLE for θ_i is found by maximizing (2). Hu (1997) shows that the theory of Wald for the classical MLE extends to the MREWLE under a suitable adaptation of Wald's assumptions. However we have not hitherto published a corresponding asymptotic theory for the nonparametric case. That theory can be developed by calling on the theory for empirical processes. We present the elements here to complete the basic development of the theory.

The focus of that theory is the error $\widehat{F}_i - F_i$, F_i denoting the cdf for population i. That error can be represented in terms of its two additive components: $\widehat{F}_i - F_i = \widehat{F}_i - \overline{F}_i + \overline{F}_i - F_i$ where $\overline{F}_i = \sum_{j=1}^n \omega_{ij} F_j$. The first of these components $\widehat{F}_i - \overline{F}_i$ represents the precision of $\widehat{F}_i$ as an unbiased estimator of $\overline{F}_i$. The second component represents the bias in $\widehat{F}_i$ as an estimator of F_i, namely $\overline{F}_i - F_i$.

The bias cannot be removed by sampling. However as the sample size increases we can reduce the bias by making the relevance weights concentrate more heavily on the populations that most closely resemble the one of interest; contextual knowledge will be needed. For example, in nonparametric regression we might know the population cdf's vary smoothly as functions of an ancillary variable like time t.

In our asymptotic theory we assume that the bias approaches zero as $n \to \infty$. In that theory, all the elements which vary with the sample size including the relevance weights have the suffix n attached. At the same time since the population of interest i has been fixed throughout we drop that suffix. Thus in the sequel, F replaces F_i while $\widehat{F}_n$ replaces $\widehat{F}_{ni}$. Finally we assume $n_j \equiv 1$

The key elements of our asymptotic theory are:

- the REWED defined by $\widehat{F}_n(y) = \sum_{j=1}^n \omega_{nj} I(Y_{nj} \leq y)$;
- the *relevance weighted average distribution function* (REWADF) defined by
$$\overline{F}_n(y) = \sum_{j=1}^{n} \omega_{nj} F_{nj}(y), \quad -\infty < y < \infty;$$
- the pth quantile of $\overline{F}_n$ defined by $\xi_{p(n)} = \inf\{y : \overline{F}_n(y) \geq p\}$, $0 < p < 1$;
- the pth *relevance weighted quantile* (REWQ) estimator defined by $\hat{\xi}_{np} = \inf\{y : F_n(y) \geq p\}$ for a sample $(Y_{n1}, \ldots, Y_{nn})$.

Letting ξ_p denote the pth quantile of the ith population distribution and using the notation just defined above, we may state our asymptotic results beginning with the strong consistency of $\widehat{F}_n$. As noted by an anonymous referee, the proofs may be omitted since they follow from standard theory (Koul, 1992).

Theorem 1. *Strong consistency of $\widehat{F}_n$:*

(a) *Suppose $\sum_{n=1}^{\infty} \exp(-\varepsilon^2 K_n) < \infty$ for all $\varepsilon > 0$, with $K_n = (\sum_{j=1}^{n} \omega_{nj}^2)^{-1}$. Then $|F_n(y) - \overline{F}_n(y)| \to 0$ a.s. for all y.*

(b) *Further, if $|F(y) - \overline{F}_n(y)| \to 0$ for all y, then $|\widehat{F}_n(y) - F(y)| \to 0$ a.s. for all y.*

Corollary 1. *If $\log(n)/K_n = o(1)$, then $|\widehat{F}_n(y) - \overline{F}_n(y)| \to 0$ a.s. for every y.*

The hypotheses of the theorem are easily satisfied for example, when $\max_j(\omega_{nj}) = o[\log(n)^{-1}]$. The assumption $|F(y) - \overline{F}_n(y)| \to 0$ for all y is essential; without this, we cannot get a consistent estimator of the cdf $F(y)$. Qualitatively this condition is the one which gives operational meaning to the notion of "relevance weights".

Even stronger conclusions are the following.

Theorem 2. *Uniformly strong consistency of $\widehat{F}_n(y)$:*

(a) *Under the hypothesis of Theorem 1(a), and the further assumptions that (i) $\sup_{y,n} \bar{f}_n(y)$ is bounded and (ii)*

$$\limsup_{M\to\infty} \sup_{n} \left[(1 - \overline{F}_n(M)), \overline{F}_n(-M)\right] \to 0,$$

where $\bar{f}_n(y)$ is the derivative of $\overline{F}_n(y)$, then $\sup_y |\widehat{F}_n(y) - \overline{F}_n(y)| \to 0$, a.s.

(b) *Further, if $\sup_y |F(y) - \overline{F}_n(y)| \to 0$, then $\sup_y |\widehat{F}_n(y) - F(y)| \to 0$, a.s.*

When the distributions underlying our investigation derive from the same family, the conditions of the last theorem are usually satisfied.

Theorem 3. *Strong consistency of $\hat{\xi}_{np}$:*

(a) *Suppose $y = \xi_{p(n)}$ solves uniquely the inequalities $\overline{F}_n(y-) \le p \le \overline{F}_n(y)$. Then $\hat{\xi}_{np} - \xi_{p(n)} \to 0$ a.s. as $n \to \infty$.*

(b) *Furthermore if $\sup_y |F(y) - \overline{F}_n(y)| \to 0$, then $\hat{\xi}_{np} - \xi_p \to 0$ a.s. for $n \to \infty$.*

The uniqueness of $\xi_{p(n)}$ required in the last theorem cannot be dropped. Next we give a useful probabilistic inequality for quantile estimators.

Theorem 4. *Suppose $x = \xi_{p(n)}$ solves uniquely, the inequalities $\overline{F}_n(x-) \leq p \leq \overline{F}_n(y)$ for any given $p \in (0,1)$. Then*

$$P\left(|\hat{\xi}_{np} - \xi_{p(n)}| > \varepsilon\right) \leq 2\exp\left[-2\delta_\varepsilon^2(n)K_n\right],$$

for every $\varepsilon > 0$ and n, where $\delta_\varepsilon(n) = \min\left[\overline{F}_n(\xi_{p(n)} + \varepsilon) - p, p - \overline{F}_n(\xi_{p(n)} - \varepsilon)\right]$.

The previous theorem shows $P\left(|\hat{\xi}_{np} - \xi_{p(n)}|\varepsilon\right)$ converges to 0 exponentially fast. The value of ε (> 0) may depend upon K_n if desired. These bounds hold for each $n = 1, 2, \ldots$, and so may be applied for any fixed n as well as for asymptotic analysis.

Except for the case of iid random variables, we cannot always find the exact distribution of $\hat{\xi}_{np}$. The asymptotic distribution of $\hat{\xi}_{np}$ given in the following theorem may therefore be useful.

Theorem 5. *Let $0 < p < 1$ and $V_n = \sum_{j=1}^n \omega_{nj}^2 F_{nj}(\xi_{p(n)})\left[1 - F_{nj}(\xi_{p(n)})\right]$. Assume $\overline{F}_n$ is differentiable at $\xi_{p(n)}$, $\inf_n \overline{F}_n'(\xi_{p(n)}) > c > 0$ while at the same time $\max_{1 \leq j \leq n}(\omega_{nj} V_n^{-1/2}) \to 0$ as $n \to \infty$. Then*

$$\lim_{n \to \infty} P\left[\bar{f}_n(\xi_{p(n)})(\hat{\xi}_{np} - \xi_{p(n)})V_n^{-1/2} \leq t\right] = \Phi(t),$$

where $\Phi(t)$ is the distribution function of $N(0,1)$.

3 Applying the NP-REWL

In this section, we use REW sample quantiles obtained from the nonparametric REWL to estimate location parameters, and compare these estimators with the weighted sample mean estimators.

Example 1. Let $\{Y_j\}$ be an independent sample with $Y_j \sim N(\mu, \sigma_j^2)$ $j = 1, \ldots, n$, the σ_j^2 being known and μ unknown. An estimate of μ is required.

Using the weighted sample mean in Example 1 to estimate μ seems natural: $\hat{\mu} = \sum_{j=1}^n c_j Y_j, \sum_{j=1}^n c_j = 1$ and $c_j \geq 0$. We easily deduce that $c_j = (1/\sigma_j^2)/\sum_{j=1}^n (1/\sigma_j^2)$ minimizes the mean squared error. Then

$$\hat{\mu} \sim \mathrm{AN}\left[\mu, \left(\sum_{j=1}^n 1/\sigma_j^2\right)^{-1}\right].$$

Now let us try using the median to estimate μ. Let F_{nj} be the distribution of Y_j and $\overline{F}_n = \sum_{j=1}^n \omega_{nj} F_{nj}$. The median of $\overline{F}_n$ is μ and we use the sample median $\hat{\xi}_{\text{med}}$ to estimate μ. By the results of the previous section, we get

$$\hat{\xi}_{\text{med}} \sim \text{AN}\left\{\mu, \frac{\sum_{j=1}^{n} \omega_{nj}^2}{4\left[\sum_{j=1}^{n} \omega_{nj} f_{nj}(\mu)\right]^2}\right\};$$

here $f_{nj}(\mu) = \left(\sqrt{2\pi}\sigma_j\right)^{-1}$.

To minimize the variance of the limit distribution subject to $\sum_{j=1}^{n} \omega_{nj} = 1$ we require $\omega_{nj} = 1/\sigma_j / \sum_{j=1}^{n} 1/\sigma_j$. The asymptotic relative efficiency of these two estimators is

$$\text{ARE}(\widehat{\mu}, \hat{\xi}_{\text{med}}) = \frac{2}{\pi}.$$

Remarks 1. 1. For the iid normal case, the ARE of the sample mean estimator relative to the sample median estimator is $2/\pi$. Here we have proved that when the samples are from normal distributions with the same mean, but different variances, the ARE of the weighted sample mean estimator relative to the weighted sample median estimator yields the same value $2/\pi$.

2. The weights used in the sample mean are different from the weights used in the sample median. We only compare the two best estimators here. If we use the same weights, the ARE can be larger or smaller than $2/\pi$.

3. The weighted sample median should be more robust than the weighted sample mean.

Example 2. Consider the double exponential family. Assume the density of Y_j to be $1/2r_j \exp(-|x - \mu|/r_j)$; the r_j are known while μ is unknown $j = 1, \ldots, n$.

We again use the weighted sample mean and weighted sample median to estimate μ. Choose $c_j = (1/r_j^2)/\sum_{j=1}^{n}(1/r_j^2)$ to minimize the mean squared error. Then

$$\widehat{\mu} \sim \text{AN}\left[\mu, 2 \Big/ \sum_{j=1}^{n}(1/r_j^2)\right].$$

By choosing $\omega_{nj} = (1/r_j)\sum_{j=1}^{n}(1/r_j)$, we get the weighted sample median $\hat{\xi}_{\text{med}}$. From the results of the previous section, $\hat{\xi}_{\text{med}} \sim \text{AN}\left[\mu, 1\sum_{j=1}^{n}(1/r_j^2)\right]$.

The asymptotic relative efficiency of these estimators is $\text{ARE}(\widehat{\mu}, \hat{\xi}_{\text{med}}) = 2$.

Remarks 2. 1. The ARE of the best sample mean estimator relative to the best sample median estimator does not depend on the $\{r_j\}$'s. The median is a more efficient estimator.

2. As in the normal case, the weights used in the sample mean do not equal the weights used in the sample median.

3. The weighted sample median should be more robust.

We have shown asymptotically that the REW quantile estimator possesses a number of desirable asymptotic properties. We now apply the estimator in simulated examples to gain insight into its performance with finite samples and compare the REW quantile estimator with the well-known Nadaraya–Watson estimator. The Gaussian kernel function is used to generate the relevance weights.

Simulation study 1

A random sample of size n is simulated from the model

$$Y = X(1 - X) + \varepsilon,$$

with $\varepsilon \sim N(0, 0.5)$ independent of $X \sim U(0,1)$. A typical realization when $n = 1000$ is shown in Figure 1. The bandwidth used here and below is $h = 0.1$. Let us next add 50 outliers from $N(2, 0.5)$ to the simulation experiment just described. The result is shown in Figure 2.

Simulation study 2

Instead of the normal error model used in Simulation 1, we now sample ε from a double exponential distribution with $r = 0.1$. Figure 3 shows the

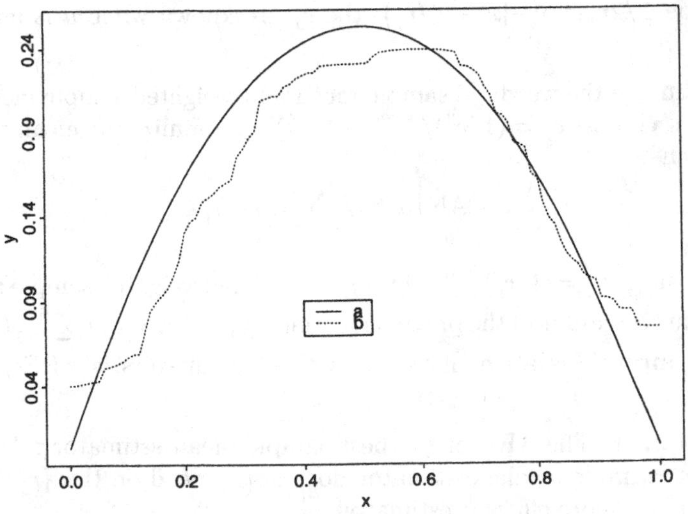

FIGURE 1. The model is $Y = X * (1 - X) + \varepsilon$, where X is uniform $(0, 1)$ and ε is $N(0, 0.5)$. The sample size $n = 1000$, the true curve is a and the realization is b.

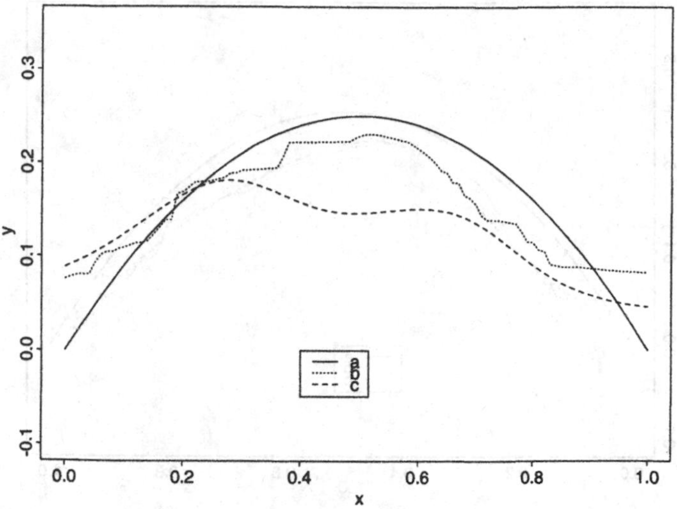

FIGURE 2. A comparison of the Nadaraya–Watson estimate with REW quantile estimator with outliers. To the data depicted in Figure 1, we add 50 ε-outliers from $N(2, 0.5)$. The true curve is a, the REW quantile estimator b, and the Nadaraya–Watson c.

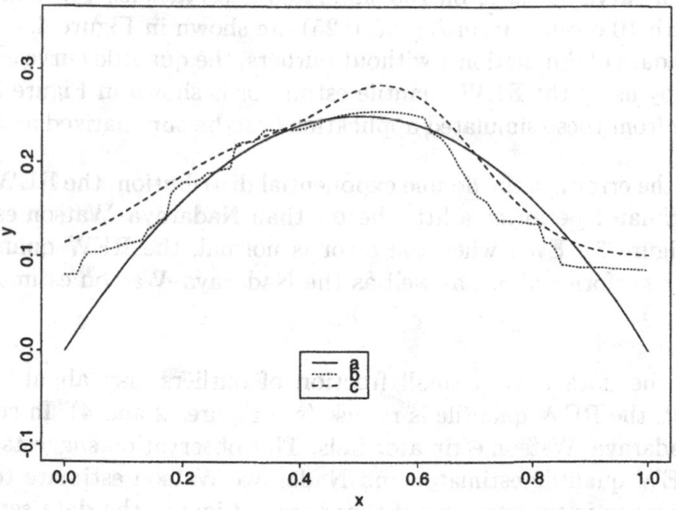

FIGURE 3. A comparison of the Nadaraya–Watson estimate with REW quantile estimator. The model is $Y = X * (1 - X) + \varepsilon$, where X is from uniform $(0, 1)$ and ε from a double exponential distribution with $r = 0.1$. The sample size is $n = 100$ and the bandwidth, $h = 0.1$. The true curve is a, the REW quantile estimator b, and the Nadaraya–Watson c.

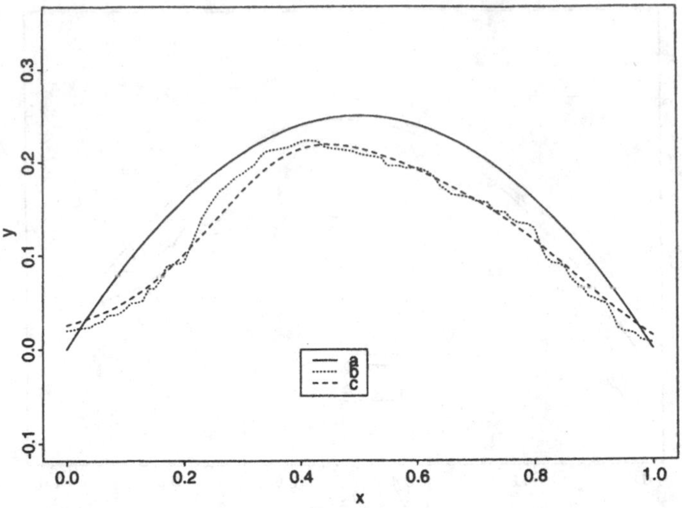

FIGURE 4. A comparison of the Nadaraya-Watson estimate with REW quantile estimator with outliers. To the data depicted in Figure 3. we add 10 ε-outliers from $N(-.5, .25)$. The true curve is a, the REW quantile estimator b, and the Nadaraya–Watson c.

results of a curve fit based on 100 simulated observations. The simulation results with 10 outliers from $N(-.5, 0.25)$ are shown in Figure 4.

For the data of Simulation 1 without outliers, the quantile curve estimate obtained by using the REW quantile estimator is shown in Figure 5.

Results from these simulated applications can be summarized as follows:

1. When the error has the double exponential distribution, the REW quantile estimator performs a little better than Nadaraya–Watson estimate (see Figure 3). Even when the error is normal, the REW quantile estimator performs about as well as the Nadaraya–Watson estimate (see Figure 1).

2. When the data have a small fraction of outliers, say about 5 or 10 percent, the REW quantile is robust (see Figures 2 and 4). In contrast, the Nadaraya–Watson estimator fails. This observation suggests we use the REW quantile estimator and Nadaraya–Watson estimate together as a diagnostic to determine if there are outliers in the data set. If the REW quantile estimator and Nadaraya–Watson estimate disagree, then we should reconsider the model and the outliers.

3. The REW quantile estimator seems promising judging from these simulation studies.

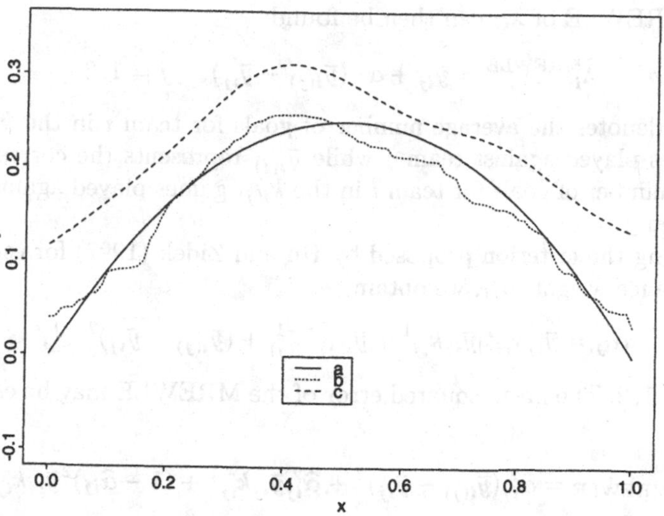

FIGURE 5. A REW quantile estimator of a quantile curve. The .25 quantile curve is estimated for the data depicted in Figure 1. The true quantile curve is a, and the REW quantile estimator is b.

4 Applying the P-REWL

In this section we demonstrate how the REWL may be used in practice, relying here on its parametric version. For this application we let Y_{ij} represent the number of goals team i scores in any one game played against team j in the National Hockey League's (NHL's) 1996–97 season. Our empirical analysis revealed a very small (formally insignificant) negative correlation between Y_{ij} and Y_{ji}. For simplicity we ignore this correlation and assume $Y_{ij} \sim$ Poisson (λ_{ij}) are independently distributed random variables.

Our aim is the prediction of (Y_{12}, Y_{21}) where "1" and "2" denote respectively the Vancouver Canucks and the Calgary Flames, a pair of teams that met five times during that season. For the predictive probability of $Y_{ij} = y_{ij}$ we use $\exp(-\hat{\lambda}_{ij})\hat{\lambda}_{ij}^{y_{ij}}/y_{ij}!$ where $\hat{\lambda}_{ij}$ is any given estimator of the Poisson mean.

Near the beginning of the season we have little "direct" information about the relative strengths of these two teams 1 and 2. The maximum likelihood estimator (MLE), the average number of goals scored in previous matches between these two teams, could not even be computed. In contrast the MREWLE would use data from games these teams played against other teams in the league.

We may now introduce our prior information through the relevance weights. To that end, we assume the scores for Vancouver in games played against all teams other than Calgary to be "exchangeable". The analogous assumption is made about the Calgary scores.

The MREWLE of λ_{ij} can then be found:

$$\hat{\lambda}_{ij}^{\text{MREWLE}} = \overline{y}_{ij} + \alpha_{ij}(\overline{y}_{i(j)} - \overline{y}_{ij}), \quad j = 1, 2. \tag{5}$$

Here $\overline{y}_{ij}$ denotes the average number of goals for team i in the k_{ij} previous games played against team j while $\overline{y}_{i(j)}$ represents the corresponding average number of goals for team i in the $k_{i(j)}$ games played against teams other than j.

Adopting the criterion proposed by Hu and Zidek (1997) for estimating the relevance weight, α_{ij}, we obtain,

$$\widehat{\alpha}_{ij} = \overline{y}_{ij} k_{ij}^{-1} [\overline{y}_{ij} k_{ij}^{-1} + \overline{y}_{i(j)} k_{i(j)}^{-1} + (\overline{y}_{i(j)} - \overline{y}_{ij})^2]^{-1}, \tag{6}$$

for i, $j = 1, 2$. The mean squared error of the MREWLE may be estimated by

$$\widehat{\text{MSE}}_{\text{MREWLE}} = \widehat{\alpha}_{ij}^2 (\overline{y}_{i(j)} - \overline{y}_{ij})^2 + \widehat{\alpha}_{ij}^2 \overline{y}_{ij} k_{ij}^{-1} + (1 - \widehat{\alpha}_{ij})^2 \overline{y}_{ij} k_{ij}^{-1}. \tag{7}$$

Notice that a lack-of-precision in the biased estimate of the population average number of goals for i when playing j $(i, j = 1, 2)$ (as indicated by large values of $\overline{y}_{ij}$ or $\overline{y}_{i(j)}$) and a large bias (as indicated by a large value of $|\overline{y}_{i(j)} - \overline{y}_{ij}|$) can both result in a diminished value for the estimated optimal relevance weight in agreement with intuition. Notice also that the value of that estimated weight will change as the season wears on and both k_{ij} and $k_{i(j)}$ increase in size. In fact as k_{ij} increases the estimated relevance weight will tend to zero so that we would tend to rely less and less on the less directly relevant data.

The estimation bias may be estimated by $\widehat{\text{bias}}_{ij} = \widehat{\alpha}_{ij}|\overline{y}_{i(j)} - \overline{y}_{ij}|$. We may use $\widehat{\text{var}}_{ij} = \widehat{\alpha}_{ij}^2 \overline{y}_{ij} k_{ij}^{-1} + (1 - \widehat{\alpha}_{ij})^2 \overline{y}_{ij} k_{ij}^{-1}$ to estimate the estimator's variance. A reasonable 95% confidence interval for the mean number of the goals based on the MREWLE would be $(\hat{\lambda}_{ij}^{\text{MREWLE}} - \widehat{\text{bias}}_{ij} - 1.96\sqrt{\widehat{\text{var}}_{ij}}, \hat{\lambda}_{ij}^{\text{MREWLE}} + \widehat{\text{bias}}_{ij} + 1.96\sqrt{\widehat{\text{var}}_{ij}})$.

In the above study, when we predict Vancouver's score, we have used the scores obtained when Vancouver played teams other than Calgary but not those from games Calgary played against teams other than Vancouver. We know that when Vancouver plays Calgary, Vancouver's score will depend on both Vancouver's offense and the Calgary's defense. So we should further consider Calgary's defense. The quality of that defense can be measured through the scores obtained by other teams playing Calgary, that is $Y_{j2} \sim \text{Poisson}(\lambda_{j2})$ $(j \neq 1)$.

The MREWLE, say MREWLE1 of λ_{ij} now becomes

$$\hat{\lambda}_{ij}^{\text{MREWLE1}} = \overline{y}_{ij} + \alpha_{ij}^{(1)}(\overline{y}_{i(j)} - \overline{y}_{ij}) + \beta_{ij}^{(1)}(\overline{y}_{(i)j} - \overline{y}_{ij}), \quad i, j = 1, 2, \tag{8}$$

where $\overline{y}_{(i)j}$ denotes the average number of goals for teams other than i in the $k_{(i)j}$ games played against team j.

Again using the criterion proposed by Hu and Zidek (1997) based on the approximate Akaike criterion, an optimal relevance weight may be estimated by

$$\widehat{\alpha}_{ij}^{(1)} = \frac{\overline{y}_{ij} k_{ij}^{-1} [\overline{y}_{(i)j} k_{(i)j}^{-1} + (\overline{y}_{(i)j} - \overline{y}_{ij})(\overline{y}_{(i)j} - \overline{y}_{i(j)})]}{A+B} \tag{9}$$

and

$$\widehat{\beta}_{ij}^{(1)} = \frac{\overline{y}_{ij} k_{ij}^{-1} [\overline{y}_{i(j)} k_{i(j)}^{-1} + (\overline{y}_{i(j)} - \overline{y}_{ij})(\overline{y}_{i(j)} - \overline{y}_{(i)j})]}{A+B}, \tag{10}$$

for i, $j = 1$, 2, where

$$A = \overline{y}_{ij} k_{ij}^{-1} [\overline{y}_{(i)j} k_{(i)j}^{-1} + \overline{y}_{i(j)} k_{i(j)}^{-1} + (\overline{y}_{(i)j} - \overline{y}_{i(j)})^2],$$

and

$$B = k_{(i)j}^{-1} \overline{y}_{(i)j} (\overline{y}_{i(j)} - \overline{y}_{ij})^2 + k_{i(j)}^{-1} \overline{y}_{i(j)} (\overline{y}_{(i)j} - \overline{y}_{ij})^2 + k_{(i)j}^{-1} k_{i(j)}^{-1} \overline{y}_{(i)j} \overline{y}_{i(j)}.$$

The mean square error of the MREWLE1 of (8) may be estimated by

$$\widehat{\text{MSE}}_{\text{MREWLE1}} = (\hat{\lambda}_{ij}^{\text{MREWLE1}} - \overline{y}_{ij})^2 + (\widehat{\alpha}_{ij}^{(1)})^2 \overline{y}_{i(j)} k_{i(j)}^{-1} + (\widehat{\beta}_{ij}^{(1)})^2 \overline{y}_{(i)j} k_{(i)j}^{-1} + (1 - \widehat{\alpha}_{ij}^{(1)} - \widehat{\beta}_{ij}^{(1)})^2 \overline{y}_{ij} k_{ij}^{-1}. \tag{11}$$

Based on the MREWLE1, the 95% predictive interval for the the number of goals would be:

$$\left(\hat{\lambda}_{ij}^{\text{MREWLE1}} - \widehat{\text{bias}}_{ij}^{(1)} - 1.96 \sqrt{\widehat{\text{var}}_{ij}^{(1)}}, \hat{\lambda}_{ij}^{\text{MREWLE1}} + \widehat{\text{bias}}_{ij}^{(1)} + 1.96 \sqrt{\widehat{\text{var}}_{ij}^{(1)}} \right),$$

where

$$\widehat{\text{bias}}_{ij}^{(1)} = |\hat{\lambda}_{ij}^{\text{MREWLE1}} - \overline{y}_{ij}|,$$

and

$$\widehat{\text{var}}_{ij}^{(1)} = (\widehat{\alpha}_{ij}^{(1)})^2 \overline{y}_{i(j)} k_{i(j)}^{-1} + (\widehat{\beta}_{ij}^{(1)})^2 \overline{y}_{(i)j} k_{(i)j}^{-1} + (1 - \widehat{\alpha}_{ij}^{(1)} - \widehat{\beta}_{ij}^{(1)})^2 \overline{y}_{ij} k_{ij}^{-1}.$$

We now compare the MLE to the point estimators MREWLE in (5) and the MREWLE1 in (8) (with estimated α_{ij}). We find the mean squared errors (squared root) of the predictors as well as the predictive interval for goals in a future game. Table 1 shows those predictions for each of the last four games and future games between Vancouver and Calgary during the season.

Notice that as a predictor of the numbers of goals both MREWLE and the MREWLE1 have smaller overall mean squared errors than the MLE for each of the two teams. Not surprisingly both tend to be more stable over these games. In the second game for example, the MLE has

TABLE 1. A Game-by-Game Comparison of the Maximum Relevance Weighted Likelihood and Maximum Likelihood Predictors (the point estimations, the mean square errors and the confidence intervals) of Vancouver and Calgary Goals for the Last Four Games and the Future Game of the 1996–97 Season.

Score	MREWLE Predictor		MREWLE1 Predictor		MLE Predictor	
(V,C)	Vancouver	Calgary	Vancouver	Calgary	Vancouver	Calgary
(4,3)	3.2 (0.57)	1.4 (0.87)	2.9 (0.42)	2 (1.14)	3.0 (1.73)	1.0 (1.0)
C.I.	[2.0,4.4]	[0,3.3]	[2.0,3.7]	[0.1,4.0]	[0.8,7.7]	[0.1,4.7]
(0,3)	3.2 (0.39)	2.4 (0.74)	3.2 (0.39)	2.7 (0.81)	3.5 (1.32)	2.0 (1.0)
C.I.	[2.4,4.0]	[1.4,3.4]	[2.4,4.0]	[1.2,4.2]	[0.9,6.1]	[0,3.9]
(5,2)	2.8 (0.59)	2.6 (0.37)	2.9 (0.70)	2.8 (0.54)	2.3 (0.88)	2.3 (0.88)
C.I.	[1.5,4.0]	[1.9,3.3]	[1.8,4.1]	[1.8,3.8]	[0.6,4.1]	[0.6,4.1]
(3,3)	3.1 (0.21)	2.6 (0.39)	3.0 (0.14)	2.8 (0.59)	3.0 (0.87)	2.3 (0.75)
C.I.	[2.6,3.6]	[1.8,3.3]	[2.7,3.3]	[1.7,3.8]	[1.3,4.7]	[0.8,3.7]
MSE	2.4		2.2		3.3	
Future	3.1 (0.23)	2.6 (0.27)	3.0 (0.14)	2.6 (0.31)	3 (0.77)	2.4 (0.69)
Game	[2.6,3.6]	[2.0,3.1]	[2.7,3.3]	[2.1,3.2]	[1.5,4.5]	[1.0,3.8]

TABLE 2. The Estimated Game-by-Game Probabilities and 90% Confidence Intervals (in %) of a Win, Lose or Draw for the Vancouver Canucks against the Calgary Flames Derived from the Maximum Relevance Weighted Likelihood and Maximum Likelihood Estimators for the Last Four Games and Future Games of the 1996-97 Season.

Game Number		2	3	4	5	Future
MREWLE	Win	72	55	43	50	50
	C.I.	[22,100]	[26,82]	[14,75]	[31,71]	[34,68]
	Tie	14	16	18	17	17
	C.I.	[1,22]	[10,19]	[12,22]	[13,19]	[14,19]
	Lose	14	29	39	33	33
	C.I.	[0,63]	[8,58]	[14,72]	[16,53]	[18,50]
MREWLE1	Win	57	50	43	45	48
	C.I.	[15,97]	[18,85]	[14,77]	[26,69]	[34,62]
	Tie	17	16	17	17	17
	C.I.	[3,22]	[9,19]	[11,22]	[15,18]	[16,18]
	Lose	26	34	40	38	35
	C.I.	[0,73]	[6,69]	[12,74]	[17,59]	[23,50]
MLE	Win	77	66	41	51	51
	C.I.	[2,100]	[4,100]	[2,93]	[9,94]	[10,91]
	Tie	13	15	19	17	17
	C.I.	[0,44]	[0,40]	[5,42]	[4,29]	[6,26]
	Lose	10	20	41	32	32
	C.I.	[0,94]	[0,88]	[2,93]	[2,81]	[3,79]

to rely on the results of just a single game, Game 1. In contrast the MREWLE and the MREWLE1 rely on all the games previously played by Vancouver and Calgary. Both the MREWLE and the MREWLE1 predictors have much smaller mean square errors and shorter confidence interval lengths. This result derives from the fact that both the MREWLE and the MREWLE1 use all the relevant information. For example, in Game 5, the MREWLE, MREWLE1 and MLE predictors of Vancouver's score (with square root of the mean square error and the confidence intervals) are 3.1 (0.21) [2.6, 3.6], 3.0 (0.28) [2.7, 3.3] and 3.0 (0.87) [1.3, 4.7], respectively. Although their point predictors are very similar, the MLE has much longer confidence intervals. The MREWLE1 and the MREWLE are quite similar, but the MREWLE1 has smaller mean squared error because the MREWLE1 also uses information about the defensive capacity of Calgary's team. The MREWLE1 predicts slightly bigger values for Calgary's score. This is because Vancouver's defense is not very good. The average score of other teams is about 3.4, when they play against Vancouver.

We can use the estimators above to construct predictive distributions for the number of goals to be scored by each of the teams and in turn to estimate the probabilities and the 90% confidence intervals of a win, lose or draw Vancouver. Those estimates appear in Table 2. Observe how the predictive probabilities obtained from the MLE, the MREWLE and the MREWLE1 converge. As the number of Vancouver—Calgary games increases we see at most small differences in the last two of the four games above. On the other hand, we see substantial differences in the first two games. The confidence intervals for the MREWLE and the MREWLE1 are much shorter than the confidence intervals of the MLE.

We have used the MREWLE-based method to predict the probabilities and confidence intervals of a Win, Lose or Draw for the Vancouver Canucks against the Calgary Flames. Further we can use those results to predict the probabilities and confidence intervals of a Win or Lose in the playoffs.

5 Discussion

In this article we have tried to show how the intuitively natural idea of the relevance weighted likelihood arises and how it may be used in both parametric and nonparametric settings. The methodology has promise although much more work remains to be done on the problem of selecting the relevance weights. Moreover, we have yet to understand why the methods we develop for finding and fitting the MREWLE leads in certain cases to known empirical Bayes estimators even though we have not explicitly formulated a prior distribution.

An extensive literature on the local likelihood illustrates the usefulness of the REWL. We believe that further development will benefit from the abstract study of the idea of relevance weighting through the formalization and extension of this notion as we have tried to indicate in this paper.

Much work remains to be done. As noted by an anonymous referee, we need to develop our asymptotic theory for the case where the relevance weights are allowed to depend on the data. Furthermore he or she emphasizes the importance of exploring linkages with empirical Bayes theory. The case of nonindependent observations has been explored and is expected to be presented elsewhere.

Acknowledgments: We are indebted to an anonymous referee for acquainting us with the monograph of Koul and the contents that relate to the asymptotic theory of the nonparametric REWL.

6 Appendix: Proofs of the Theorems

Lemma 1 (Marcus and Zinn (1984)). *Let $\{c_n\}$, $n = 1, \ldots, \infty$, be a sequence of real numbers and $\{Y_n\}$, $n = 1, \ldots, \infty$, a sequence of independent random variables. Define $U_n(t)$ by*

$$U_n(t) = \sum_{j=1}^{n} c_j \left[I(Y_j \le t) - P(Y_j \le t) \right].$$

Then

$$P\left[|U_n(t)| \left(\sum_{j=1}^{n} {c_j}^2 \right)^{-1/2} > \lambda \right] \le \exp(-\lambda^2/8)(1 + 2\sqrt{2\pi}\lambda),$$

for all $\lambda > 0$.

Proof of Theorem 1.

$$\left|\widehat{F}_n(y) - \overline{F}_n(y)\right| = \left| \sum_{j=1}^{n} \omega_{nj} \left[I(Y_{nj} \le x) - F_{nj}(x) \right] \right| \stackrel{\text{def}}{=} |V_n(x)|,$$

say. So on applying Lemma 1 with $\lambda = \varepsilon K_n^{1/2}$,

$$\begin{aligned} P(|\widehat{F}_n(y) - \overline{F}_n(y)| > \varepsilon) &= P\left[|V_n(x)| \left(\sum_{j=1}^{n} \omega_{nj}^2 \right)^{-1/2} > \varepsilon \left(\sum_{j=1}^{n} \omega_{nj}^2 \right)^{-1/2} \right] \\ &\le (1 + 2\sqrt{2\pi\varepsilon K_n^{1/2}}) \exp(-\varepsilon^2 K_n/8), \end{aligned}$$

for every $\varepsilon > 0$. The assumption $\sum_{n=1}^{\infty} \exp(-\varepsilon^2 K_n) < \infty$, implies that $K_n \to \infty$ when $n \to \infty$. It follows that for every $\varepsilon > 0$, there exists N, such that for every $n > N$

$$(1 + 2\sqrt{2\pi\varepsilon K_n^{1/2}}) < \exp\left(\frac{\varepsilon^2 K_n}{16}\right).$$

Consequently

$$(1 + 2\sqrt{2\pi\varepsilon K_n^{1/2}}) \exp\left(-\frac{\varepsilon^2 K_n}{8}\right) \leq \exp\left(-\frac{\varepsilon^2 K_n}{16}\right).$$

But $\sum_{n=1}^{\infty} \exp(-\varepsilon^2 K_n) < \infty$ for all $\varepsilon > 0$. So $\sum_{n=1}^{\infty} \exp(-\varepsilon^2 K_n/16) < \infty$ for every $\varepsilon > 0$. Hence

$$\sum_{n=1}^{\infty} P\left[\left|\widehat{F}_n(y) - \overline{F}_n(y)\right| > \varepsilon\right] < \infty \text{ for all } \varepsilon > 0.$$

The Borel–Cantelli Lemma then implies $\left|\widehat{F}_n(y) - \overline{F}_n(y)\right| \to 0$ a.s. for every x. □

Proof of Theorem 2. Let M be a large positive integer and

$$u_n = \max_{-M^2 \leq j \leq M^2} \left|F_n(j/M) - \overline{F}_n(j/M)\right|.$$

By Theorem 1, $u_n \to 0$ a.s. Also monotonicity implies that for $(j-1)/M < t \leq j/M$

$$\begin{aligned} F_n(t) - \overline{F}_n(t) &\leq F_n(j/M) - \overline{F}_n[(j-1)/M] \\ &= [F_n(j/M) - \overline{F}_n(j/M)] + \{\overline{F}_n(j/M) - \overline{F}_n[(j-1)/M]\}. \end{aligned}$$

By similar reasoning,

$$\begin{aligned} &F_n(t) - \overline{F}_n(t) \\ &\geq \{F_n[(j-1)/M] - \overline{F}_n[(j-1)/M]\} - \{\overline{F}_n(j/M) - \overline{F}_n[(j-1)/M]\}. \end{aligned}$$

So

$$\begin{aligned} &\limsup_{n\to\infty} \left|\widehat{F}_n(y) - \overline{F}_n(y)\right| \\ &\leq \limsup_{n\to\infty} u_n \\ &\quad + \limsup_{n\to\infty} \max_{-M^2 \leq j \leq M^2} \left[\overline{F}_n\left(\frac{j}{M}\right) - \overline{F}_n\left(\frac{j-1}{M}\right), 1 - \overline{F}_n(M), \overline{F}_n(-M)\right] \\ &\leq \limsup_{n\to\infty} u_n + \limsup_{n\to\infty} M^{-1} \sup_{x,n} \overline{F}_n(y) \\ &\quad + \limsup_{n\to\infty} [(1 - \overline{F}_n(M)), \overline{F}_n(-M)], \end{aligned}$$

under the assumptions. Since M is arbitrary, the result follows. □

Proof of Theorem 3. Let $\varepsilon > 0$. By the uniqueness condition and the definition of $\xi_{p(n)}$,

$$\overline{F}_n(\xi_{p(n)} - \varepsilon) < p < \overline{F}_n(\xi_{p(n)} + \varepsilon).$$

By Theorem 2, $F_n(\xi_{p(n)} - \varepsilon) - \overline{F}_n(\xi_{p(n)} - \varepsilon) \to 0$ a.s. and $F_n(\xi_{p(n)} + \varepsilon) - \overline{F}_n(\xi_{p(n)} + \varepsilon) \to 0$ a.s. Hence $P\big[F_m(\xi_{p(m)} - \varepsilon) < p < F_m(\xi_{p(m)} + \varepsilon)$, for $m \geq n\big] \to 1$ as $n \to \infty$. That is, $P\big(\sup_{m>n} |\hat{\xi}_{pm} - \xi_{p(m)}| > \varepsilon\big) \to 0$ as $n \to \infty$. This completes the proof. □

To prove Theorem 4, we need the following useful result of Hoeffding Hoeffding (1963).

Lemma 2. *Let $Y_1, \ldots, Y_n$ be independent random variables satisfying*

$$P(a_j \leq Y_j \leq b_j) = 1,$$

for each i, where $a_j < b_j$. Then for $t > 0$,

$$P\left\{\sum_{j=1}^{n}[Y_i - E(Y_j)] \geq t\right\} \leq \exp\left[-2t^2 \Big/ \sum_{j=1}^{n}(b_j - a_j)^2\right].$$

Proof of Theorem 4. Fix $\varepsilon > 0$. Then

$$P\big(|\hat{\xi}_{np} - \xi_{p(n)}| > \varepsilon\big) \leq P(\hat{\xi}_{np} \geq \xi_{p(n)} + \varepsilon) + P(\hat{\xi}_{np} \leq \xi_{p(n)} - \varepsilon).$$

But with $Y_j = \omega_{nj} I(Y_{nj} > \xi_{p(n)} + \varepsilon)$,

$$\begin{aligned} P(\hat{\xi}_{np} \geq \xi_{p(n)} + \varepsilon) &= P\big[p > F_n(\xi_{p(n)} + \varepsilon)\big] \\ &= P\left[\sum_{j=1}^{n} \omega_{nj} I(Y_{nj} > \xi_{p(n)} + \varepsilon) > 1 - p\right] \\ &= P\Bigg\{\sum_{j=1}^{n}[Y_j - E(Y_j)] > 1 - p \\ &\qquad\qquad - \sum_{j=1}^{n} \omega_{nj}\left[1 - F_{nj}(\xi_{p(n)} + \varepsilon)\right]\Bigg\} \\ &= P\left\{\sum_{j=1}^{n}[Y_j - E(Y_j)] > \overline{F}_n(\xi_{p(n)} + \varepsilon) - p\right\}. \end{aligned}$$

Because $P(0 \leq Y_j \leq \omega_{nj}) = 1$ for each j, by Lemma 2, we have

$$P(\hat{\xi}_{np} \geq \xi_{p(n)} + \varepsilon) \leq \exp\left(-2\delta_1^2 \Big/ \sum_{j=1}^{n} \omega_{nj}^2\right) = \exp(-2\delta_1^2 K_n);$$

here $\delta_1 = \overline{F}_n(\xi_{p(n)} + \varepsilon) - p$. Similarly,

$$P(\hat{\xi}_{np} \leq \xi_{p(n)} - \varepsilon) \leq \exp\left(-2\delta_2^2 \Big/ \sum_{j=1}^{n} \omega_{nj}^2\right) = \exp(-2\delta_2^2 K_n),$$

where $\delta_2 = p - \bar{F}_n(\xi_{p(n)} - \varepsilon)$.

Putting $\delta_\varepsilon(n) = \min(\delta_1, \delta_2)$, completes the proof. □

Proof of Theorem 5. Fix t, and put

$$G_n(t) = P[\bar{f}_n(\xi_{p(n)})(\hat{\xi}_{np} - \xi_{p(n)})V_n^{-1/2} \leq t] = P(\hat{\xi}_{np} \leq a_n),$$

where $a_n = \xi_{p(n)} + tV_n^{1/2}/\bar{f}_n(\xi_{p(n)})$ Then by the definition of $\hat{\xi}_{np}$

$$G_n(t) = P\big[\widehat{F}_n(a_n) \geq p\big].$$

Thus

$$\begin{aligned} G_n(t) &= P\Bigg[\sum_{j=1}^{n} \omega_{nj} I(Y_{nj} \leq a_n) \geq p\Bigg] \\ &= P\Bigg\{V_n^{-1/2} \sum_{j=1}^{n} \omega_{nj}\big[I(Y_{nj} \leq a_n)\big] - E\big[I(Y_{nj} \leq a_n)\big]\Bigg\} \\ &\qquad\qquad \geq V_n^{-1/2}\Bigg\{p - \sum_{j=1}^{n} \omega_{nj} E\big[I(Y_{nj} \leq a_n)\big]\Bigg\}\Bigg) \\ &= P(Z_n \geq c_n); \end{aligned}$$

here

$$Z_n = V_n^{-1/2} \sum_{j=1}^{n} \omega_{nj}\big\{I(Y_{nj} \leq a_n) - E\big[I(Y_{nj} \leq a_n)\big]\big\},$$

and

$$c_n = V_n^{-1/2}\Bigg\{p - \sum_{j=1}^{n} \omega_{nj} E\big[I(Y_{nj} \leq a_n)\big]\Bigg\}.$$

We first prove $Z_n \to N(0,1)$ in distribution and then that $c_n \to -t$ as $n \to \infty$ to complete the proof. To this end

$$Z_n = \sum_{j=1}^{n} \omega_{nj} V_n^{-1/2}\big[I(Y_{nj} \leq a_n) - F_{nj}(a_n)\big] = \sum_{j=1}^{n} \eta_{nj},$$

where $\eta_{nj} = \omega_{nj} V_n^{-1/2}\big[I(Y_{nj} \leq a_n) - F_{nj}(a_n)\big]$.

From the condition $\max_{1 \leq j \leq n}(\omega_{nj} V_n^{-1/2}) \to 0$, we get $\max_{1 \leq j \leq n} \omega_{nj} \to 0$ and $V_n \to 0$. We then easily obtain for every $\varepsilon > 0$ and $\tau > 0$:

1. $\sum_{j=1}^{n} P\big(|\eta_{nj}| \geq \varepsilon\big) \to 0$. (Since $|\eta_{nj}| \leq 2\max_{1 \leq j \leq n}(\omega_{nj} V_n^{-1/2}) \to 0$);
2. $\sum_{j=1}^{n} \Big(E\big[\eta_{nj}^2 I\big(|\eta_{nj}| < \tau\big)\big] - \big\{E\big[\eta_{nj} I\big(|\eta_{nj}| < \tau\big)\big]\big\}^2\Big) \to 1$. (For n large enough, $I\big(|\eta_{nj}| < \tau\big) = 1$);
3. $\sum_{j=1}^{n} E\big[\eta_{nj} I\big(|\eta_{nj}| < \tau\big)\big] \to 0$.

So $Z_n \to N(0,1)$ in distribution by the CLT (see (Chung, 1968, p. 191) for triangular independent random variables.

Next we prove $c_n \to -t$.

$$\begin{aligned} c_n &= V_n^{-1/2}\Big\{p - \sum_{j=1}^{n} \omega_{nj} E\big[I(Y_{nj} \le a_n)\big]\Big\} \\ &= V_n^{-1/2} \sum_{j=1}^{n} \omega_{nj}\big[F_{nj}(\xi_{p(n)}) - F_{nj}(a_n)\big] \\ &= -V_n^{-1/2} \sum_{j=1}^{n} \omega_{nj}\big[F_{nj}(\xi_{p(n)})(a_n - \xi_{p(n)}) + o(a_n - \xi_{p(n)})\big] \\ &\qquad\qquad \to -t \text{ as } n \to \infty. \end{aligned}$$

The proof is now complete. □

To prove Theorem 5, we need the following results, see Shorack and Wellner (1986, p. 855).

Lemma 3 (Bernstein). *Let $Y_1, Y_2, \ldots, Y_n$ be independent random variables satisfying $P(|Y_j - E(Y_j)| \le m) = 1$, for each j, where $m < \infty$. Then, for $\varepsilon > 0$,*

$$P\Big\{\Big|\sum_{j=1}^{n}[Y_j - E(Y_j)]\Big| \ge \varepsilon\Big\} \le 2\exp\Big[-\frac{\varepsilon^2}{2\sum_{j=1}^{n}\mathrm{Var}(Y_j) + 2/3m\varepsilon}\Big],$$

for all $n = 1, 2, \ldots$.

Lemma 4. *Let $0 < p < 1$. Suppose conditions 1–3 of Theorem 5 hold. Then with probability 1 (hereafter wp1)*

$$|\hat{\xi}_{np} - \xi_{p(n)}| \le \frac{(\sqrt{c^*/2} + 1)K_n^{-1/2}(\log K_n)^{1/2}}{\overline{F}_n(\xi_{p(n)})},$$

for all sufficiently large n.

Proof. Since $\overline{F}_n$ is continuous at $\xi_{p(n)}$ with $\overline{F}_n'(\xi_{p(n)}) > 0$, $\xi_{p(n)}$ solves uniquely $\overline{F}_n(x-) \le p \le \overline{F}_n(y)$ and $p = \overline{F}_n(\xi_{p(n)})$. Put

$$\varepsilon_n = (\sqrt{c^*/2} + 1)K_n^{-1/2}(\log K_n)^{1/2}/\bar{f}_n(\xi_{p(n)}).$$

We then have

$$\begin{aligned} \overline{F}_n(\xi_{p(n)} + \varepsilon_n) - p &= \overline{F}_n(\xi_{p(n)} + \varepsilon_n) - \overline{F}_n(\xi_{p(n)}) \\ &= \bar{f}_n(\xi_{p(n)})\varepsilon_n + o(\varepsilon_n) \\ &\ge \sqrt{c^*/2}(\log K_n)^{1/2}/K_n^{1/2}, \end{aligned}$$

for all sufficiently large n.

Likewise we may show that $p - \overline{F}_n(\xi_{p(n)} - \varepsilon_n)$ satisfies a similar inequality. Thus, with $\delta_\varepsilon(n)$ as defined in Theorem 4, we have

$$2K_n\delta_\varepsilon(n)^2 \geq c^* \log K_n,$$

for all n sufficiently large. Hence by Theorem 4,

$$P\big(|\hat{\xi}_{np} - \xi_{p(n)}| > \varepsilon_n\big) \leq \frac{2}{K_n^{c^*}},$$

for all sufficiently large n.

This last result, hypothesis 3 of this theorem and the Borel–Cantelli Lemma imply that wp1 $|\hat{\xi}_{np} - \xi_{p(n)}| > \varepsilon_n$ holds for only finitely many n. This completes the proof. □

Lemma 5. *Let $0 < p < 1$ and T_n be any estimator of $\xi_{p(n)}$ for which $T_n - \xi_{p(n)} \to 0$ wp1. Suppose $\overline{F}_n$ has a bounded second derivative in the neighborhood of $\xi_{p(n)}$. Then wp1*

$$\overline{F}_n(T_n) - \overline{F}_n(\xi_{p(n)}) = \overline{F}'_n(\xi_{p(n)})(T_n - \xi_{p(n)}) + O\big((T_n - \xi_{p(n)})^2\big),$$

as $n \to \infty$.

Proof. The proof is an immediate consequence of the Taylor expansion.

For convenience in presenting the next result, we set

$$D_n(x) = \big[F_n(\xi_{p(n)} + x) - F_n(\xi_{p(n)})\big] - \big[\overline{F}_n(\xi_{p(n)} + x) - \overline{F}_n(\xi_{p(n)})\big].$$

Lemma 6. *Let $\{a_n\}$ be a sequence of positive constants such that*

$$a_n \sim c_0 K_n^{-1/2}(\log K_n)^q,$$

as $n \to \infty$, for constants $c_0 > 0$ and $q \geq 1/2$. Let $m_n = \max_{1\leq j\leq n}\{\omega_{nj}\}$ and

$$H_{pn} = \sup_{|x|\leq a_n} \big|D_n(x)\big|.$$

If $m_n = o\big[K_n^{-3/4}(\log K_n)^{(q-1)/2}\big]$, then under the hypothesis of Theorem 5, wp1

$$H_{pn} = O\big[K_n^{-3/4}(\log K_n)^{1/2(q+1)}\big]. \quad \square$$

Proof. Let $\{b_n\}$ be any positive integers such that $b_n \sim c_0 K_n^{1/4}(\log K_n)^q$ as $n \to \infty$. For successive integers $r = -b_n, \ldots, b_n$, put $\eta_{r,n} = a_n b_n^{-1} r$ and $\alpha_{r,n} = \overline{F}_n(\xi_{p(n)} + \eta_{r+1,n}) - \overline{F}_n(\xi_{p(n)} + \eta_{r,n})$. The monotonicity of F_n and $\overline{F}_n$ implies that for $\eta_{r,n} \leq x \leq \eta_{r+1,n}$,

$$\begin{aligned} D_n(x) &\leq \big[F_n(\xi_{p(n)} + \eta_{r+1,n}) - F_n(\xi_{p(n)})\big] - \big[\overline{F}_n(\xi_{p(n)} + \eta_{r,n}) - \overline{F}_n(\xi_{p(n)})\big] \\ &\leq D_n(\eta_{r+1,n}) + \big[\overline{F}_n(\xi_{p(n)} + \eta_{r+1,n}) - \overline{F}_n(\xi_{p(n)} + \eta_{r,n})\big]. \end{aligned}$$

Similarly,

$$D_n(x) \geq D_n(\eta_{r,n}) - [\overline{F}_n(\xi_{p(n)} + \eta_{r+1,n}) - \overline{F}_n(\xi_{p(n)} + \eta_{r,n})].$$

So

$$H_{pn} \leq A_n + \beta_n,$$

where $A_n = \max\{|D_n(\eta_{r,n})|: -b_n \leq r \leq b_n\}$ and $\beta_n = \max\{\alpha_{r,n}: -b_n \leq r \leq b_n - 1\}$. Since $\eta_{r+1,n} - \eta_{r,n} = a_n b_n^{-1} \sim K_n^{-3/4}$, $-b_n \leq r \leq b_n - 1$, we have by the Mean Value Theorem (MVT) that

$$\alpha_{r,n} \leq \left[\sup_{|x|\leq a_n} \overline{F}'_n(\xi_{p(n)} + x)\right](\eta_{r+1,n} - \eta_{r,n}) \sim \left[\sup_{|x|\leq a_n} \overline{F}'_n(\xi_{p(n)} + x)\right] K_n^{-3/4},$$

$-b_n \leq r \leq b_n - 1$. Thus

$$\beta_n = O(K_n^{-3/4}), \quad n \to \infty.$$

We now establish that wp1

$$A_n = O\left[K_n^{-3/4}(\log K_n)^{1/2(q+1)}\right] \text{ as } n \to \infty.$$

By the Borel–Cantelli Lemma it suffices to show that

$$\sum_{n=1}^{\infty} P(A_n \geq \gamma_n) < \infty,$$

where $\gamma_n = c_1 K_n^{-3/4}(\log K_n)^{1/2(q+1)}$ for some constant $c_1 > 0$. Now

$$P(A_n \geq \gamma_n) \leq \sum_{r=-b_n}^{b_n} P(|D_n(\eta_{r,n})| \geq \gamma_n),$$

and

$$|D_n(\eta_{r,n})| = \left|\sum_{j=1}^{n} \omega_{nj}\left([Y_{nj} \in (\xi_{p(n)}, \xi_{p(n)} + \eta_{r,n})] - E\{I[Y_{nj} \in (\xi_{p(n)}, \xi_{p(n)} + \eta_{r,n})]\}\right)\right|,$$

by definition. With $Y_j = \omega_{nj} I(Y_{nj} \in (\xi_{p(n)}, \xi_{p(n)} + \eta_{r,n}))$, Bernstein's Lemma (see Lemma 3) implies

$$P(|D_n(\eta_{r,n})| \geq \gamma_n) \leq 2\exp(-\gamma_n^2/D_n),$$

where $D_n = 2\sum_{j=1}^{n} \mathrm{Var}(Y_j) + 2/3 m_n \gamma_n$.

Choose $c_2 > \sup_{n,i} F_{nj}(\xi_{p(n)})$. Then there exists an integer N such that

$$F_{nj}(\xi_{p(n)} + a_n) - F_{nj}(\xi_{p(n)}) < c_2 a_n,$$

and

$$F_{nj}(\xi_{p(n)}) - F_{nj}(\xi_{p(n)} - a_n) < c_2 a_n,$$

both of the above inequalities being for all $n > N$ and $j = 1, \ldots, n$. Then

$$\sum_{j=1}^{n} \mathrm{Var}(Y_j) \leq \sum_{j=1}^{n} \omega_{nj}^2 c_2 a_n = K_n^{-1} c_2 a_n.$$

Hence

$$\gamma_n^2 / D_n \geq \gamma_n^2 / \{2K_n^{-1} c_2 a_n + 2/3 m_n \gamma_n\} \geq c_1^2 \log K_n / (4 c_2 c_0),$$

for all sufficiently large n. The last result obtains because of the condition $m_n = o[K_n^{-3/4} (\log K_n)^{(q-1)/2}]$.

Given c_0 and c_2, we may choose c_1 large enough that $c_1^2 (4 c_2 c_0)^{-1} > c^* + 1$. It then follows that there exists N^* such that

$$P(|D_n(\eta_{r,n})| \geq \gamma_n) \leq 2K_n^{-(c^*+1)},$$

for all $|r| \leq b_n$ and $n > N^*$. Consequently, for $n > N^*$

$$P(A_n \geq \gamma_n) \leq 8 b_n K_n^{-(c^*+1)}.$$

In turn this implies

$$P(A_n \geq \gamma_n) \leq 8 K_n^{-c^*}.$$

Hence $\sum_{n=1}^{\infty} P(A_n \geq \gamma_n) < \infty$, and the proof is complete. □

Proof of Theorem 5. Under the assumptions, we may apply Lemma 4. This means Lemma 5 becomes applicable with $T_n = \hat{\xi}_{np}$ and we have *wp*1,

$$\overline{F}_n(\hat{\xi}_{np}) - \overline{F}_n(\xi_{p(n)}) = \bar{f}_n(\xi_{p(n)})(\hat{\xi}_{np} - \xi_{p(n)}) + O(K_n^{-1} \log K_n), \text{ as } n \to \infty.$$

Now using Lemma 6 with $q = 1/2$, and appealing to Lemma 4 again, we may pass from the last conclusion to: *wp*1

$$F_n(\hat{\xi}_{np}) - F_n(\xi_{p(n)}) = \bar{f}_n(\xi_{p(n)})(\hat{\xi}_{np} - \xi_{p(n)}) + O[K_n^{-3/4} (\log K_n)^{3/4}], \text{ as } n \to \infty.$$

Finally, since *wp*1: $F_n(\hat{\xi}_{np}) = p + O(m_n)$, as $n \to \infty$, we have wp1

$$p - F_n(\xi_{p(n)}) = \bar{f}_n(\xi_{p(n)})(\hat{\xi}_{np} - \xi_{p(n)}) + O(K_n^{-3/4} (\log K_n)^{3/4}), \text{ as } n \to \infty.$$

This completes the proof. □

7 References

Akaike, H. (1973). Information theory and an extension of entropy maximization principle. In B. Petrov and F. Csak (Eds.), *2nd International Symposium on Information Theory*, pp. 276–281. Kiado: Akademia.

Akaike, H. (1977). On entropy maximization principle. In P. Krishnaiah (Ed.), *Applications of Statistics*, pp. 27–41. Amsterdam: North-Holland.

Akaike, H. (1978). A Bayesian analysis if the minimum aic procedure. *Ann. Inst. Statist. Math. 30A*, 9–14.

Akaike, H. (1982). On the fallacy of the likelihood principle. *Statist. Probab. Lett. 1*, 75–78.

Akaike, H. (1983). Information measures and model selection. *Bull. Inst. Internat. Statist. 50*, 277–291.

Akaike, H. (1985). Prediction and entropy. In *A celebration of statistics*, pp. 1–24. New York-Berlin: Springer.

Berger, J.O. (1985). *Statistical Decision Theory and Bayesian Analysis, Second edition.* New York: Springer-Verlag.

Chung, K.L. (1968). *A Course in Probability Theory.* New York: Harcourt Brace and World, Inc.

Efron, B. (1996). Empirical Bayes methods for combining likelihoods (with discussion). *J. Amer. Statist. Assoc. 91*, 538–565.

Fan, J., N.E. Heckman, and W.P. Wand (1995). Local polynomial kernel regression for generalized linear models and quasi-likelihood functions. *J. Amer. Statist. Assoc. 90*, 826–838.

Hoeffding, W. (1963). Probability inequalities for sum of bounded random variables. *J. Amer. Statist. Assoc. 58*, 13–30.

Hu, F. (1994). Relevance weighted smoothing and a new bootstrap method. Technical report, Department of Statistics, University of British Columbia.

Hu, F. (1997). Asymptotic properties of relevance weighted likelihood estimations. *Canad. J. Statist. 25*, 45–60.

Hu, F. and J.V. Zidek (1993). A relevance weighted quantile estimator. Technical report, Department of Statistics, University of British Columbia.

Hu, F. and J.V. Zidek (1997). The relevance weighted likelihood. Unpublished.

Koul, H.L. (1992). *Weighted Empiricals and Linear Models*, Volume 21 of *IMS Lecture Notes*. Hayward, CA: Inst. Math. Statist.

Marcus, M.B. and J. Zinn (1984). The bounded law of the iterated logarithm for the weighted empirical distribution process in the non-iid case. *Ann. Prob. 12*, 335–360.

Shorack, G.R. and J.A. Wellner (1986). *Empirical Processes with Applications to Statistics*. New York: Wiley.

Staniswalis, J.G. (1989). The kernel estimate of a regression function in likelihood—based models. *J. Amer. Statist. Assoc. 84*, 276–283.

Tibshirani, R. and T. Hastie (1987). Local likelihood estimation. *J. Amer. Statist. Assoc. 82*, 559–567.

Lecture Notes in Statistics

For information about Volumes 1 to 74, please contact Springer-Verlag

Vol. 75: P.K. Goel, N.S. Iyengar (Eds.), Bayesian Analysis in Statistics and Econometrics. xi, 410 pages, 1992.

Vol. 76: L. Bondesson, Generalized Gamma Convolutions and Related Classes of Distributions and Densities. viii, 173 pages, 1992.

Vol. 77: E. Mammen, When Does Bootstrap Work? Asymptotic Results and Simulations. vi, 196 pages, 1992.

Vol. 78: L. Fahrmeir, B. Francis, R. Gilchrist, G. Tutz (Eds.), Advances in GLIM and Statistical Modelling: Proceedings of the GLIM92 Conference and the 7th International Workshop on Statistical Modelling, Munich, 13-17 July 1992. ix, 225 pages, 1992.

Vol. 79: N. Schmitz, Optimal Sequentially Planned Decision Procedures. xii, 209 pages, 1992.

Vol. 80: M. Fligner, J. Verducci (Eds.), Probability Models and Statistical Analyses for Ranking Data. xxii, 306 pages, 1992.

Vol. 81: P. Spirtes, C. Glymour, R. Scheines, Causation, Prediction, and Search. xxiii, 526 pages, 1993.

Vol. 82: A. Korostelev and A. Tsybakov, Minimax Theory of Image Reconstruction. xii, 268 pages, 1993.

Vol. 83: C. Gatsonis, J. Hodges, R. Kass, N. Singpurwalla (Editors), Case Studies in Bayesian Statistics. xii, 437 pages, 1993.

Vol. 84: S. Yamada, Pivotal Measures in Statistical Experiments and Sufficiency. vii, 129 pages, 1994.

Vol. 85: P. Doukhan, Mixing: Properties and Examples. xi, 142 pages, 1994.

Vol. 86: W. Vach, Logistic Regression with Missing Values in the Covariates. xi, 139 pages, 1994.

Vol. 87: J. Müller, Lectures on Random Voronoi Tessellations.vii, 134 pages, 1994.

Vol. 88: J. E. Kolassa, Series Approximation Methods in Statistics. Second Edition, ix, 183 pages, 1997.

Vol. 89: P. Cheeseman, R.W. Oldford (Editors), Selecting Models From Data: AI and Statistics IV. xii, 487 pages, 1994.

Vol. 90: A. Csenki, Dependability for Systems with a Partitioned State Space: Markov and Semi-Markov Theory and Computational Implementation. x, 241 pages, 1994.

Vol. 91: J.D. Malley, Statistical Applications of Jordan Algebras. viii, 101 pages, 1994.

Vol. 92: M. Eerola, Probabilistic Causality in Longitudinal Studies. vii, 133 pages, 1994.

Vol. 93: Bernard Van Cutsem (Editor), Classification and Dissimilarity Analysis. xiv, 238 pages, 1994.

Vol. 94: Jane F. Gentleman and G.A. Whitmore (Editors), Case Studies in Data Analysis. viii, 262 pages, 1994.

Vol. 95: Shelemyahu Zacks, Stochastic Visibility in Random Fields. x, 175 pages, 1994.

Vol. 96: Ibrahim Rahimov, Random Sums and Branching Stochastic Processes. viii, 195 pages, 1995.

Vol. 97: R. Szekli, Stochastic Ordering and Dependence in Applied Probability. viii, 194 pages, 1995.

Vol. 98: Philippe Barbe and Patrice Bertail, The Weighted Bootstrap. viii, 230 pages, 1995.

Vol. 99: C.C. Heyde (Editor), Branching Processes: Proceedings of the First World Congress. viii, 185 pages, 1995.

Vol. 100: Wlodzimierz Bryc, The Normal Distribution: Characterizations with Applications. viii, 139 pages, 1995.

Vol. 101: H.H. Andersen, M.Højbjerre, D. Sørensen, P.S.Eriksen, Linear and Graphical Models: for the Multivariate Complex Normal Distribution. x, 184 pages, 1995.

Vol. 102: A.M. Mathai, Serge B. Provost, Takesi Hayakawa, Bilinear Forms and Zonal Polynomials. x, 378 pages, 1995.

Vol. 103: Anestis Antoniadis and Georges Oppenheim (Editors), Wavelets and Statistics. vi, 411 pages, 1995.

Vol. 104: Gilg U.H. Seeber, Brian J. Francis, Reinhold Hatzinger, Gabriele Steckel-Berger (Editors), Statistical Modelling: 10th International Workshop, Innsbruck, July 10-14th 1995. x, 327 pages, 1995.

Vol. 105: Constantine Gatsonis, James S. Hodges, Robert E. Kass, Nozer D. Singpurwalla(Editors), Case Studies in Bayesian Statistics, Volume II. x, 354 pages, 1995.

Vol. 106: Harald Niederreiter, Peter Jau-Shyong Shiue (Editors), Monte Carlo and Quasi-Monte Carlo Methods in Scientific Computing. xiv, 372 pages, 1995.

Vol. 107: Masafumi Akahira, Kei Takeuchi, Non-Regular Statistical Estimation. vii, 183 pages, 1995.

Vol. 108: Wesley L. Schaible (Editor), Indirect Estimators in U.S. Federal Programs. viii, 195 pages, 1995.

Vol. 109: Helmut Rieder (Editor), Robust Statistics, Data Analysis, and Computer Intensive Methods. xiv, 427 pages, 1996.

Vol. 110: D. Bosq, Nonparametric Statistics for Stochastic Processes. xii, 169 pages, 1996.

Vol. 111: Leon Willenborg, Ton de Waal, Statistical Disclosure Control in Practice. xiv, 152 pages, 1996.

Vol. 112: Doug Fischer, Hans-J. Lenz (Editors), Learning from Data. xii, 450 pages, 1996.

Vol. 113: Rainer Schwabe, Optimum Designs for Multi-Factor Models. viii, 124 pages, 1996.

Vol. 114: C.C. Heyde, Yu. V. Prohorov, R. Pyke, and S. T. Rachev (Editors), Athens Conference on Applied Probability and Time Series Analysis Volume I: Applied Probability In Honor of J.M. Gani. viii, 424 pages, 1996.

Vol. 115: P.M. Robinson, M. Rosenblatt (Editors), Athens Conference on Applied Probability and Time Series Analysis Volume II: Time Series Analysis In Memory of E.J. Hannan. viii, 448 pages, 1996.

Vol. 116: Genshiro Kitagawa and Will Gersch, Smoothness Priors Analysis of Time Series. x, 261 pages, 1996.

Vol. 117: Paul Glasserman, Karl Sigman, David D. Yao (Editors), Stochastic Networks. xii, 298, 1996.

Vol. 118: Radford M. Neal, Bayesian Learning for Neural Networks. xv, 183, 1996.

Vol. 119: Masanao Aoki, Arthur M. Havenner, Applications of Computer Aided Time Series Modeling. ix, 329 pages, 1997.

Vol. 120: Maia Berkane, Latent Variable Modeling and Applications to Causality. vi, 288 pages, 1997.

Vol. 121: Constantine Gatsonis, James S. Hodges, Robert E. Kass, Robert McCulloch, Peter Rossi, Nozer D. Singpurwalla (Editors), Case Studies in Bayesian Statistics, Volume III. xvi, 487 pages, 1997.

Vol. 122: Timothy G. Gregoire, David R. Brillinger, Peter J. Diggle, Estelle Russek-Cohen, William G. Warren, Russell D. Wolfinger (Editors), Modeling Longitudinal and Spatially Correlated Data. x, 402 pages, 1997.

Vol. 123: D. Y. Lin and T. R. Fleming (Editors), Proceedings of the First Seattle Symposium in Biostatistics: Survival Analysis. xiii, 308 pages, 1997.

Vol. 124: Christine H. Müller, Robust Planning and Analysis of Experiments. x, 234 pages, 1997.

Vol. 125: Valerii V. Fedorov and Peter Hackl, Model-oriented Design of Experiments. viii, 117 pages, 1997.

Vol. 126: Geert Verbeke and Geert Molenberghs, Linear Mixed Models in Practice: A SAS-Oriented Approach. xiii, 306 pages, 1997.

Vol. 127: Harald Niederreiter, Peter Hellekalek, Gerhard Larcher, and Peter Zinterhof (Editors), Monte Carlo and Quasi-Monte Carlo Methods 1996, xii, 448 pages, 1997.

Vol. 128: L. Accardi and C.C. Heyde (Editors), Probability Towards 2000, x, 356 pages, 1998.

Vol. 129: Wolfgang Härdle, Gerard Kerkyacharian, Dominique Picard, and Alexander Tsybakov, Wavelets, Approximation, and Statistical Applications, xvi, 265 pages, 1998.

Vol. 130: Bo-Cheng Wei, Exponential Family Nonlinear Models, ix, 240 pages, 1998.

Vol. 131: Joel L. Horowitz, Semiparametric Methods in Econometrics, ix, 204 pages, 1998.

Vol. 132: Douglas Nychka, Walter W. Piegorsch, and Lawrence H. Cox (Editors), Case Studies in Environmental Statistics, viii, 200 pages, 1998.

Vol. 133: Dipak Dey, Peter Müller, and Debajyoti Sinha (Editors), Practical Nonparametric and Semiparametric Bayesian Statistics, xv, 408 pages, 1998.

Vol. 134: Yu. A. Kutoyants, Statistical Inference For Spatial Poisson Processes, vii, 284 pages, 1998.

Vol. 135: Christian P. Robert, Discretization and MCMC Convergence Assessment, x, 192 pages, 1998.

Vol. 136: Gregory C. Reinsel, Raja P. Velu, Multivariate Reduced-Rank Regression, xiii, 272 pages, 1998.

Vol. 137: V. Seshadri, The Inverse Gaussian Distribution: Statistical Theory and Applications, xi, 360 pages, 1998.

Vol. 138: Peter Hellekalek, Gerhard Larcher (Editors), Random and Quasi-Random Point Sets, xi, 352 pages, 1998.

Vol. 139: Roger B. Nelsen, An Introduction to Copulas, xi, 232 pages, 1999.

Vol. 140: Constantine Gatsonis, Robert E. Kass, Bradley Carlin, Alicia Carriquiry, Andrew Gelman, Isabella Verdinelli, Mike West (Editors), Case Studies in Bayesian Statistics, Volume IV, xvi, 456 pages, 1999.

Vol. 141: Peter Müller, Brani Vidakovic (Editors), Bayesian Inference in Wavelet Based Models, xi, 394 pages, 1999.

Vol. 142: György Terdik, Bilinear Stochastic Models and Related Problems of Nonlinear Time Series Analysis: A Frequency Domain Approach, xi, 258 pages, 1999.

Vol. 143: Russell Barton, Graphical Methods for the Design of Experiments, x, 208 pages, 1999.

Vol. 144: L. Mark Berliner, Douglas Nychka, and Timothy Hoar (Editors), Case Studies in Statistics and the Atmospheric Sciences, x, 208 pages, 2000.

Vol. 145: James H. Matis and Thomas R. Kiffe, Stochastic Population Models, viii, 220 pages, 2000.

Vol. 146: Wim Schoutens, Stochastic Processes and Orthogonal Polynomials, xiv, 163 pages, 2000.

Vol. 147: Jürgen Franke, Wolfgang Härdle, and Gerhard Stahl, Measuring Risk in Complex Stochastic Systems, xvi, 272 pages, 2000.

Vol. 148: S.E. Ahmed and N. Reid, Empirical Bayes and Likelihood Inference, xiv, 256 pages, 2000.